Social Butterflies

MONOGRAPHS IN POPULATION BIOLOGY
SIMON LEVIN, ROB PRINGLE, AND CORINA TARNITA, SERIES EDITORS

A complete series list follows the index.

# Social Butterflies

HENRY S. HORN

PRINCETON UNIVERSITY PRESS
Princeton and Oxford

Copyright © 2021 by Princeton University Press

Princeton University Press is committed to the protection of copyright and the intellectual property our authors entrust to us. Copyright promotes the progress and integrity of knowledge. Thank you for supporting free speech and the global exchange of ideas by purchasing an authorized edition of this book. If you wish to reproduce or distribute any part of it in any form, please obtain permission.

Requests for permission to reproduce material from this work
should be sent to permissions@press.princeton.edu

Published by Princeton University Press
41 William Street, Princeton, New Jersey 08540
6 Oxford Street, Woodstock, Oxfordshire OX20 1TR

press.princeton.edu

All Rights Reserved

Library of Congress Cataloging-in-Publication Data

Names: Horn, Henry S., 1941–2019, author.
Title: Social butterflies / Henry S. Horn.
Other titles: Monographs in population biology.
Description: Princeton, New Jersey : Princeton University Press, 2021. | Series: Monographs in population biology | Includes bibliographical references and index.
Identifiers: LCCN 2020040068 (print) | LCCN 2020040069 (ebook) | ISBN 9780691206301 (paperback) | ISBN 9780691206295 (hardback) | ISBN 9780691212685 (ebook)
Subjects: LCSH: Butterflies—Identification. | Butterflies—Behavior.
Classification: LCC QL562.4 .H67 2021 (print) | LCC QL562.4 (ebook) | DDC 595.78/9156—dc23
LC record available at https://lccn.loc.gov/2020040068
LC ebook record available at https://lccn.loc.gov/2020040069

British Library Cataloging-in-Publication Data is available

Editorial: Alison Kalett and Whitney Rauenhorst
Production Editorial: Mark Bellis
Text and Cover Design: C. Alvarez-Gaffin
Production: Jacqueline Poirier
Publicity: Matthew Taylor and Amy Stewart
Copyeditors: Lucinda Treadwell and Daniel Rubenstein

This book has been composed in Times

Printed on acid-free paper. ∞

Printed in the United States of America

10 9 8 7 6 5 4 3 2 1

# Contents

| | |
|---|---|
| Foreword | vii |
| Preface | ix |
| 1. Introduction: Mysteries and Methods | 1 |
| 2. Morphological and Perceptual Adaptations to Habitat and Society: Bases for Cues and Rules of Behavior | 27 |
| 3. Tactical Forward Vagrancy: The Ringlet *Coenonympha tullia inornata* | 74 |
| 4. Fortuitous Site-Fidelity? The Eyed Brown *Satyrodes eurydice* | 96 |
| 5. Setting and Running a Trapline: The Great Spangled Fritillary *Speyeria cybele* | 110 |
| 6. Defining and Defending a Territory: The Viceroy *Limenitis archippus* | 125 |
| 7. Sociology at a Singles Bar: The Pearly Eye *Enodia anthedon* | 138 |
| 8. Do Butterflies Make Decisions? | 160 |
| 9. Life History Consequences of Individual Behavior | 187 |
| 10. Summary *and Speculations* | 202 |
| Appendix A. Taxonomy | 213 |
| Appendix B. More Natural History | 216 |
| Appendix C. Machinery | 221 |
| Appendix D. NetLogo Programs | 230 |
| Notes | 237 |
| Bibliography | 249 |
| Index | 265 |

# Foreword

This book represents a milestone in the long and influential history of the Monographs in Population Biology. Henry Stainken Horn (1941–2019) was for all of us a scientific inspiration for his unmatched ability to seamlessly interweave theory, natural history, and empiricism. He was a great teacher, beloved by both students and colleagues alike for his generosity in sharing his expertise and encouragement with others. For him, the Monographs represented the perfect vehicle to combine these talents, and to help build a conceptual foundation for the subject. Henry edited or coedited the MPB series from 1989 to 2019—almost its complete history—and had much to do with encouraging and improving what was published, from the earliest volumes. His influence from the start stemmed from the degree to which the founding editor, Robert MacArthur, trusted his judgment, and because Henry was extremely smart and equally passionate about coaxing the best out of authors. He read every proposal and manuscript thoroughly and selflessly in terms of the time he invested in each, and the quality of the eventual products owes more to Henry Horn than to any other person. The MPB series was one of his great loves.

This book is a milestone for two reasons. First, because Henry's first book—*The Adaptive Geometry of Trees*, a brilliant collection of insights into what shaped the observed morphological patterns—appeared almost exactly 50 years ago as the third book in the series. Second, because it is Henry's swan song—Henry passed away shortly after completing his penultimate draft (the version published here represents the version submitted by Henry with only light editing undertaken subsequently). These two monographs are appropriate bookends on a remarkable career. As will be evident from the current volume, Henry was a masterful writer and storyteller, able to combine an unparalleled command of natural history and ecology with powerful mathematical insights, and to still make the product fun and captivating to read. His mastery is evident in the breadth of his writings in the series, from vegetation to butterflies, but his erudition touched everything he encountered. His philosophy of scientific inquiry is perhaps best summarized in this short quote from the book you are holding:

> I get great joy from discovering the facts of natural history, but my most profound satisfaction comes when those facts are organized conceptually. That conceptual organization is at first less a formal testing of theory against fact

than an interplay between fact and fancy until the fancy becomes refined into theory, and the theory can then be tested with further facts.

*Social Butterflies* is a realization of this philosophy. Henry Horn produced seminal work on plant biology and he constructed crystal-clear visual models of life history evolution, but butterflies were his first love. When he summered in Whitehall, NY with his family, he would spend all day, every day, outside watching how butterflies navigated the fields, woods, and forests surrounding his house in search of food, mates, and egg-laying sites. As was his style, Henry always looked for the simplest explanation for the patterns he witnessed—movements in the case of butterflies. He always wanted to understand the mechanisms underlying behavioral, physiological, or morphological traits, as well as their adaptive or evolutionary function. For Henry, butterflies revealed an array of different mating systems across the landscape. Thus, depending on whether females aggregated or traveled along predictable routes, males were forced to respond in different ways—sometimes establishing territories, sometimes defending key locations that females favored, and sometimes simply wandering in search of females with which they would consort only briefly. But understanding the "why" was not enough. Henry also had to know the "how," and this is where his ability to draw on so many biological and physical first principles made him unique. By "putting himself inside the bodies" of the butterflies, he sought to make sense of their world from their perspective; and from this unique vantage point, he identified how the structure of the landscape influenced how each butterfly species gathered and used information to shape its movements. Using very creative and nontraditional approaches—erecting large arrays of mirrors, or building a rotating "rotisserie" on a fishing pole—he was able to alter their behaviors to test, and often validate, his novel ideas. This monograph illustrates these and many more scientific escapades. Beyond revealing the wonders of butterfly behavioral and sensory ecology, it also shows how one of the greatest natural historians to bridge the 20th and 21st centuries unraveled complex problems by identifying their essence, cleverly solving them with simple backyard tools, and sharing them with the world in lucid and engaging prose.

Henry Horn had an unbridled love for Nature, for extracting pattern from masses of observations, and for the sheer joy of communicating what one had learned. We will all miss him as a scholar, as a colleague, and as a shepherd of this series. But he will live on in his writings, which will continue to transfer his wisdom, creativity, and nonconformity to us and to future generations.

Alison Kalett, Simon Levin, Robert Pringle, Daniel Rubenstein, and Corina Tarnita
Princeton, New Jersey
March 2020

# Preface

> None but those who were deprived of their Senses, would go in Pursuit of Butterflies.
> –Moses Harris, 1766, *The Aurelian*; attributed arguably to "Some Relations" of "Lady Glanvil."

Lady Eleanor Glanville, a 17th-century entomologist, deserved more sensitive relatives, who could see that her passion for butterflies honed her senses, rather than taking them away. My own study of butterflies has been organized in the logical and analytic fashion that I shall present in the Introduction, but it started with a youthful passion that has continued to interact with historical accidents and personal proclivities that ultimately underlie the outline of this book.

You are welcome to skip right to the Introduction for the spurious impression that this book reports a project that was brilliantly designed from the outset. But you would miss a deep implicit theme, namely that joyous exploration of outright natural history interacts constructively with conceptual biology.

## 0.1 ANCIENT HISTORY

Several blocks from my childhood home in Augusta, Georgia, the street lost its pavement and continued as a dirt track through an abandoned field that was growing to woodland. I was not supposed to go there, especially alone, but I did. What drew me were birds, trees, and butterflies. The butterflies collected in prodigious numbers at the edges of evaporating mud puddles. In my memory they were mostly Tiger Swallowtails *Papilio glaucus* and Zebra Swallowtails *Eurytides marcellus* and large Cloudless Sulphurs *Phoebis sennae*, and they sat in little groups according to their kind, all pointing in the same direction. At the time, I wanted to learn more about birds and trees, but I just wanted to enjoy the

FIGURE 0.1. Brothers Charles M., William M., and David J. Horn, observing water striders at Sheldrake Creek, near Cayuga Lake, NY, July 1956. Our only joint publication to date is Horn, Horn, Horn, and Horn 1976.

butterflies, creeping close and lying on my belly very carefully to avoid the stains of red clay that would betray where I had been.

My father, the late Reverend Henry E. Horn, laid the groundwork for my formal study of butterflies in the early 1950s, when he taught my brothers and me to make butterfly nets out of broom handles, coat hangers, and cheese cloth (see Figure 0.1). We established an insect collection drawn from our backyard and vacant lots in Cambridge, Massachusetts,[1] from vacation spots near Ithaca, New York, and East Arlington, Vermont, and from our jobs at Boy Scout camps in Waltham, Massachusetts, and West Rindge, New Hampshire. Brother David was the most compulsively organized, and until recently he continued to curate the bulk of our childhood collection as Professor of Entomology at Ohio State University.

In 1967 my wife, Elizabeth, and I discovered that we could not afford the down payment on a house in Princeton, New Jersey, so we took the money that she had earned while we were living on my National Science Foundation Graduate Fellowship, and we bought a piece of abandoned farmland near Whitehall, New

PREFACE  XI

FIGURE 0.2. Field assistants Eric and Jennifer Horn, central campus of the study site, near Whitehall, NY, August 1982.

York. We began to spend our summers there, and over several years built a cabin and a work-shed from salvaged barn boards (see Figure 0.2). We lived without electricity, telephone, or plumbing.

I had originally intended to use the land for long-term manipulative experiments on forest succession, about which I had some biologically deterministic ideas (Horn 1971, 1975, 1976a). However, I soon discovered that the most interesting features of vegetation dynamics on this site were crucially dependent on historical accidents of the state of the land when initially fallowed. This meant that understanding the dynamics depended more on history than on biological insight. I set out to understand why this situation was different from what I was used to back in New Jersey (Horn 1981; Horn et al. 1989). However, I was soon attracted to an entirely different set of problems, involving butterflies.

## 0.2 HISTORICAL OUTLINE

In 1973, I encountered a mysterious species that I had never seen before in all my youthful collecting in New England. It turned out to be the Inornate Ringlet *Coenonympha tullia inornata*.[2] I knew from the work of a former graduate student, G. Scott Anthony, that this species was in the process of newly expanding its range into New England (Keji 1963, Ferris 1970, Shapiro 1974, Iftner 1997). I thought that here might be a chance to study the development of a population from the moment of its establishment. So, I made a little clamp to hold butterflies while I marked them with a small number that I could later read without recapturing them (Appendix C.1, and Horn 1976b). The clamp also let me release butterflies immediately right where I had caught them.

As long as I was catching Ringlets, I caught and marked any other species that I could, eventually marking 26 species of butterflies. I began to watch them, concentrating on those species whose marked individuals I encountered often enough to be exciting. I soon found five species whose social behaviors, at least in caricature, spanned nearly the range of behaviors observed in vertebrates (Brown 1964, Bradbury and Vehrencamp 1977, Rutowski and Alcock 1989). The Inornate Ringlet showed tactical vagrancy, its males wandering with no detectable pattern, and encountering females rarely and at random. The Eyed Brown *Satyrodes eurydice* was habitat-restricted in its ranging, following linear depressions in the vegetation that further restricted the wanderings of each individual to a portion of the available habitat. The Great Spangled Fritillary *Speyeria cybele* exemplified traplining, repeatedly circulating among sites where females were likely to appear. Male Viceroys *Limenitis archippus* appeared to defend territories, individualistically bounded real estate that females were likely to traverse. The Northern Pearly Eye *Enodia anthedon* aggregated at a food source, yeasty slime fluxes on wounded woody vegetation, where females went for the nutrients, and males went to access both nutrients and females.

Two species were common and easy to study, the Inornate Ringlet and the Northern Pearly Eye. Two were uncommon, but still easy to study, the Viceroy and the Northern Eyed Brown. One was uncommon and very difficult to study, the Great Spangled Fritillary, but as the apparent trapliner, it was so intriguing that I persevered.

I have yet to put together all of my information on the developing population of the Ringlet, but its range extension and population genetics became the

PhD dissertation of Diane Wiernasz (Wiernasz 1983, 1989). The rest of my work on butterflies exploits my interest in biological problems with geometrical overtones, and my lifelong penchant for mechanical, optical, and electronic tinkering (cf. Hoban 1974). Having been strongly influenced by an article by John Rader Platt on the strategy of "Strong Inference" in science (Platt 1964), I usually try to test specific hypotheses with planned comparisons or with manipulative experiments. However, this is a recurring theme, rather than a litany (Horn and Farnsworth 1986), and some of my hypotheses remain speculative.

## 0.3 PERSONAL BIASES

The timing of field seasons, both within and between years, compromised optimal design with family schedule and teaching obligations. But two additional constraints from my own behavior deserve pedagogical notes. Early on I experienced "data-generated narcosis," and later, an increasing reluctance to slaughter butterflies.

Data-generated narcosis has the following character. Having once decided that it was a good idea to gather a particular type of data, you gather more and more without ever reevaluating whether those data are appropriate to the questions that you are asking. I always caution students about this affliction; so there is no excuse whatsoever for my not recognizing my own case. Nevertheless, for season after season I laboriously marked individual butterflies, logged their day-to-day locations, compiled the data, and analyzed them to find barely detectable differences due to variations in ranging behavior. Each year I planned to increase the sample size next year, to explore the statistical significance of those tiny differences. I could not understand why dramatic differences in the observed behavior of individuals of the different species were generating such similar patterns of day-to-day ranging. Of course, in retrospect the answer is obvious. Over the course of a day or two, any individual butterfly, using whatever pattern of behavior, could conceivably wander across my entire study site. So individual behavior is less of a constraint on aggregated day-to-day ranging than is the geographic configuration of boundaries between appropriate and inappropriate habitat. The behaviors differed among species from moment to moment, and documenting them was a matter of mapping the movements of individuals on a scale of seconds, not of days. I did not discover this and amend my ways until 1981, seven years into the project.

I began the formal work on this project with a reference collection of hundreds of dead butterflies. The collection was so beautiful that I had no qualms about murdering butterflies at the time, but I did not have a specific purpose either. Accordingly, I missed a unique opportunity to document an interesting pattern that is firmly in my mind, but that may or may not be real, namely that the Ringlet population was uniform at first, but became increasingly sexually dimorphic in size and color. So my collection just sits there as a monument to early insensitivity.

As I began marking and observing butterflies in the field, I quickly found that several species are injured by the marking process,[3] some obviously, some in ways that subtly changed their behavior. I removed those species from my fieldwork, in part because I could not trust their data, but largely because I felt sorry and ashamed to see the changed and maladaptive behavior that I had caused. As the project developed, I spent less time massively marking and recapturing, and more time simply observing a few known individuals. I got to know the idiosyncrasies of, e.g., Viceroys #74.4, #77.10, #81.3, #81.8, #81.10, #81.17, and #83.1; Pearly Eyes #74:13, #80.102•, #83.36, and #83.20; Eyed Brown #83.3; and Fritillaries #74.56, #82.0 and #82:1. I even named them, respectively: Viceroys *Methuselah, Adam, John, Charlie I, Arthur, Charlie II,* and *Narcissus*; Pearly Eyes *Nick, Susannah, Robert E.,* and *Ulysses S.*; Eyed Brown *Traveler*; and Fritillaries *Don Juan, Main Man,* and *Pierre*. On the advice of colleagues, I shall repress some of the names henceforth, but Bekoff (1994) has eloquently defended the use of such names for individual research subjects.

Once I learned something of how butterflies live their lives, I began to have difficulty with the notion of taking their lives from them. This happened despite my craving data that can only be gotten from dead butterflies. For example, how many mates a female will accept is shown by dissection to see how many spermatophores she has accumulated (after Burns 1968). If I want to establish that males are at a particular place for the purpose of, or to the effect of, encountering potential mates, . . . then I must show the presence of potential mates. If all the females that I collect in that place have one and only one spermatophore, then I have no evidence that those females are potential mates. However, if I find any virgins or any polyandrous females, I have strong evidence that the females present are at least potential mates.

Here is a specific instance of the problem. Male Viceroys defend territories where wandering females are concentrated (Chapter 6). So I collected a sample of fresh females to look for virgins and a sample of worn and experienced females to look for multiply mated females. All twelve female Viceroys contained one and only one spermatophore. Twelve is a pitifully small sample size. I could gather more, but I simply can't do it to a small population that I have grown to know and respect. Furthermore, for one of the most thoroughly studied butterfly populations in the world, the Bay Checkerspot (*Euphydryas editha bayensis*), Harrison (et al. 1991) has argued that the removal of females during her study appreciably increased the likelihood of local extirpation, even though the numbers involved were small relative to natural fluctuations in the population. So there is an important scientific gap in my work that I am not temperamentally suited to fill, and I have to argue on indirect evidence, i.e., the published behavior of congeneric species, among which free-flying females are found, albeit uncommonly, to carry no or multiple spermatophores (Chapter 1, Table 1.3). Fortunately, I can shift the blame for my reticence to sacrifice butterflies onto the scientific literature, where there has been a surge of excellent papers on ethical considerations in fieldwork (For précis and bibliographies, see, e.g., Bekoff et al. 1992, Farnsworth and Rosovsky 1993, Putnam 1995, Farnsworth 1995, and Bekoff 2007).

## 0.4 EMERGENT QUESTIONS, AND MY INTENT

That is how the research behind this book came about. Major elements of the formal design were serendipity and idiosyncratic enthusiasms. Serendipity allowed the butterflies to pose their own mysteries. In particular, how can creatures with such simple perceptual and integrative abilities engage in such apparently complex behavior? What simple environmental cues and internal rules of behavior might different species use to organize behaviors like vagrancy, site-fidelity, traplining, territoriality, and mating aggregation? And do the different behaviors carry different consequences for successful propagation in the local population? This book is a systematic exploration of these mysteries. Their solutions are incomplete, in the sense that from every tentative solution, new mysteries emerge.

I hope that these solutions will be provisionally convincing for specialists in butterfly physiology, behavior, and ecology, and that lepidopterists will treat the new mysteries, and the inevitable shortcomings of my work, as opportunities for more research. But I also hope that those with broader interests in animal behavior will find intriguing comparisons and contrasts between the behaviors of their beloved organisms and those of butterflies. Finally, and most fervently, I hope that a wide range of students early in their scientific careers will see that first-rate science can be pursued with joyous passion, even with limited resources, . . . because outright natural history always provides novel observations that stimulate and respond to conceptual thought.

## 0.5 ACKNOWLEDGMENTS

I owe deep thanks to many people for help over the decades of research behind this book. Foremost is my wife Elizabeth (Betty) Gates Horn, whose main contributions have been intellectual discussion during the best of times and emotional support during the worst. The bulk of the fieldwork has been solo, but I had extensive help in the 1977 season from Jeff Georgia with the Fritillary and the Eyed Brown, and in the 1978 and 1980 seasons from Diane Wiernasz with the Ringlet. Jeff's work resulted in an undergraduate senior thesis (Georgia 1978), and Diane's work was a small part of a far more extensive and independent PhD thesis (Wiernasz 1983). I also had episodic help in the field from Elizabeth Horn, daughter Jennifer Horn, son Eric Horn, brother David Horn, and summer neighbor Eileen Hart.

For extensive conversations that changed the way I think about *Social Butterflies*, I thank Diane Wiernasz, Dan Rubenstein, Art Shapiro, Phil DeVries, Helen Dunlap-Pianka, Mike Singer, Ron Rutowski, John Endler, Larry Gilbert, Lord Robert May (né Bob), Nick Davies, Nick Haddad, James Brown, and Jim Gould.

For specific encouragement during difficult times, I am very grateful to Bob Lederhouse, Paul Ehrlich, Gary Bernard, the late Charles Remington, the late Sir Richard Southwood, Egbert Leigh, Naomi Pierce, John Hoogland, Ed Wilson, Judith May, Emily Wilkinson, Bernd Heinrich, James Marden, John Bonner, Simon Levin, Christie Riehl, and Cassie Stoddard.

The budget for my study of butterflies was modest and came mostly from Princeton University's Eugene Higgins Trust Fund, administered by the Department of Biology, plus my own seminar honoraria that I had set aside in what I called the "Fund for Woodsy Lore."

A narrative outline of this book was written, and the bulk of the data was initially analyzed, during a sabbatical term in the spring of 1989. I had an office at the Institute for Advanced Study in Princeton, New Jersey. I am grateful to its then director, the late Marvin Goldberger, and its chief administrator Rachel Gray for the rare tranquility and the superb tea and cookies.

Writing the book itself started during a sabbatical term in the spring of 1996 at Stanford University. I thank my official hosts Marcus Feldman and Deborah Gordon, as well as my unofficial hosts Carol Boggs, Ward Watt, Stu Weiss, Paul Ehrlich, and Diane Wagner. For help in the library I thank Jill Otto.[4]

Several fully developed but amorphous ideas crystallized during the Fourth International Conference on the Biology of Butterflies in March 2002 in Leeuwenhorst, The Netherlands (Lewis and Bryant 2002). I thank Paul Brakefield for engineering my presence there, no small feat given my reclusiveness at the time. I recall with particular joy discussing my poster with three personal heroes simultaneously: Christer Wiklund, Dick Vane-Wright, and Robert Pyle. I also had very encouraging discussions with Doekele Stavenga, the late Ilkka Hanski, and Phil DeVries.

The first thorough draft, in the form of a narrative outline, was written during a third sabbatical term in the spring of 2006 at the University of Arizona in Tucson.

I thank my official hosts, Michael and Carole Rosenzweig, and my unofficial host, Brian Enquist, for just the right proportions of intellectual stimulation and privacy to get the job done.

In 2010, I adopted a strategy of signing up to give a formal talk at least once a year, based on a chapter that was almost, but not quite, ready for a detailed draft. At the same time, I began to develop graphical models for the behavior of my studied species, using the software package NetLogo (Wilensky1999). This exercise forced me to be very explicit about the mechanisms of behavior that I was proposing, and it confirmed most of my previous inferences of their results, but most importantly, it sharpened my overview of the essential similarities and differences among species. I am thankful for episodic discussions with Colin Twomey on biological realities and how to convert them into computer algorithms. And I appreciate the forbearance of the many others who attended my talks, and whose comments helped me to clarify ideas, evidence, and rhetoric.

Why has it taken so long to write? Perhaps the noblest excuse is that I am attracted to broad concepts of ecology and behavior, but I am exceedingly detail-oriented in matters of natural history, and I have trouble bridging the gap between these attitudes. The same problem arises in my teaching, which until recently has taken precedence over research in my attempts to solve it. Since formal retirement in June 2011, I have gradually shed much of my formal teaching, and I now have the freedom of mind to address the larger research projects that have long been in the background while I have published smaller projects.

Retirement from formal mentoring has also turned my thoughts to my own mentors, both teachers and colleagues, to all of whom I am indebted for insights that I have brought to *Social Butterflies*. My early mentors and heroes for contagiously enthusiastic scholarship and engaging writing include the late Thomas Patrick Burns, Thomas Shinagel, the late Sir Arthur Loveridge, the late Ernest E. Williams, Edward O. Wilson, and the late Martin H. Zimmerman. I profited greatly from discussions about graphical rhetoric with the late Amy Bordvik. My companion in learning statistics and computer programming was Eric R. Pianka. What other mathematics I have is due mostly to osmosis from the late Robert H. MacArthur, Egbert G. Leigh, Lord Robert May, and Simon A. Levin. Robert MacArthur made an unusual contribution to this project as my exemplar and mentor for cabin-building and the Thoreauvian summer lifestyle. And although *Social Butterflies* is my first project whose substantive conception and design were entirely my own, it builds on the behavioral style of my graduate mentor Gordon Orians, and the ecological style of my early colleague Robert MacArthur, both of whom championed the search for interpretable patterns via continual interplay between conceptual thought and empirical natural history.

*Social Butterflies* is dedicated to the memory of my late mother, Catherine Hedwig Stainken Horn, Master of Science, Cornell 1939, who taught me more about the rudiments of the scientific method than anything that I have read since. Not the least wise of her observations was, "You learn something new every day, if you are not careful." She died in 2007, but she had a chance to read and criticize my draft of 1996.

Social Butterflies

CHAPTER ONE

# Introduction

## *Mysteries and Methods*

> "We have got to the deductions and the inferences," said Lestrade, winking at me. "I find it hard enough to tackle the facts, Holmes, without flying away after theories and fancies." "You are right," said Holmes demurely; "you do find it very hard to tackle the facts."
> –Inspector Lestrade and Sherlock Holmes, Sir Arthur Conan Doyle, 1892. "The Boscombe Valley Mystery," in *The Adventures of Sherlock Holmes.*

I get great joy from discovering the facts of natural history, but my most profound satisfaction comes when those facts are organized conceptually. That conceptual organization is at first less a formal testing of theory against fact than an interplay between fact and fancy until the fancy becomes refined into theory, and the theory can then be tested with further facts.

### 1.1. THE GOALS OF THIS BOOK

My initial fancy is that butterflies have patterns of ranging and mate-finding that seem to imitate the more complex behaviors seen in vertebrates. In this book, I hope to present empirical evidence to convince you that five species show five kinds of interactive behavior that are sufficiently complex to qualify as "social." I shall argue that the social behavior of butterflies uses cues and rules that are so simple that they are amenable to experimental manipulation in the field, and that the behavior has significant consequences for life history, which in turn have implications for local population biology. The cues and rules are aided by aspects of physiology that may allow behavioral automata to generate seemingly complex behavior. So in a sense, this book explores emergent patterns through three levels of biological organization: physiology, behavior, and ecology. As such, it responds to a call by Nathan (et al. 2008) for an integrated study of movement at successive scales: from the modes and mechanisms of orientation and movement, through momentary tracks in the favored habitat, to consequences for life history with implications for the demographics of the local population.

This chapter begins with a caution that mate-finding is an interaction between male and female. When I use language that describes males looking for females, I continue a long tradition of bias in human observational skills, and I shall try to correct this bias when I interpret my observations. Next, I describe why I did not continue intensive study of some species, partly as a context for the five species that I chose, and additionally for the potential interest of others who might encounter any of the full complement of species in other places. The bulk of the chapter sets the stage with description of the habitats and mating behavior of the five species that I have studied most intensively. I then introduce the males, their patterns of ranging, and the mysteries they pose. There is the obligatory parade of study site, methods, analyses, and statistics. Then I return to a road map of the rest of the chapters and a discussion of how they lead to my goal of beginning to understand emergent patterns of natural history.

## 1.2. WHO IS SEARCHING FOR WHOM?

I am concerned primarily with the behavior of males as they search for females, and only secondarily with the behavior of females in quest of suitable sites for egg-laying. This androcentric bias is more observational than ecological. I am a visual creature, and males are more active in quests for females than vice versa, at least for the butterfly species that I have observed. Furthermore, I can easily see the goal of a male's search, namely a female. But it is much more difficult for me to distinguish the visual characteristics of a potential site for oviposition, and the ultimate laying of eggs is almost certainly triggered by chemical cues that I cannot sense, but which may play a part in the butterfly's search as well. Nevertheless, many of the analyzed details of male behavior are only interpretable in terms of observed or strongly inferred aspects of female behavior.

A traditional classification of the behaviors of males looking for mates divides them between "patroller" and "percher," with some species exercising both (Scott 1974). Scott defines patrollers as "flying almost constantly" all day or all afternoon in search of females (Scott 1986, pp. 46–47). Such flight requires energetically efficient morphology and movement of wings. The flying male must detect a female that is either sitting or flying, depending on the species and the status of the female. Scott's perchers "dart out from vantage points," inspect an object that may be a female, and then return to that perch or another nearby. Such flight requires rapid acceleration toward and maneuvering around the object inspected, as well as escape should the object turn out to be a dragonfly or other predator. In general, the perching male detects and responds to a moving object. Patrollers sometimes

INTRODUCTION

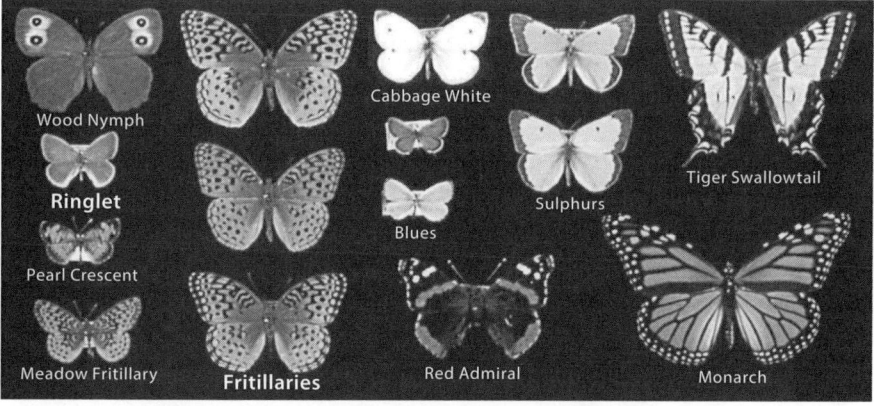

A. Patrollers search "by flying almost constantly" all day or all afternoon

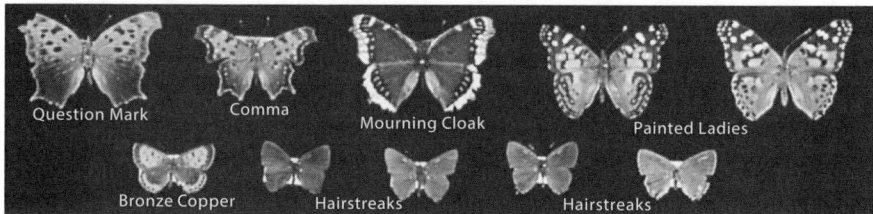

B. Perchers "dart out from vantage points" to maneuver around objects and return

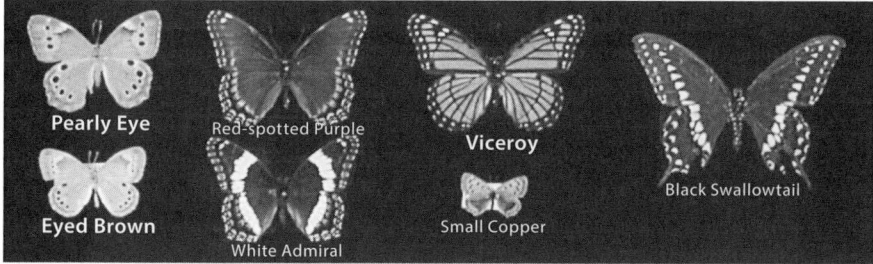

C. Some Perchers also Patrol (and vice versa)

FIGURE 1.1. Common species of the study area by their male mate-finding behavior. Behavioral categories are taken from Scott (1986). The wings of Patrollers (A) tend to be more convex, especially the forewing and its outer edge. The forewings of Perchers (B) tend to be more pointed, with a straight, or even concave outer edge. These flight adaptations, especially the exceptions, are discussed in Chapter 2 (Section 2.2.1).

perch in potentially productive sites, and perchers may patrol a bit as they return to perching. Figure 1.1 arranges the species of my study site according to Scott's classification of patrollers versus perchers.

With this tradition in mind,[1] I initially gathered data under the assumption that males are searching for females. However, my analyses and modeling in some cases suggest that the actively moving male is not searching, but rather is putting himself in a better position to be found by a more sedentary female. Indeed, when mate-finding is an intrinsically interactive process, the conventional logical distinctions between searcher and quarry, and even between cause and effect, are moot.[2]

## 1.3 SETTING THE STAGE: SPECIES, HABITATS, AND MATING

### *1.3.1 Choosing which Species to Study*

I started by marking any species that I could see and capture easily. Deciding which to continue studying in detail was a compromise between intrinsic interest in their behavior and feasibility of studying that behavior in my locale and using my techniques. Bulk re-sighting statistics for my first intensive field season (1974) are reported in Table 1.1. The table separates species by the general reasons for studying them or not.

Ten species that were marked (and an additional fifteen that were not marked) were rare, irregular in presence, and/or high-and-wide-flying (Table 1.1A). Of these, the Clouded Sulphur was common, but flew warily and rapidly enough to avoid capture, and high enough to traverse extensive hedgerows, so that marking seemed futile. The Green Comma, White Admiral, Red Spotted Purple, and Tiger Swallowtail were highly active in heavily wooded areas, where they were unobservable in and above the canopy. The Baltimore was tempting, given the ecological and evolutionary studies of Bowers, and the potential comparisons with its congeners (Bowers, Ehrlich, Singer, et al., as later reviewed by Ehrlich and Hanski 2004). However, I found it only near patches of its larval food plant, turtlehead (*Chelone glabra*), which is itself so patchy and ephemeral in my study area that I decided not to put any more stress on the population of Baltimore butterflies.

Six species were difficult to work with (Table 1.1B). The Bronze and American Coppers were patchy in their distribution, both seemingly associated with drier patches of soil, the Bronze in otherwise moist meadow, the American in already relatively dry upland. Both Coppers sallied out from perches and apparently chased other butterflies. Both are small and hard to transfer from net to marking clip (see the clip in use in Figure 1.6). In addition, the somewhat larger Bronze Coppers power their delicate wings with strong muscles, so I could not mark them without a risk of their damaging themselves. The Hairstreaks are more robust, are often placid when approached from behind, and tend to return to a nearby perch after disturbance. So they can often be captured directly in the marking clip. However, they typically perch on vegetation at the edge of a woodsy opening, and are almost impossible to follow in flight. Accordingly, finding Hairstreaks, and especially re-finding individuals, would be far more dependent on vagaries in my behavior than on regularity in theirs.[3] The Meadow Fritillary is the most promising species in this difficult category. It is small but sturdy, easy to mark but hard to read. However, the Meadow Fritillary was uncommon in 1974, and it turned out to be even less abundant in subsequent years.

Two species were manifestly vagrant over distances that extended beyond my study area (Table 1.1C). This was obvious from the behavior of the Cabbage Butterfly, which, though easy to capture, often flew well above the vegetation to traverse hedgerows as the Clouded Sulphur does. It took more work to show the vagrancy of the Pearl Crescent. They are small but robust and easy to mark. However, they tend to rest directly on the ground, with their wings spread out so that marks on the underside are not visible. So I often had to recapture Pearl Crescents to read their numbers, and my recaptures may have artificially extended their intrinsic vagrancy.

It was difficult to read the identifying numbers on marked individuals of two species, the Little Wood Satyr and the Wood Nymph (Table 1.1D). The Satyr, like the Pearl Crescent, often perches with its wings outspread, obscuring the marks. Furthermore, it frequents patchy woodland, where it can readily disappear into the vegetation. The undersides of the wings of the Wood Nymph are often so dark that the black marks show little contrast. The 1974 re-sighting records for both species are consistent with those of vagrants among the other species, though I often observed Wood Nymphs interacting at aggregations on patches of wild marjoram *Origanum vulgare*, where they were sipping nectar, an unusual behavior for their subfamily, the satyrines.

Although the species above were not marked and re-sighted intensively after 1974, some of them will reappear briefly for exemplary behavior or simple convenience, especially in Chapter 2, showing potential adaptations of flight and vision.

The 1974 season offered promising re-sighting statistics for the five species whose intensive study gave most of the material for this book (Table 1.1E). The Ringlet was common enough for many to be marked initially, and it had a high percentage of recaptures among the prospective vagrants. The Eyed Brown, though not so common, was sufficiently local in its behavior to ensure many recaptures. The bulk re-sighting statistics were not so promising for the large Fritillaries (mainly the Great Spangled), but the 1974 re-sightings included long flights away from a particular location and precise returns at long intervals, and one instance of an individual travelling a complex closed circuit of 200 meters, and then repeating the same circuit two more times (mapped in Figure 5.3). These observations reinforced the classical story of Fritillaries' repeating long, closed paths (e.g., Opler and Krizek 1984). The Viceroy was uncommon in 1974, but a

TABLE 1.1. Bulk Re-sighting Statistics for All Marked Butterflies in 1974 Field Season.

| Species | Marked | Re-sightings | % Re-sightings | Male Behavior* |
|---|---|---|---|---|
| *A. Rare and/or Irregular* | | | | |
| Black Swallowtail | 1 | 0 | 0 | Patroller/Percher |
| Clouded Sulphur | 1 | 0 | 0 | Patroller |
| Green Comma | 2 | 1 | 50 | Percher |
| Monarch | 2 | 0 | 0 | Patroller |
| Tailed Blue | 2 | 0 | 0 | Patroller |
| Coral Hairstreak | 3 | 0 | 0 | Percher |
| White Admiral | 4 | 0 | 0 | Percher/Patroller |
| Red Spotted Purple | 5 | 0 | 0 | Percher/Patroller |
| Tiger Swallowtail | 10 | 1 | 10 | Patroller |
| Baltimore | 13 | 4 | 31 | Percher |
| *B. Difficult to Work with* | | | | |
| Bronze Copper | 7 | 0 | 0 | Territorial? |
| American Copper | 8 | 7 | 87 | Territorial? |
| Hairstreaks (3 species)** | 44 | 13 | 30 | Percher |
| Meadow Fritillary | 20 | 22 | 110 | Trapliner? |
| *C. Widely Vagrant* | | | | |
| Cabbage Butterfly | 28 | 1 | 4 | Vagrant |
| Pearl Crescent | 125 | 9 | 7 | Vagrant |
| *D. Identifying #s Obscure* | | | | |
| Little Wood Satyr | 48 | 13 | 27 | Vagrant? |
| Wood Nymph | 101 | 35 | 35 | Vagrant/Percher? |
| *E. Intensely Studied* | | | | |
| Ringlet | 88 | 56 | 64 | Vagrant |
| Eyed Brown | 30 | 44 | 147 | Site-faithful |
| Large Fritillary (2 species)** | 73 | 20 | 27 | Trapliner |
| Viceroy | 16 | 17 | 106 | Territorial |
| Pearly Eye | 113 | 113 | 100 | Singles' Bar |

  * "Patroller" and "Percher" are characterizations from Scott 1986, and correspond to my observations. Other terms and queries are my own inferences.
  ** Species that were hard to distinguish without recapture were numbered sequentially, but are pooled in this compilation: Striped, Banded, and Hickory Hairstreaks; Aphrodite and Great Spangled Fritillary.

few individual males circulated among the same neighboring perches over several days, and the sets of perches differed for different males. Male Pearly Eyes were common, and individuals were frequently re-sighted sipping fermented sap at particular wounds on woody plants. Accordingly, these five species deserve the following detailed descriptions.

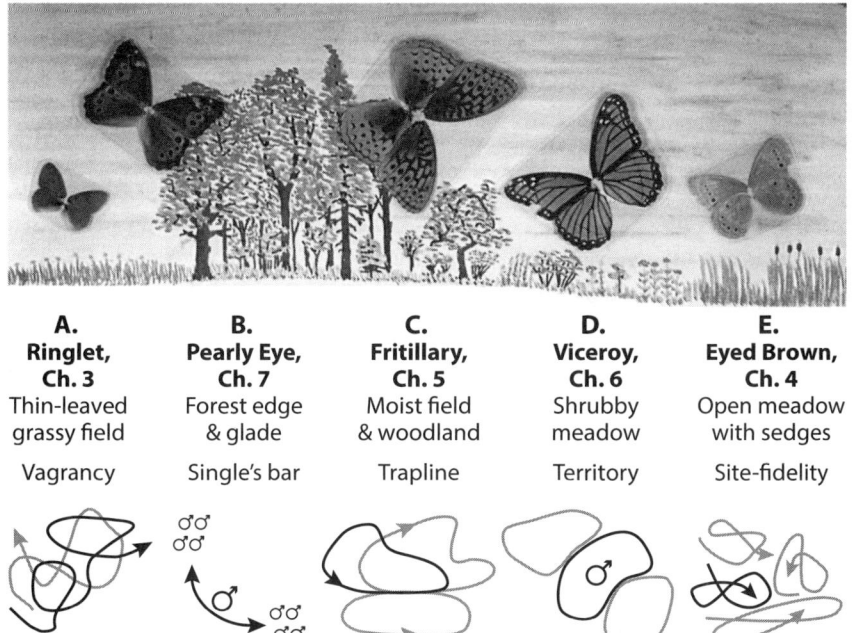

FIGURE 1.2. Habitats and behavior of my intensively studied species.

*1.3.2 The Species Studied Intensively*

The five species that I studied most intensively are the (Inornate = Plain = Common) Ringlet (*Coenonympha tullia inornata*), the Eyed Brown (*Satyrodes eurydice*), the Great Spangled Fritillary (*Speyeria cybele*), the Viceroy (*Limenitis archippus*), and the (Northern) Pearly Eye (*Enodia anthdon*). They are illustrated in Figure 1.2, along with characterizations to be discussed in the following text.

My usage of names follows Cech and Tudor (2005) and Glassberg (2017), but I shall shamelessly use shortened vernacular names in the text, specifically: Ringlet, Eyed Brown, Fritillary, Viceroy, and Pearly Eye. The only species for which this might cause confusion is the Ringlet, which is called the Large Heath in Europe (where the name "Ringlet" is given to species of the genus *Erebia*), and the Fritillary. Slight ambiguity is appropriate for the Fritillary because some of my observations of unmarked individuals may be Aphrodite (*Speyeria aphrodite*) or very rarely Atlantis Fritillary (*Speyeria atlantis*). The behavior of marked Aphrodite was indistinguishable from that of marked Great Spangled Fritillary; nevertheless, only Great Spangled Fritillaries were tallied in all analyses of marked "Fritillaries" of known individual identity. Detailed notes on Latin names, vernacular names, and behaviorally analogous European species are given in Appendix A.

## 1.3.3 Habitats

All sightings of the five studies species are tabulated by habitat in Table 1.2 and later diagrammed in Figure 8.1. "Field" means dry upland like that in Figure 1.4. "Meadow" denotes wet meadow like that in Figure 1.5. "Woodland" has a discontinuous canopy of trees above a grassy or shrubby understory. "Forest" has a closed canopy of trees.

The Ringlet favors dry upland fields, the Eyed Brown and Viceroy favor wet meadows, and the Pearly Eye favors open woodland. The Fritillary ranges over all the semi-open habitats. These are the characterizations diagrammed in Figure 1.2.

The spatial configuration of favored and avoided landscape elements interacts with the movements of butterflies in complex ways that will be explored separately for each species in Chapters 3 through 7, and then compared in Chapters 8 and 9. The four major landscape elements differ in their characteristic patchiness and contiguity. Forest with its continuously contiguous canopy is generally an extensive matrix in areas long abandoned from agriculture. Woodland is generally found along the edges of all the other landscape elements, and so it is usually composed of long narrow strips, but there may be isolated woodland glades in forest. Wet meadows are found in small patches along watercourses, and they are usually bounded and often connected by woodland. Dry upland fields are found

TABLE 1.2. Habitat of Day-to-Day Sightings of Studied Species. These data are also diagrammed in Figure 8.1.

| Species and sex | Sightings | %Field | %Meadow | %Woodland | %Forest |
|---|---|---|---|---|---|
| Ringlet | | | | | |
| male | 1,284 | **77** | 2 | 20 | 1 |
| female | 824 | **79** | 2 | 18 | 1 |
| Eyed Brown | | | | | |
| male | 298 | 9 | **72** | 19 | 0 |
| Fritillary | | | | | |
| male | 206 | **44** | 23 | **29** | 4 |
| Viceroy | | | | | |
| male | 179 | 16 | **70** | 14 | 0 |
| female | 35 | 17 | **57** | 26 | 0 |
| Pearly Eye | | | | | |
| male | 1,716 | 21 | 1 | **74** | 4 |
| female | 471 | 15 | 1 | **83** | 1 |

*Note:* Boldface numbers accumulate 50% or more of the sightings.

INTRODUCTION

in areas recently abandoned from agriculture; they may be small or large patches, variously broken up by hedgerows and wider strips of woodland. The effective contiguity of the favored habitat depends not only on its spatial distribution, but also on how a given species of butterfly behaves at the boundary with another habitat, and within that other habitat. The interactions between butterfly behavior and habitat configuration are discussed in Chapter 8 (especially Section 8.1 and Figure 8.1).

## 1.3.4 Mating

When a male encounters a free-flying female, she is a potential mate only if she is a virgin or if she is a mated female that will accept another mating. So it is important to know how frequently virgins and multiply mated females are found among free-flying butterflies, and species tend to segregate into those whose females accept only a single mating versus those whose females will mate two or more times (Burns 1968). A mating male usually passes a packet of sperm in a structure called a spermatophore, and the female dispenses sperm from these packets stored in her bursa copulatrix. Dissecting a female and counting the spermatophores in her bursa copulatrix estimates the minimum number times she has mated, subject to cautions about null matings, multiple spermatophores, and sperm competition raised by, e.g., Cordero (2002) and Platt and Allen (2001). Spermatophore counts for my intensively studied species are presented in Table 1.3.

The frequency of free-flying apparent virgins in the Ringlet suggests that mate-finding may be less efficient in this species than the others, with consequences discussed in Chapter 3.

Female Ringlets, Fritillaries, and Viceroys tend strongly to mate only once, though female Viceroys will accept a second mating in the laboratory (Platt and Brower 1968). Given the paucity of data for the Viceroy, I have compiled data from congeners (*Limenitis* spp.) that suggest that multiple matings may occur, albeit rarely. The importance of monandry for the Ringlet and Fritillary are discussed in Sections 3.4.2 and 5.2. Conversely the importance of potential multiple matings for female Viceroys in discussed in Sections 6.1 and 6.2.2.

Multiple matings are more common in the Eyed Brown and the Pearly Eye. Accordingly, for these species, any female that a male encounters may be a potential mate.

Field observations show that Ringlets and Fritillaries may mate throughout the day, but matings of Pearly Eyes may be restricted to the afternoon, with consequences discussed in Section 7.5.

TABLE 1.3. Mating Statistics

| Species | Spermatophores per female | | | Source | Number in copula | Time of day found (EST) |
| --- | --- | --- | --- | --- | --- | --- |
| | *0* | *1* | *2–3* | | | |
| Ringlet | 1 | 9 | 0 | My study | 9 | 10:00–15:30 |
| | 8 | 56 | 4 | Shields 1967 | | |
| | 2 | 40 | 2 | Ehrlich & Ehrlich 1978, W.US | | |
| Eyed Brown | 0 | 12 | 3 | My study | 1 | 14:00 |
| Fritillary | 0 | 4 | 1 | My study | 3 | 10:00–14:00 |
| | 0 | 65 | 3 | Burns 1968, VA & MD | | |
| Viceroy | 0 | 12 | 0 | My study | 1 | 10:30 |
| L. arthemis + astyanax | 8 | 169 | 36 | Platt & Allen 2001, MA* | | |
| L. lorquini | 0 | 1 | 0 | Ehrlich & Ehrlich 1978, W.US | | |
| L. weidemeyeri | 0 | 1 | 0 | Ehrlich & Ehrlich 1978, W.US | | |
| L. camilus | – | – | some | Lederer 1960, Europe | | |
| Pearly Eye | 1 | 29 | 3 | My study | 4 | 13:30–15:00 |

\* Spermatophore counts for *Limenitis arthemis* + *L. astyanax* were calculated from total sample size and percentages reported by Platt and Allen (2001) as "A.P.P. unpublished data"; the 8 virgins might actually be 9. Data are reported for other *Limenitis* species in addition to the Viceroy because the data for the Viceroy *Limenitis archippus* are so sparse.

## 1.4 FIVE SPECIES HAVE FIVE KINDS OF RANGING BEHAVIOR

Quantitative details of sample sizes, lifespans, ranging, and oviposition are recorded and discussed in Chapter 9 and Appendix B.

One needs to be careful about applying concepts like "territory" and "trapline" to butterfly behavior, especially if the behavior is to be compared with that of vertebrates. "Territory" implies recognition and active defense of a particular piece of real estate, not simply sitting at a particular place, flying away from time to time, and returning (Scott 1986). "Trapline" implies repeated visits to a number of specific sites in some recognizable sequence, and perhaps even implies that the sites have initially been recognized as promising some kind of reward. Many workers, following Scott (1974, 1986), prefer a terminology that is more neutrally descriptive of initial observations, contrasting the aforementioned "perchers," species that watch for cues of potential mates and sally out to investigate them from a fixed perch, versus "patrollers" that fly continually over the landscape, responding to such cues as fall within their radius of detection.[4] I ask you to accept the following précis with charitable faith, pending critical support in the individual species' descriptions of Chapters 3 through 7.

*The Common Ringlet* (Figure 1.2A) wanders about in dry upland fields, without any discernible pattern. Females appear to encounter sites for oviposition at random, and males to encounter females at random. Females may have difficulty finding a mate, or being found by a mate, and they tend to accept only a single mating. Both sexes are short-lived as adults, typically living less than a week.

*The Northern Eyed Brown* (Figure 1.2E) behaves like the Ringlet when at low density in sedgy meadows. However, when their density is high, male Eyed Browns appear to restrict their wanderings, and perhaps they even defend territories against each other. Females accept more than one mating. Both sexes are long-lived as adults, with some individuals observed for several weeks. This surprised me, given their structural and behavioral delicacy.

*Male Great Spangled Fritillaries* (Figure 1.2C) establish fixed circuits, which they patrol on a regular schedule, hunting for freshly emerged and virgin females. Females tend to mate only once. Mated female Fritillaries may patrol similar traplines when they lay their eggs. Males emerge long before most of the females, and both sexes live long adult lives, up to a month or so for the males. My teaching duties precluded field seasons long enough to measure the local lifespan of females.

*Male Viceroys* (Figure 1.2D) defend particular pieces of real estate that lie along paths to which females converge as they follow streams to lay eggs on young willows and aspens. There is something of a paradox in that egg-laying females seem already to have mated. This will be discussed further in Chapter 6 (especially Section 6.2.2). Both sexes live long adult lives, up to a month for the males.

*Female Northern Pearly Eyes* (Figure 1.2B) feed at tree wounds, where the leaking sap has been infected by yeast to form a nutritive brew. Male Pearly Eyes gather at such sites, sometimes tolerating each other's presence in a tight aggregation, sometimes appearing to vie for exclusive access to a given wound and the attracted females. Because females will accept more than a single mating, the females that come to a tree wound are potential mates for the resident males. Both sexes live long adult lives, several weeks to a month or so.

Thus the social behavior of these butterflies appears, at least superficially, to span nearly the full range of social systems seen in vertebrates: asociality, adaptively adjusted home range, traplining, territoriality, and resource-based coloniality.

## 1.5 MYSTERIES

I am, heart and soul, a natural historian. I revel in simple observations of the mysteries of nature. But I am also an analytical and mechanical tinkerer. I like to solve mysteries. I am often asked if my analytical understanding undermines mystery and spoils my joy in nature. The answer is an emphatic, "No!" In fact, solving a mystery at one level simply poses more mysteries for me at other levels. So here is the first mystery, followed by some of the solutions and the mysteries that they generate.

How do butterflies, with their limited sensory and integrative capabilities, achieve such complex social behavior?

A first step in answering this question is to examine the sensory and motor capabilities of butterflies (Chapter 2). There are some surprises concerning the sophistication of some of their interactions with their environment, albeit these interactions usually involve a simple response to a message that has penetrated a very specialized filter.

A second step is to describe the behavior and to probe its complexity. Simple rules of individual behavior can produce rich interactive behavior. Is complexity a direct observation or a first-order inference, . . . or is it an anthropocentric imposition? Do butterflies make decisions? Do they know what they are doing? These questions are explored for each of the five most studied species in Chapters 3 through 7. Each exploration is aided by a simple model of behavior, based on the sensory, especially visual, capabilities inferred in Chapter 2. Tentative answers are compiled in Chapter 8.

What are the effects of individual behavior on life history, and hence on local population dynamics? Conversely, do butterflies show adaptive patterns related to the dynamics of their habitats? These particular questions are posed in Chapter 9, but attempts to answer them lead beyond the purview of this book, to questions of how the behavior evolved.

If you can't wait to find more answers and questions, move right to Section 1.7, Road Map of the Chapters.

## 1.6 STUDY SITE AND METHODS

### *1.6.1 Study Site*

The study site is on 72 hectares (178 acres) of abandoned farmland 7.5 km ESE of Whitehall, New York, in the middle of the little nubbin of New York that sticks into Vermont just southeast of Lake Champlain. The habitat is (or "was" during my fieldwork) a mosaic of dry upland fields and wet meadows, varying in size

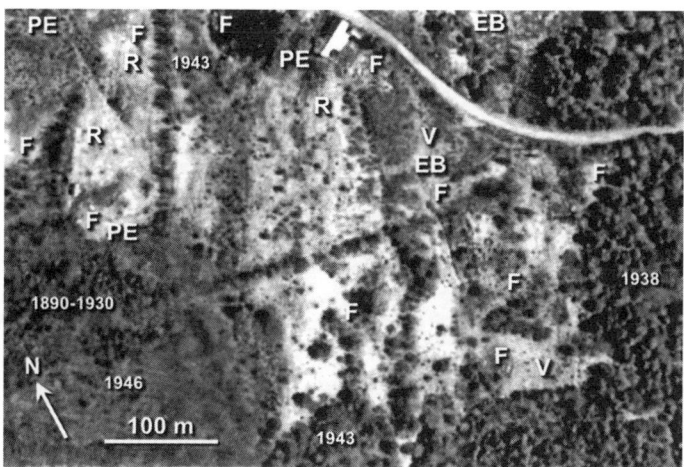

FIGURE 1.3. Aerial photograph shows the configuration of field and forest in 1973. The gray-scale photo, courtesy of the USDA's Agricultural Stabilization and Conservation Service (#8-2-90, 5 July 1973), has been hand-colored in green to emphasize forest, copses, and large trees. Also shown are birthdates of local large trees, estimated by counting the annual rings in a sample of wood taken with an increment borer. The most intensive study sites for each species are also shown as: R = Ringlet, EB = Eyed Brown, F = Fritillary (Great Spangled), V = Viceroy, and PE = Pearly Eye.

from 0.2 to 1.5 hectares and last farmed for hay in about 1963, set in a matrix of hedgerow and forest that dates from between ca. 1830 and 1935. Figure 1.3 is an aerial photograph that shows the configuration of field and forest in 1973. Also shown are birthdates of trees, estimated by counting the annual rings in a sample of wood taken with an increment borer (Stokes and Smiley 1968). Figures 1.4 and 1.5 show respectively a typical dry upland field in gently rolling terrain, and a wet meadow in a local valley.[5] Thus there are four fundamentally different habitats each with discrete boundaries: dry upland field, wet meadow, woodland edge and glade, and forests of varied ages. Butterflies are common and accessible in the fields, meadows, and woodland edges.

### *1.6.2 Methods*

Here I describe my general methods, especially those that were used uniformly for all species. Some methods are described in other chapters if their details are important to interpret results, or if they were used mainly with a single species. Technical details of physical machinery are relegated to Appendix C.

FIGURE 1.4. Typical dry upland field in gently rolling terrain (August 1974). This is the canonical habitat for the Ringlet. Pearly Eyes frequent the woodland edges of such fields, e.g., beyond the tent in the upper right. Fritillaries pass through wet and sheltered patches wherever there are appropriate flowers with nectar, or violets (*Viola* spp.), the larval food plant, below the other vegetation, e.g., a swale just beyond the goldenrods (*Solidago* spp.) on the left. The dominant plants are grasses, especially poverty grass *Danthonia spicata*, and the notorious star thistle *Centaurea maculata*.

Most of the butterflies were captured with a net. I started with the standard techniques of my youth, lying in wait near flowers or other attractants, and occasionally sprinting across open fields. However, once I began to watch the behavior of individuals more closely, and to learn the cues by which their flight was guided, I could routinely place my net where they were going before they got there.[6] I also exploited their escape behavior to their disadvantage. For example, most satyrines escape with a powerful initial upstroke, putting the butterfly into a brief "power dive" (Scott 1986); so a low sweep of the net is more likely to catch them than a high sweep. Viceroys sometimes escape with the slow flat-winged, highly maneuverable float typical of their genus; but they sometimes adopt the rapid, dihedral "V"-winged sail more characteristic of the Monarch *Danaus plexippus*, seemingly in behavioral mimicry. In the former case, netting from behind is more successful; in the latter, from in front, for the aerodynamic reasons of relative

INTRODUCTION

FIGURE 1.5. Wet meadow in a local valley (August 1981). The Eyed Brown inhabits the wettest, sedgy sites (left foreground). Territories of male Viceroys divide areas of mixed vegetation (center). Fritillaries refuel here, especially at flowers like *Asclepias* (milkweed and butterfly weed) and *Eupatorium* (Joe-Pye weed and boneset, lower left).

maneuverability discussed by Dalton (1977). On occasion I caught sitting butterflies by hand. I also used a baited trap designed by Platt (1969), with modifications for easier access suggested by Diane Wiernasz. However, the trap could not be used when and where hornets (Vespulidae) were common, because they made short work of the trapped butterflies. Learning to handle butterflies so that they are not damaged also requires close observation and exploitation of the varieties of escape behavior, and their differences among species and between sexes within a species.

My basic technique depends on marking butterflies with individual numbers, using the technique described in Horn (1976b; see also Figure 1.6 and Appendix C.1). The numbers on these butterflies can be read with close-focusing binoculars,[7] so that it is not necessary to disturb the creatures by recapturing when they are encountered at a later date (see Figure 1.7). Whenever I encountered a marked butterfly, I recorded its location to an accuracy of about ±6 meters on a gridded map. The resulting tables of "who-was-where-when" provided data for analyses of ranging patterns from day to day. Movements on a finer scale were recorded by following individual butterflies for up to an hour at a time, tracing their movements

FIGURE 1.6. Male Viceroy #83:1 "Narcissus" (5 Aug. 1983) in holding clip (Horn 1976b). A number is carefully written through a hole in the clip, by use of a black Sanford Sharpie pen. In the background are net, notebook, close-focusing binoculars (my current favorite is the *Papilio* model by Pentax), and clipboard with detailed maps of each field. These are used to record day-to-day or minute-to-minute ranging of the marked butterfly. See Appendix C for technical details of physical machinery that I used.

FIGURE 1.7. Male Ringlet #75:289 from the August 1975 flight. Even for a Ringlet, whose forewing length is only 20 mm, the number is easy to read without recapturing the butterfly.

INTRODUCTION

to the nearest meter on a fine-scale map and recording time signals every 10 or 30 seconds (see Appendix C.2 for the antique circuitry of the timer).

There are potentially complex biases introduced by the effects of capture on subsequent behavior (Singer and Wedlake 1981, Morton 1982, Orive and Baughman 1989, Smith 1995), and changes in behavior may affect population biology (Harrison et al. 1991, Ehrlich and Hanski 2004). Accordingly, I watched each captured individual after release, and I recorded any difference in behavior from that before capture. These observations help me to address biases in ways that are appropriate for each species and for each analysis. Details for individual species are in Chapters 3–7. Moreover, whenever it made sense to do so, I considered the first encounter with a marked butterfly to be the first free-ranging encounter, hours after marking, rather than the marking episode itself, in accordance with the advice of Smith (1995).

Female butterflies tend to fall into mating categories by species: strongly monandrous, mostly monandrous, and potentially polyandrous. To categorize my species, I counted the number of spermatophores that males had left in the reproductive tract of a female, adding information from my own dissections to that in the literature (Table 1.3). Many authors have urged caution when interpreting the number of spermatophores found in a dissection, and Cordero (2002) even cites and discusses instances of "mating" without transfer of a spermatophore, and of transfer of two spermatophores in a single mating. Accordingly, I treat my spermatophore counts as strongly suggestive, rather than as definitive evidence of the number of separate matings by a given female.

To view the world through a butterfly's eye, I used a fake compound eye. Details are described in Appendix C.4 and defended in Chapter 2, Section 2.3.3. It suffices here to say that this exercise emphasized the extraordinary importance of visual edges, especially the configuration of near and far horizons. So while following butterflies in the field, I would often put my eye in the place where a butterfly had been when it dramatically changed its behavior, and I would look for edges that might have relevance to the new behavior. Tentative insights were tested by creating manicured environments with new edges, and testing butterflies for predicted behavior (especially for the Eyed Brown in Section 4.3 and Viceroy in Section 6.3.1).

Some behavior was recorded on videotape. This was most successful with the Pearly Eye, some of whose interactions could be engineered to be within a focused field of view, though rapid movements and poor to highly varied lighting in their woodland edge habitat made recording difficult, especially in late afternoon and early evening. The fanciest machine that I used was a Sony 8 mm Handicam 10x Model CCD–FX411. Tedious frame-by-frame analyses were adequate for generating insights, but too feeble for public documentation and definitive testing of ideas. Accordingly, I designed and built a portable machine inspired by Dalton's (1975) laboratory-bound stroboscopic flash. Mechanical and electronic details are in Appendix C.3. The resulting photographs of butterflies in flight provide important insights in Chapter 2 and Chapter 7. Over the course of my fieldwork, and especially afterward, consumer video recorders improved steadily, and analytical techniques using microcomputers have improved dramatically in power and accessibility. Some of my currently open questions could likely be answered definitively by making and analyzing digital videos.

To analyze the cues that influence butterflies' behavior, various lures were placed in front of them (after Tinbergen et al. 1942, Crane 1955, and Magnus 1958). Butterflies themselves were used as lures by tethering them with a loop of fine thread about the thorax between forewings and hindwings (Figure 1.8). The tethered butterfly, dangling from the end of a stick, was placed in front of the subject under observation. An artificial lure was constructed from a fisherman's

FIGURE 1.8 Tethered Banded Hairstreak *Satyrium calanus*, July 1977. The hairstreaks are tough for their size and survive and behave well when tethered. So do the Fritillary, Viceroy, and Pearly Eye. Nevertheless, the procedure requires delicacy and careful monitoring. I never felt comfortable tethering a given butterfly for more than a few trials before releasing it, and I ignored data from formerly tethered individuals seen later. Worth (1980) makes ingenious use of a tiny fisherman's toggle to prevent tangling of an animal in its tether, but I had no such trouble. The Ringlet and Eyed Brown sometimes became lethargic during and after tethering. All data using such individuals were ignored.

INTRODUCTION

spinning reel that turned a small piece of cardboard at the end of an automobile speedometer cable (christened the "Butterfly Rotisserie" of Figure 7.8 and Appendix C.6). Some results are in Chapter 2, and more results and details are in Chapter 7. The paper lure could be of any color, and could be stationary or spun at various speeds. Olfactory lures were made by soaking cellulose sponges in sugar solutions and ferments (Appendix C.8 and Chapter 7). Finally, visual cues for orientation in the habitat were distorted by large mirrors that were placed so as to reflect portions on the horizon into the wrong places (Appendix C.7 and Chapter 6). Further details of these techniques are in appropriate chapters.

Mapping of movements was done at two scales of time and space, with corresponding differences in resolution. For hour-to-hour and day-to-day observations of marked individuals, an aerial photograph (Figure 1.3) was given a grid system at 36 meter intervals, oriented to hedgerows and other prominent features.[8] Additional orienting features were mapped onto the photo, either by triangulation with a plane-table or by polar-coordinate plotting, using a lensatic compass and either a 20 meter tape or a split-image rangefinder. The location of an observation was recorded in my field notebook as a letter-number for the grid-square, followed by a code for one of nine sections within the grid-square: N, E, S, or W middle third of edge; NE, SE, SW, or NW corner, or M for middle. Hence the resolution of locations of observations by hour or by day is about ±6 meters (adding my own uncertainties relative to landmarks in the field, certainly never worse than ±10 m).

Flight paths of particular individuals that were followed continuously for minutes at a time were traced on pseudo-vellum paper clamped to maps of particular fields or meadows. These maps were made with the same techniques as above, but at a much finer scale, and with much more landscape detail measured directly on the ground. I could record movements near me to ±1 meter, and more distant movements to ±5 meters; any movements beyond my confident estimation were recorded as dotted lines.

*1.6.3 Methods I Would Use Today*

None of the following technologies were available for most or all of the time spanning my fieldwork. Today I would record far more with digital camera and high-definition video.[9] I would make my maps as I originally did, but aided by satellite imagery, global positioning systems (GPS), and geographical information systems (GIS). I would take my own low-altitude photos to fill in fine-scale maps of fields and meadows, using a drone mini-helicopter. Of course, the maps would still be checked for scale, orientation, parallax, etcetera with measurements on the ground of distances between mapped landmarks, including my own survey flags.[10] I would record the positions of hour-by-hour and day-by-day captures as GPS coordinates, and I would record my own movements as continuous GPS tracks. I would trace moment-to-moment movements of individual butterflies onto a map on the touch-sensitive display of a digital tablet or laptop computer, carried on the mini-table described in Appendix C.1.

### 1.6.4 Analyses and Statistics

Quantitative data are the foundation of modern scientific orthodoxy. Accordingly, here is a statement of personal doctrine. My main analyses of population parameters are ad hoc inventions. It is not that I am lazy, or ignorant, or disdainful of other approaches. I have tried to follow individual butterflies as well as populations, and so I am acutely aware that the space that they occupy is highly patchy, and that even the patches are internally inhomogeneous (see Figures 1.3, 1.4, 1.5, and 8.1). Yet all the traditional mark-and-recapture analyses make a fundamental assumption of homogeneous mixing of marked and unmarked individuals, even those analyses that admit variations in behavior within the lifespan of individuals, among individuals, or among trapping sites (Wilbur and Landwehr 1974, Pollock et al. 1990, Amstrup et al. 2005). So heterogeneity of habitat usually precludes my use of traditional analyses.

The most nearly homogeneous habitats are the dry upland fields that are home to the Ringlet, though even here there are differences in slope, exposure, shape of field, and types and densities of grasses, forbs, shrubs, and saplings (Figure 1.4). Accordingly, I initially calculated population sizes for the Ringlet using the Manley-Parr version of the Jolly-Seber index, from the recipes given by Southwood (1978). The calculated variance is so large, however, that the mean minus the standard deviation is often lower than the number of marked individuals known to be alive, which must perforce be an absolute lower bound on the population size at a given time (see also a similar warning in Amstrup et al. 2005, pp. 46–55). Wiernasz (1983), facing the same problem with data from the same population, found more realistic seasonal patterns in the number of individuals known to be alive than in numbers from Manley-Parr and especially Jolly-Seber calculations. I follow Wiernasz's lead, and present and interpret patterns in the number of individuals known to be alive, rather than mark-recapture calculations. Similar arguments are given for ad hoc analyses of ranging and local survivorship in Chapters 3 and 9 respectively, as the data do not support assumptions needed for fancier analyses (such as those reviewed, e.g., by Turchin 1998). In particular, for a population that is intensely sampled, aging from initial capture may be preferable to correcting with guessed age by wear, because the disruption of the life-plan by capture may start a disturbed butterfly anew (Smith 1995). Of course, I have excluded from demographic calculations individuals that were initially captured in highly worn condition. Demography is based on numbers of individuals known to be alive, i.e., marked butterflies observed on a given date or a later date. Losses before the next sampling date include both deaths and permanent emigrations, so demographic patterns are literally changes in local residency rather than survivorship. Nevertheless in Chapter 9, I often use the language of "survival" for four reasons: (1) Individuals that disappear from the local population are lost to its subsequent dynamics whether they died or emigrated, (2) I suspect that many of the losses are indeed due to death, (3) I use traditional demographics of survival to estimate parameters of local residency, and (4) Phrases like "local survival" and "local lifespan" are rhetorically useful and technically correct.

My attitude toward statistics is like that toward general analysis. I would rather do a simple test with a full understanding of its weaknesses than a fancy test that fits many appropriate special conditions, and yet makes a fundamental assumption that just isn't so. In many cases I am describing a new empirical observation, rather than testing an a priori hypothesis. Accordingly, the statistical question that I am asking is whether the data that I have gathered are sufficiently copious and sufficiently concentrated that I am unlikely to have been fooled by a statistical fluke. Many tests simply ask whether my observations depart further from a 50:50 outcome than would an appropriate number of tosses of an unbiased coin 95% of the time (the "binomial test," e.g., Bailey 1995). Binary categorical data are tested for independence to the same criterion by a 2×2 $\chi 2$ test if the data are copious, or a Fisher (1958) exact test if the data are few (Gotelli and Ellison 2013). I often plot or report the mean of a variable ± the standard error of its mean, as recommended by Finney and Harper (1993). The point at which two ranges reported in that way would overlap is very near the point at which they would just be judged not to differ by a one-tailed t-test at the 0.05 level of significance, given a large number of measurements, say 30. I also report proportions based on 30 or more observations with ± their standard error (Bailey 1995). My data fail nearly all the tests that James and McCulloch (1990) propose for the valid use of multivariate analysis in ecology. The bivariate relations that I have found are empirical descriptions, rather than tests of hypotheses. Their statistics have all the usual strengths and weaknesses associated with inductive ad hoc hypotheses (cf. Mentis 1988, and Gotelli and Ellison 2013).

Of course, where I am testing a specific hypothesis with a designed experiment, I interpret my statistics in the standard way. And I try whenever possible to plan the sequence of experiments with a table of random numbers. Fisher (1935, p. 13ff.) makes an elegant case for the power and necessity of randomization, describing a test of a lady's assertion that she can tell whether milk has been added to hot tea or the tea has been added to the milk.[11]

A further problem arises when trying to accumulate a large number of behavioral observations in nature from a relatively small number of butterflies, not all of them marked. Some individuals may be observed several times, while others are observed once. The repeated data from one individual are not "independent" in the same sense as data from separate individuals. This problem, a kind of "pseudo-replication" (Hurlbert 1984), is a recondite one, with ramifications from the behavioral level (Catchpole 1989) to the landscape level (Hargrove and Pickering 1992). My approach to the problem is perhaps more facile than it deserves, but the practical limitations on my observations leave little choice. I treat each observation as independent, even for some repeated observations on a single animal. For example, I may treat repeated observations on a single animal as independent when the behaviors are separated by a different kind of behavior, so that they are recurrences rather than continuations. So movements are generally accumulated for an individual, and a standard error of the mean (SEM) will count the number of individuals (see, e.g., the "velocities" of Table 9.1). But daily survival is treated as independent from day to day, so that its SEM counts the number of individual-days (see, e.g., Section 9.3). Technically my inferences do not extend to a population of organisms such as I have observed, but rather to a population of observations such as I have gathered. In a formal sense my inferences are limited to a small universe of discourse, namely my observations, which are already biased by the patchiness of the environment. I make tactical interpretations based on the empirical structure of my observations as raw data, rather than on a calculation that postulates an idealized variance based on theoretical assumptions that I know to be false. I prefer this to the easier course of using fancy conventional statistics, giving appropriate citations, and having no idea of the effect of false fundamental assumptions on their meaning.[12]

Some inferences are drawn from observations of very few individuals, perhaps even only a single one. Statistical inference when $N = 1$ can be valid for a statement of the form: "This behavior CAN happen, because it DID."

### 1.6.5 Models

Verbal descriptions of some behaviors were made more explicit as models, using the NetLogo language (Wilensky 1999, http://ccl.northwestern.edu/netlogo/), which represents individual animals as agents which can move over distances and in directions that are specified functions of the distances and directions to other agents, which can be other animals or objects, or properties of the local environment.[13] NetLogo models are proposed to simulate the search paths of Ringlets (Tactical Vagrancy Model in Chapter 3), to caricature the local movements of

INTRODUCTION 23

Eyed Browns (Fortuitous Site-Fidelity Model in Chapter 4), to contrast with the ranging of Fritillaries (contra the Fortuitous Site-Fidelity Model from Chapter 5), and to illustrate the results of proposed rules of interaction between and among Pearly Eyes (Fight-or-Follow Model in Chapter 7). Where the models mimic real behavior, they lend plausibility to interpretations; where they depart from reality, they suggest that something else is going on.

Some of my physical machinery, specifically the artificial compound eye, the artificial horizon, and rotating artificial visual lures, are "models" in the same logical sense. They can also either make an interpretation more plausible, or suggest that more work is needed.

My primary data are intrinsically spatial, maps of individual movement from day to day or from moment to moment, and my interpretations involve outright geometry. Accordingly, my models and conceptual demonstrations are geometrical, rather than algebraic.

## 1.7 ROAD MAP OF THE CHAPTERS

These methods and others are used in Chapter 2 to explore some physiological and behavioral adaptations for orientation, movement, and mate-finding. One speculative discovery is that complex environmental and social cues can be encoded and filtered into potentially simple messages. Another is that some aerial interactions might be mediated by puffs of air that butterflies throw at each other. Chapter 2 explores the physical and physiological underpinnings of observed behavior. The behavior itself is observed fact, the physics and physiology are interpretation. Physicists and physiologists would have it the other way round, and might view my incursion into their fields skeptically, especially given my primitive and outdated machinery, rudimentary statistics, and naive extrapolations.

So particularly in the Summary of Chapter 2, but in the subsequent summaries as well, I try to distinguish results in plain text versus *speculations in italics*. The *speculations* are presented as topics for further research. Each chapter summary also includes a retrospective reference to the sections of that chapter in which each result *or speculation* is introduced or discussed, along with key words from the subtitle of the book to highlight Cues, Rules, Behavior, and Life History.

Chapters 3–7 are studies of the five species, in an order that corresponds roughly to increasing perceived complexity of their ranging and social interaction. Each chapter starts with a descriptive account of the species' natural history and behavioral idiosyncrasies, particularly as I have observed them in my study area. A description of ranging behavior and interactions, emphasizing males, is illustrated with a map of the behavior of one or more individuals. For each species, a minimal model is proposed to explain a characteristic aspect of its behavior, and the search is on to find aspects of behavior that go beyond the model, be they empirical result *or speculation.*

In Chapter 3 the Plain Ringlet is found to have a large component of randomness in its ranging. A simple Tactical Forward Vagrancy Model, based on Skellam's (1951) classical work, confirms this, and prompts a theoretical calculation of how many mateable females a male will encounter if he spends all his thermally permissible adult lifespan flying around and looking. The number is surprisingly small. Adult male Ringlets have a short and demanding life, and it is conceivable that they are not so much searching for females as actively putting themselves in positions to be found by stationary females.

Site-fidelity of the Eyed Brown is described in Chapter 4. Males fly along distinct channels in the vegetation, so that their site-fidelity is partly a direct response to the network of such channels in their habitat. A Fortuitous Site-Fidelity Model shows that even a random network of channels might produce locally confining domains, and there would be no need to postulate territoriality as a mechanism to produce site-fidelity. Nevertheless, male Eyed Browns may recognize and defend individual pieces of real estate.

Chapter 5 argues that Great Spangled Fritillaries establish and patrol traplines. Although no individual has been followed around a large circuit, one male was watched during the initial stages of establishing a regular route. Many males have been observed in scheduled local repeats of segments of wider travels. Their behavior is compared to some simple but elegant insights of Anderson (1983) into the famous "traveling salesman problem." The females are "selling" their eggs to dispersed larval food plants, and the males are "selling" their sperm to females, which apparently wait near their natal sites to mate monogamously. The generalized traveling salesman problem, finding an optimally efficient tour among sites, is notorious for being intractable, even insoluble in reasonable time (Cook 2012). And adding more salesmen further complicates the problem (Fekete et al. 2004). However, competition can favor specialization on a local neighborhood, decreasing the number of sales-sites to allow a tractable near-optimization. Whether Fritillaries achieve such optimization is an open question, but I can at least estimate plausible parameters for the task. The fact that different males follow different but overlapping circuits violates an explicit prediction of the Fortuitous Site-Fidelity Model of Chapter 4, strongly implying that the individual circuits are more than fortuitous.

Territoriality of the Viceroy is described in Chapter 6. Males use aerial tactics, like those of World War I fighter pilots, to defend territories on flyways frequented by ovipositing females. This behavior is examined with mirrors, rather than with mathematical models. The mirrors are used to create artificial flyways, to which males then orient their observations.

The Pearly Eye of Chapter 7 is the most extensively studied species. Its habit of congregating at slime fluxes on trees and shrubs made it easy for me to foster interactions with artificial baits and lures. The Pearly Eye displays all the rich details of social interaction shown by Davies (1978) for the Speckled Wood Butterfly: male defense with home court advantage against other males, escalation in even contests, and sexual chases of female by males, including conga lines of males pursuing a single female. Davies' study rapidly became a classic example of game-theoretic assessment and optimal behavior (Maynard Smith 1978). Davies' observations and experiments have been given alternative interpretations by Wickman and Wiklund (1983), based on the thermal effects of basking in their favored forest sunflecks. I have found that much of the superficially complex behavior, at least for the Pearly Eye, can also be interpreted with an extension of simple principles proposed by Scott (1974)—namely, males investigate butterfly-like objects, but females avoid or ignore them. Like Davies, I believe that much more is going on, and I try to give plausible, albeit not definitive, evidence supporting this belief in Chapter 7. Scott's simple principles, codified as a NetLogo Fight-or-Follow Model, generate visual output that is frighteningly similar to videotapes of real interactions between Pearly Eyes.

Chapter 4, Chapter 5, and Chapter 7 present models in which superficially complex behavior derives from slavishly mechanical responses to specific stimuli. Conversely, Chapter 2 emphasizes how complex stimuli might be encoded as simple messages.

In Chapter 8, I emphasize departures from the models of the preceding chapters, to provide strong but indirect evidence that some butterflies make decisions. I also open, but do not altogether resolve, the question of whether butterflies know what they are doing. Chapter 8 uncovers interactions of butterflies with the landscape as well as with each other. Accordingly, behaviors of my species may be different in landscapes whose spatial configuration of habitats is different from my study area.

Chapter 9 compares and contrasts all five species to explore how individual behavior produces important demographic consequences at the scale of the local population. An important feature of Chapter 9 is the use of the same analytic machinations for all five species, an attempt to show that the differences among them, documented in Chapters 3–7, are not simply consequences of my telling qualitatively different stories.

Finally, Chapter 10 speculates about the adaptive significance of the differences among the species, and adds to previous cautions about purposive interpretation of butterfly behavior. Comparisons and contrasts may provide an example of Southwood's (1977) suggestion that localized behavior is adaptive to stable habitats and vagrancy to ephemeral habitats. I end with a summary of what my whole study has firmly established, and a list of intriguing but unanswered questions.

## 1.8 PROSPECT

Here is what I intend to show in this book. Some aspects of the physiology of butterflies underlie simple cues and simple behaviors (Chapter 2) that can generate the appearance of complex interactions (Chapter 8). Five species show variations in behavior that span much of the range of social behavior seen in animals with greater sensory and integrative sophistication. These variations are directly apparent when appropriately different kinds of data are gathered for the different species (Chapters 3–7). Gathering consistent data and treating them in a common way for all species confirms their differences, albeit indirectly (Chapter 8). The question of whether butterflies actually make decisions is exposed to highly skeptical analysis. Some strong, circumstantial evidence of residual complexity survives the skeptical mechanical analysis, but also suggests further mechanical inquiry (Chapter 8). And the differences in ranging behavior have important consequences for life history and thus for local population biology (Chapter 9).

No one of these points is totally original. What is original is the combination of them in a single study that uses similar methods on several species. A particular novel feature of my study is that, while it is firmly centered in what might be called behavioral ecology, it is solidly bookended by physiological ethology at the beginning and by population biology at the end.

CHAPTER TWO

# Morphological and Perceptual Adaptations to Habitat and Society

*Bases for Cues and Rules of Behavior*

> Nature-study is, despite all discussions and perversions, a study of nature; it consists of simple, truthful observations that may, like beads on a string, finally be threaded upon the understanding and thus held together as a logical and harmonious whole.
> –Anna Botsford Comstock, 1911, p. 1,
> *Handbook of Nature Study*

My overall goal is to observe and record the natural history of five species of butterflies, and to explore how simple cues and rules can help to understand complex behavior that has consequences for the local population. This chapter explores motor and sensory capabilities of butterflies that adapt them to appropriate habitats, help in their choice of those habitats, and facilitate their finding potential mates. In this chapter, I try to signal those "results" that are speculative by expressing them in permissive and subjunctive language. More direct and indicative language suggests confirmed or newly established facts.

I start by redescribing the mate-finding behavior of males from Chapter 1 in terms of the information and maneuvers needed to carry out that behavior. I then set my own observations, measurements, and experiments into the context of the past and current work of others on mechanisms of flight and vision, with brief notes on olfaction and hearing. Many of my insights are derivative, but some are new. The new insights may not be definitive, but I hope that they are sufficiently plausible to prompt further inquiry. And they are consistent with details presented in the following chapters; so they might even be true. I make no claim that my sensory observations and interpretations represent the sole modalities behind the cues and rules of behavior proposed in other chapters.[1] I only argue that they are plausible, sufficient, and within the demonstrated capabilities of butterflies. Accordingly, they are worth further exploration.

## 2.1 BEHAVIOR NEEDED FOR MATE-FINDING

### 2.1.1 Choice of Habitat Interacts with Mate-Finding.

The physical constraints on mate-finding are complicated. The habitat may be structurally and visually simple OR complex. A moving OR stationary male must find a moving OR stationary mate, with high OR low visual (or olfactory) contrast, against a simple OR complex background. And male and female must inspect each other, first to detect whether they are indeed male and female of matching species, and then to determine whether to mate. Note that all aspects of the mate-finding task involve an interaction of both flight and vision. Furthermore the five main binary alternatives, the "OR"s just listed, could produce a factorial total of 32 potential combinations—or more realistically, 24, because if neither male nor female moves, nobody finds a mate. The only alternatives that can be intuitively evaluated are contrast and background; it is easier to detect a high-contrast object against a simple background than a low-contrast object against a complex background. The rest require some details of movement and perception, i.e., details of interaction between flight and vision.

The male Ringlet is a patroller, spending most of its thermally permissible lifespan on the wing, with brief stops to tank up on nectar for energy. It is restricted to open, grassy habitats that are clearly distinguishable by their luminance from shady woods and shrubland. As will be shown in Chapter 3, a Ringlet can find its way to and around its grassy habitat by the simple rule of "run to daylight" (cf. Lombardi and Heinz 1963). Chapter 3 also suggests theoretical models by which a male can efficiently search its habitat, with the different models incorporating different mechanisms of visual orientation. When it comes to actual encounters, however, the male is "searching" for a stationary low-contrast target against a complex (grass) background, whereas a perched female might detect a moving high-contrast male against a simple (sky) background. In fact, a sitting female rose to encounter a flying male in all three of the completed courtships that Diane Wiernasz or I actually observed from encounter to copulation. So the male may not be "searching" so much as coursing over the habitat to put himself in a position to be "found" by a stationary female. This nontraditional interpretation is expanded in Chapter 3 (Section 3.5).

The male Eyed Brown is also mainly a local patroller, along channels in sedges, with some perching along the channels. Such channels are characterized visually by V-notches in the very near horizon. As will be shown by a simple Fortuitous Site-Fidelity Model in Chapter 4, following such channels can automatically result in individual site-fidelity, without the need to recognize particular pieces of real estate.

The male Fritillary also patrols, but over much greater distances; individuals establish routes that they repeat with greater or lesser fidelity. Chapter 5 shows that parts of the routes of several individuals overlap in commonly used "alleys," which are characterized by V-notches in the distant horizon.

During the afternoon and sometimes in the morning, the male Viceroy perches on emergent vegetation from which it orients toward low spots (effectively V-notches) in the distant horizon. It darts out in response to laterally moving objects that contrast sharply with the background sky. In Chapter 6 I argue that intruding males are actively driven from a particular piece of real estate, to which the "territorial" male returns to perch on the original spot or on some other emergent plant. During its return, and sometimes spontaneously from a favored perch, the male may briefly patrol, visiting emergent plants, and settling either briefly or for a long period. In the morning if not on "territory," males can often be found on fermenting slime fluxes, especially on willow (*Salix*). Many details of Viceroy behavior that are described here are explicitly tested in Chapter 6.

The male Pearly Eye has the most complex behavior, perching mostly on tree trunks at or near fermenting slime fluxes, darting out to inspect fluttering objects, and returning, sometimes with local patrolling characterized by helical inspections of nearby tree trunks. The male Pearly Eye also shows individually variable behavior that could be caricatured as "inquisitive," flying out to inspect any aspect of its visual environment that changes, and potentially "skittish," launching additionally in seeming response to vibrations in nearby air or substrate. Inferred aspects of visual ability in this chapter are used in Chapter 7 to argue that much of the Pearly Eye's behavior is interpretable in terms of mechanical responses to simple cues, with no need to posit social strategies and tactics. However, the present chapter also suggests that the male Pearly Eye has sufficient visual information to make "strategic decisions" about the superficial equivalents of chasing females versus brawling with other males at singles bars.

*2.1.2 Visual Cues Can Induce Inspection by Males.*

The initial stages of encounter between male and female appear to be mediated by vision. The cues vary from species to species, and may include simple aspects of flutter, color, and relative motion (cf. Tinbergen et al. 1942, Crane 1955, Silberglied 1977). My own evidence for this comes from observations of perched butterflies flying directly out to an appropriate object, and from casually flying butterflies turning suddenly to fly directly toward an appropriate object. Many instances were ad hoc observations of natural encounters, mostly with conspecifics and other butterflies, but some with potential predators like dragonflies and birds, and some with natural objects that resembled the species in size and color. I induced some interactions by presenting a tethered conspecific (see Figure 1.8 in Chapter 1), or a rotating artificial lure in what I call the "Butterfly Rotisserie" (after Magnus 1958; see Appendix C.6 and Figure 7.8). Females of all species typically ignored or avoided tethered butterflies, moving lures in the Butterfly Rotisserie, and other moving objects. Males usually responded with an approach, but sometimes seemed to enter an unresponsive state. Males of the Pearly Eye were the most varied in this respect, with identified individuals being generally more or less responsive, and with most individuals becoming habituated to the Butterfly Rotisserie after a few responses in a given session (see Sections 7.5 and 7.7 for more about their individualistic behaviors). Otherwise, reaction distances of perched butterflies to artificial lures were clear and representative. Reaction distances of flying butterflies are problematic if the lure is presented along an already established flight path; accordingly, I counted turning reactions only when a lure was presented to the side of the flight path. This will underestimate the true reaction distances if reaction is forward-biased.[2]

Statistics of the distances at which such interactions began are tabulated for males of the study species in Table 2.1, and shown graphically in Figure 2.1. The table and figure compile spontaneous interactions and those induced experimentally by presenting an actively fluttering living lure or a rotating artificial lure in the Butterfly Rotisserie. "Crucial cues" were inferred by noticing that responses

TABLE 2.1. Interaction Distances for Male Butterflies

| Species and Number of Observations | Type(s) (Details in text) | Mean interaction distance ± standard error (cm) | Maximum observed interaction distance (cm) | Typical wingspan of potential mate (cm) | Mean angle subtended by potential mate (minimum angle) | Crucial cues (Details in text) |
|---|---|---|---|---|---|---|
| Ringlet 38 | natural, tether | 26 ± 1 | 75 | 3 | 7° (2°) | flutter |
| Eyed Brown 55 flutter | tether, rotisserie | 96 ± 6 | 300 | 5 | 3° (1.1°) | reflectance, color, |
| Fritillary 30 | natural | 147 ± 10 | 300 | 8 | 3° (2°) | color |
| Viceroy 55 | natural | 167 ± 13 | 700 | 7 | 2° (0.8°) | lateral movement |
| Pearly Eye 115 flutter | natural at baits, tether | 72 ± 4 | 250 | 5 | 4° (1.4°) | reflectance, color, |

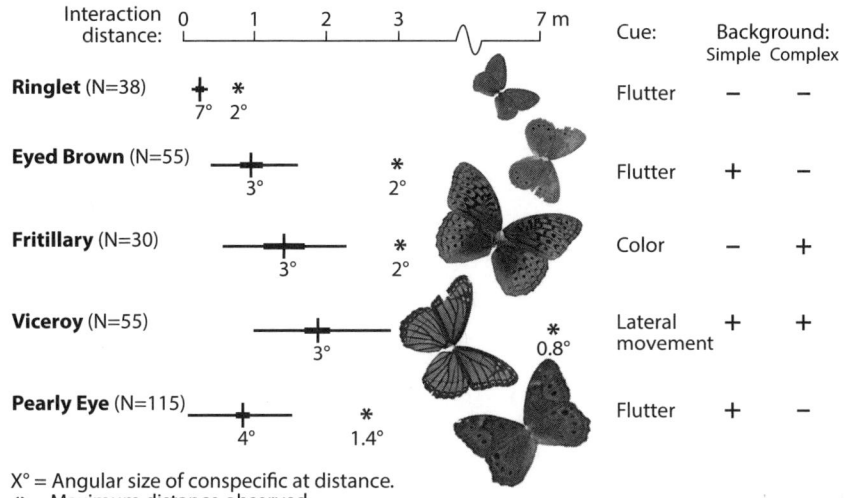

X° = Angular size of conspecific at distance.
✱ = Maximum distance observed.

FIGURE 2.1. Distances at which perched males respond to a conspecific-like object. Whisker plots give the mean distances at which perched males respond to a conspecific-like object, ± the standard error of the mean, and ± the standard deviation of the distance. Asterisks show the most extreme response for each species, and the angle subtended by the object in the visual field of the subject is given at each plotted distance. The poorest performer is the Ringlet; in addition to being small, flying males are looking for a low-contrast, grass-brown female against a complex background of patchy grass. The champion performer is the Viceroy; perched males are looking for a laterally moving female, contrastingly backlit, against a simple background of sky (see also Figure 6.4, which shows that the extreme for the Viceroy is on a long, thin tail of the distribution of response distances).

of males to a stimulus were associated most often with the tabulated factor: "flutter" means an object that was active at about the frequency of the wingbeat of a conspecific; "reflectance" means that the object was about the shade of a conspecific; "color" means an approximation of orange for the Fritillary and of light brown for the Eyed Brown and the Pearly Eye; and "lateral movement" means moving across the field of view for the Viceroy.[3] The distance of response was less than a meter for the smallest species, the Ringlet, in the order of 1–2 meters for the Eyed Brown, Fritillary, and Pearly Eye, and 2–7 meters for the Viceroy. The Ringlet fits into the pattern of lesser visual acuity for a smaller-sized eye, documented by Rutowski (2000), but it is also affected by the potential need to distinguish an object of low contrast (a grass-brown female) against a complex background (sunlit brown grass). The prodigious performance of the Viceroy is in part due to its choice of vantage points so as to perceive the contrasting darkness of a moving conspecific against the uniform brightness of the sky. Dividing the typical wingspan of a conspecific by the reaction distance allows estimation of the angle subtended by the quarry when detected. This is as little as a degree or two, which as I shall argue below (Section 2.3.1) is the angle spanned by one or very few ommatidia of the compound eye, very near a butterfly's absolute limit of visual acuity. Again the lower mean angular performance of the Ringlet is due in part to its small size and in part to the complexity of its task. The performance of the Viceroy is perhaps enhanced as its quarry moves laterally past the fields of view of several ommatidia. Insights from Table 2.1 will be augmented in Section 2.3.4 below.

### 2.1.3 Skills Needed for Mate-Finding

The mate-finding tasks described separately above for each species can be summarized as a list of behavioral skills required of flight:

- Different abilities to accelerate and maneuver.
- Efficient flight in the open and/or in cluttered habitats.

... and of rules requiring visual cues:

- Ability to choose appropriate habitat.
- Adequate vision over the range of habitats occupied.
- Orientation toward horizons, especially V-notches.
- Enhanced detection of objects of interest, and movement toward them.
- Relative ability to distinguish objects:
  near vs far, moving vs stationary, while moving vs staying stationary.

So the rest of this chapter will consider flight, vision, and briefly olfaction as ways of attaining those behaviors, rules, and cues, and then summarize the insights that will be applied in later chapters to support cues and rules proposed to underlie the behavior, and ultimately the life history of each species.

## 2.2 FLIGHT

### 2.2.1 Wing Morphology

I confess at the outset that the following analysis is unsatisfying. It is partly consistent with previous literature on mechanisms of butterfly flight and how they can adapt different species to different habitats and behavior, but there is considerable ambiguity in just what the proper adaptations might be. In particular, some very different behaviors might entail similar morphologies for achieving different modes of flight. On the positive side, some very primitive photos and videotapes of free-flying butterflies suggest potential mechanisms of flight that have been underexplored, and that do seem to differ adaptively between species that behave differently in their characteristic habitats.

Several studies have determined that wing morphology differs adaptively for constantly flying patrollers versus episodically accelerating perchers (Betts and Wooton 1988, Wickman 1992).[4] In general the wings of patrollers are broader in relation to length and more rounded than the narrower and more pointed wings of perchers. The adaptive interpretation invokes sailing, or even "slithering" (q.v. in Section 2.2.2 below), through relatively still air for the patrollers, versus decreasing aerodynamic drag and efficiently shedding vortices in accelerating maneuvers as perchers dart out and inspect objects. However, there is considerable overlap in the morphologies of the two behavioral categories, even for multivariate analyses of aerodynamically nuanced parameters such as first and second moments of wing length and of wing area about the fulcrum on the body (Betts and Wooton 1988). Part of the overlap is among species that mix patrolling and perching, and perhaps part is due to phylogenetically conservative morphology in related species that differ in behavior (Wickman 1992). But there is some ambiguity in aerodynamic requisites as well, because in order to fly in mildly turbulent air, a sailing patroller may need the same kind of narrow pointed wing for low drag and efficient vortex-shedding, as is required for the rapid maneuvers of a darting percher.

When I separate the patrollers and perchers among the common species at Whitehall, NY, the above differences are a trend, but the exceptions are equally interesting. Most of the larger patrollers have broad, rounded wings (Figure 1.1A in Chapter 1). The wings of the Sulfurs, the Tiger Swallowtail, and the Monarch are narrower and more pointed, and these species tend to continue patrolling in windier conditions. The forewings of the Red Admiral are mildly concave, and it flies rapidly and erratically in its patrolling. Conversely, the wings of the larger perchers, especially the forewings, are relatively narrow and pointed, with a concave outer edge (Figure 1.1B in Chapter 1). The wing shapes of small species do not parse cleanly between patrollers and perchers (Figures 1.1A–B in Chapter 1). The species that Scott (1986) characterizes as mixing patrolling with perching have a variety of intermediate wing shapes that correspond to my own characterization of their major activity (Figure 1.1C in Chapter 1). In particular, the Pearly Eye, which I would call a percher, because its patrolling is confined to short explorations on the way back to its perch, has the putatively adaptive narrower and more pointed wings, compared with the other satyrines (i.e., the Eyed Brown, Ringlet, and Wood Nymph).

In addition to the variable requirements that mate-finding places on flight, avoidance of aerial predators requires rapid maneuvers of all species, and hence limits the extent to which patrollers can be ideally adapted to efficient patrolling. Chai (and Srygley 1990, and Marden and Chai 1991) found that palatable species had relatively more mass in thoracic muscles than unpalatable species. I have not looked for such important differences among my species, but note only that aerial predation is a constant threat to all of them.[5]

In summary, the wings, especially the forewings, of accelerators, rapid fliers, and gliders are narrower in relation to their length than those of slow flutterers. These differences have extensive implications for behavior. Strong anecdotal examples are given in Sections 2.2.4–2.2.6 and in later chapters for searching, interacting, avoiding predation, mimicry, and the effect of damaged wings on survival.

*2.2.2 Dynamics of the Wingbeat Cycle, Vortices, and Postulated "Slithering"*

Although there is a general pattern of broad rounded wings for slow-flying patrollers and narrow pointed forewings with concave outer edges for accelerating perchers, a closer look at the aerodynamic mechanics of flapping flight is needed to gain insights into more detailed behavior. Details of wingbeat patterns were recorded with stroboscopic photography, using a homemade, portable version of the pioneering machinery of Dalton (1975). Technical aspects are described in Appendix C.3. Butterflies are released from a dark box and fly toward the light, breaking a beam of light to fire 5 flashguns in sequence, usually at intervals of 20 milliseconds.

The following description caricatures initial mechanisms proposed by Weis-Fogh (1973), further mechanisms and elegant experiments from Ellington (1984a, 1984b, 1995; see also Kingsolver 1985b), and summaries, photographs, and other graphics in Brackenbury (1992). Figure 2.2 portrays a cartoon butterfly launching into flapping flight and a stroboscope photo of a female Cabbage White (*Pieris rapae*)[6] in successive stages of its launch. As the butterfly launches, its wings, which are initially together vertically, peel apart; air rushes in and down between them, and the butterfly rises in reaction to the air movement (Figure 2.2A). As the wings flap downward, a rolling vortex slides down the leading edge to be shed at the wing tip; a ring vortex is shed down and back, and the butterfly moves up and forward in reaction (Figure 2.2B). As the wings clap together at the end of the downstroke, air is expelled from between them in a ring vortex and the butterfly is buoyed up in reaction (Figure 2.2C). The following upstroke generates a ring vortex, shed backward and slightly upward, propelling the butterfly mainly forward

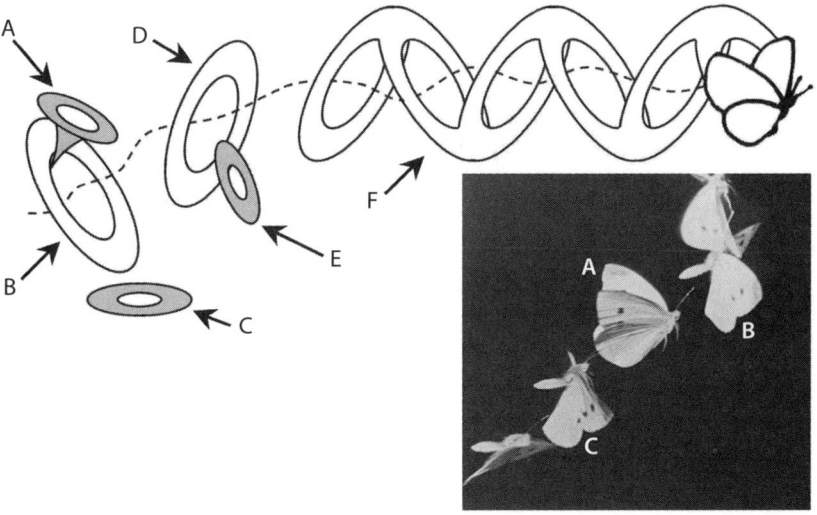

FIGURE 2.2. Idealized vortex ring wake of a talented butterfly (after Ellington 1980 and Brackenbury 1992) and strobe photo of a launching female Cabbage White. The internal flows of all illustrated vortices are directed more or less downward (generating lift) and/or backward (generating forward propulsion). For launch (A–B), lift and propulsion are generated as wings peel and/or fling apart to funnel air between them (A); then on the downstroke a rolling vortex is generated at the leading edge of the wings and slips along that edge to be shed continuously at the wingtips (B); and the downstroke ends in a clap that ejects the remaining air between them as a downward-directed vortex ring (C). The following power upstroke (D–E) sheds another vortex (D) from the wingtips; and may end in a clap ejecting an additional ring vortex (E). A change in body orientation and stroke plane ensures lift and propulsion on both downstroke and upstroke, but with different balances of lift versus propulsion. Forward flight (F) may be maintained by shedding a connected series of ring vortices from the wingtips, essentially "fanning" the air backward, or "slithering" through it. See Sections 2.2.2 and 2.2.4 for more details, including variation in the extent to which different species use the different modes of vortex ring production in different settings.

(Figure 2.2D). If the upstroke ends powerfully enough, air between the wings can be expelled in another ring vortex, which, if the body is tilted sufficiently upward, can be directed backward and down, sending the butterfly forward and up (Figure 2.2E). This up-clap has been noted in passing by others, but I think that it is a crucial feature of the flight of many satyrines, and I shall comment further on it below. Finally, the butterfly may enter a cyclic flapping in which the wings do not clap on either the downstroke or the upstroke (Figure 2.2F). This cyclic flapping sheds a connected set of vortex rings, two for each cycle.[7] This vortex wake is generally directed backward, and the butterfly moves forward in momentum-conserving reaction. Srygley and Thomas (2002) have physically visualized the vortices produced by a Red Admiral (*Vanessa atalanta*) in free forward flight. In particular they have shown the potential power in the clap vortices, on both the downstroke and the upstroke, and they give strongly suggestive evidence that vortices of successive claps can interact to produce a less energetic wake, and hence to require less power from the butterfly.

Ring vortices, especially those produced by the clap, can remain coherent and powerful over surprising distances (see Fraenkel 1972 for relevant theory). My favorite demonstration of this is the air vortex cannon, commercially available as a toy under the trade name Airzooka. It fires ring vortices that remain powerful for distances an order of magnitude larger than their diameter. So it is plausible to suggest that close interactions between individual butterflies can be mediated by buffeting ring vortices thrown toward each other, and that some of the observed feats of following at distances of several body lengths could be fostered by entrainment of the follower in the leader's vortex wake, possibly with increased efficiency due to slipstreaming (Bouferrouk 2014). Furthermore, it is likely that any pheromones that are released, even in small amounts, during interactions will remain concentrated in the vortex wake and hence potentially effective over long distances within the aerial path of the releaser.

These suggestions are a very promising topic for future exploration, following the lead of Srygley and Thomas (2002), but the venture is beyond my current technical talents and machinery. For now, I merely caution against active and "strategic" interpretation of seemingly complex interactive behavior that might be mediated through passive responses of one individual to vortices shed by another. This caution is strengthened in Sections 7.6 and 10.6.4.

The tight coupling between wing and air is shown in strobe photos of the Clouded Sulphur (Figure 2.3). In addition to the peel, clap, and flap already referred to, the wings show dramatic flexion as the stiff veins on the leading edge are powered downward, but the air beneath the more flexible parts of the wing is accelerated more gradually, probably due in part to its inertia and in part to its viscosity. Especially in slow flapping flight this flexion may provide an additional mechanism that has received little technical attention, namely "slithering" through the air as the wing carves a sinusoidal path, accelerating air backward in flow that is more nearly laminar than turbulent. Indeed, one of the added surprises of Srygley and Thomas' (2002) study is that, even in rapid flight, the flow over much of the surface of the Red Admiral's wing is more nearly attached and laminar than expected, and vortices at the trailing edge and especially at the leading edge are much smaller than expected. This slithering motion is made more plausible by a consideration of the aerodynamic role of scales and their placement on the wing.

# ADAPTATIONS TO HABITAT AND SOCIETY

**A.** Peel Stage in Context

**B.** Open-Wing Flexion in Context

**C.** Wings as a Unit     **D.** Wings Partially Slotted     **E.** Viscous Flexion

FIGURE 2.3 Male Clouded Sulphurs showing all the actions (except down-clap) described in Figure 2.2. Note the extreme flexure of the wings, due to interactions among forces applied by the butterfly, heterogeneous elasticity of the wings, and viscous, turbulent, and inertial properties of the air. Such flexure is variable among photos taken at the same stages of the wingbeat cycle.

### 2.2.3 How Scales May Aid Slow Flight, especially for Patrollers

Nachtigall (1965) classically asserted that "Butterflies with scales removed may fly poorly or not at all."[8] This was a decade before the unsteady aerodynamics of flapping flight were beginning to be understood (Weis-Fogh 1973), but Nachtigall confirmed part of the poor performance by putting a dried butterfly specimen in a small wind-tunnel, where the de-scaled wing had a lower lift to drag ratio than an intact wing, especially at realistically intermediate angles of attack. Potential explanations of the effect of scales (e.g., Dudley 2000, pp. 117–118) have suggested that they induce a lower aerodynamic drag by analogy to the "sharkskin effect" (Bechert et al. 1985), in which microscale ribs parallel to water flow prevent small boundary vortices from coalescing laterally, and hence keep them small.

Something else or additional may be going on with butterfly scales, which are arrayed in neat rows like lapped shingles (see Figure 2.4). The following aerodynamic caricature starts with an assumption of nearly laminar flow at a negligible angle of attack. This assumption is realistic for the slithering motion proposed above (Section 2.2.2), in which the wing deforms to follow the air around it. But in addition, the result of the caricature will be seen to extend the range of velocities under which airflow will be more nearly laminar.

Starting with nearly laminar flow, I suggest that the rows of scales induce parallel coherent micro-turbulence, one tubular vortex-cell for each row. Because such cells will be stationary at contact with the scales (the "no-slip condition," e.g., Vogel 1994, pp. 18–20), their internal rotational velocity will be about the airspeed of the wing, and the external velocity will be about the same, but directed backward, favoring coherence. One might think of the tubular vortex-cells heuristically as tiny roller bearings, each about the diameter of the scales and the length of the row of scales. Such bearings could dramatically decrease drag in the direction normal to the rows of scales. Indeed, the frictional forces would be those generated by accelerating only the air within about a scale's length of the wing, rather than air at the much greater distances of turbulence generated by macro-separation of airflow from the wing. This mechanism is likely to break down at a sufficiently higher airspeed, as faster angular rotation of the tubular vortices could enlarge their diameters so that they are no longer constrained to the distance between scale rows, and lose their coherence. However, up to that airspeed, the "roller bearing hypothesis" has several important interrelated consequences:

- Friction is dramatically reduced, with consequent reduction in drag coefficient.
- The effect is anisotropic: lower friction for flow normal to the scale rows.
- The airspeed of the transition from laminar to turbulent flow is increased.
- Lift/drag ratio is improved, especially at low angles of attack.

Note that all of these effects contribute to increasing the range of airspeeds for which airflow over the main surface of the wing is more nearly laminar than expected. Thus they feed back to strengthen the assumption of laminar flow made at the outset. They would also produce unexpectedly small leading and trailing separation vortices, which is what Srygley and Thomas (2002) observed.

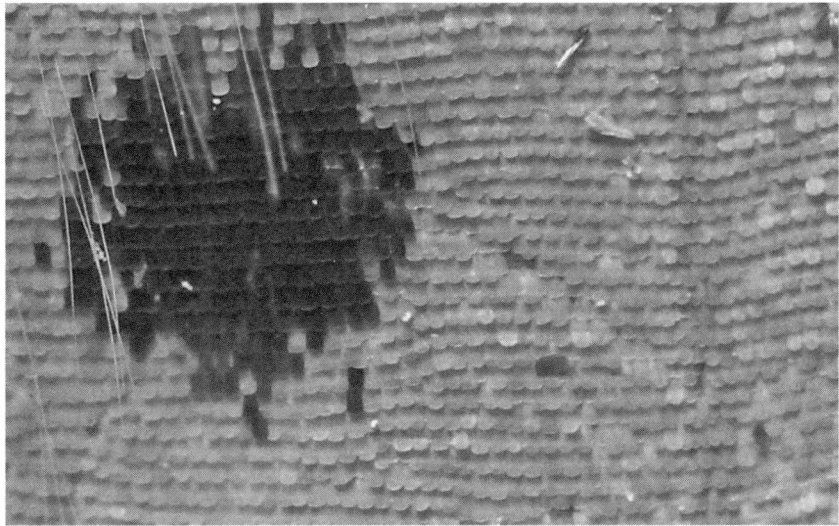

**A.** Great Spangled Fritillary.

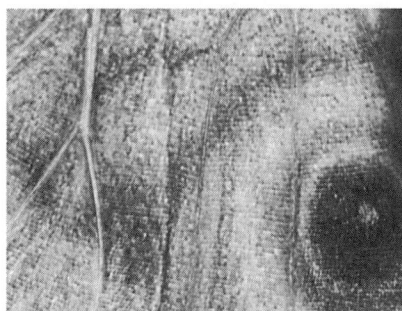

**B.** Pearly Eye, fresh female.

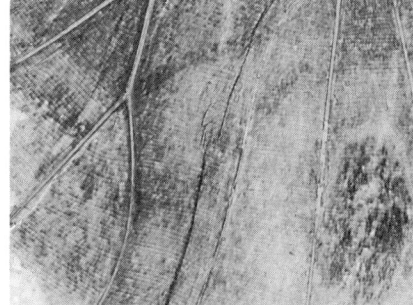

**C.** Pearly Eye, tattered male after ca. 2 weeks.

FIGURE 2.4. Scales on the hindwing of Great Spangled Fritillary (A) and Pearly Eye (B and C). The tiny ribbed scales are shingled in rows that are roughly perpendicular to the relatively stiffer veins (one near right edge of A) that reinforce the membranous wing. This reduces frictional drag specifically to airflow along the regions bordered by veins, which may increase the airspeed at which laminar flow transitions to turbulent, and may direct the formation and coalescence of turbulent vortices (see Section 2.2.3 for details). Normal wear and tear tatters wing edges and strips away scales, decreasing efficiency of flight, especially for male Pearly Eyes.

The detailed aerodynamics are of course more complex than the heuristic caricature above. My ethereal tubular "roller bearing" cells are "trapped vortices" in the airplane and ground vehicle literature (Yeung 2006, Bouferrouk 2014), and the conditions that allow the "trapping" to be self-stabilizing are stringent (Riddle el al. 1999, Tutty et al. 2013). Nevertheless, Hu and Tamai (2008) have visualized such trapped vortices and confirmed their aerodynamic effects for corrugated plates at a scale of about 10 mm, and the effects themselves have been shown for the corrugated wings of dragonflies at a scale of about 1 mm (review in Hu and Tamai). Bixler and Bhushan (2013 and refs.) have shown the anisotropy of drag for both a Morpho (*Morpho didius*) butterfly wing, and a bioinspired artificial surface designed to match the geometric surface characteristics of that wing, with the corrugations due to rows of scales being at a physical separation of about 100 μm. And the improved lift/drag ratio of a scale-textured wing was discovered and documented by Nachtigall (1965) long ago.

Figure 2.4 also shows that the rows of scales are at right angles to the wing veins that flank them, just as they are between appreciably separated veins over the wings of almost every butterfly that I have observed closely. Here I compound my previous speculation with the suggestion that the lowered induced friction parallel to the veins should favor airflow along the axis of the veins. As the wing flaps downward, the membrane between veins may be bowed upward, and vice versa, channeling that flow. Such cupped channels are barely visible as ripples in the wings of the Clouded Sulphur in Figure 2.3. The orientation of most veins is such that these channels might concentrate airflow in a pivot-fugal direction toward the trailing edge of the wing. That flow could favor the initiation, coalescence, and separation of coherent trailing edge vortices, in a macroscopic version of the "sharkskin effect."[9]

Recent studies of the flight of butterflies have combined visualization of vortices shed by real butterflies, behavior of simulated wing surfaces, and fluid dynamic calculations (Zheng et al. 2013, Yokoyama et al. 2013, and Fei and Yang 2015). These studies share several results:

- confirmation that wing deformation is important in real butterflies and in models,
- smaller than expected vortices shed by real butterflies,
- better performance of the real butterflies than the models,
- the notion that something is missing from the analysis.

Potential candidates for missing components are the mechanisms described above: nearly laminar slithering of the wing through the air, and the contribution of scales to low drag and perhaps to favoring nearly laminar flow at appreciable airspeeds. Putting together laminar with turbulent fluid dynamics is an extreme challenge, particularly given that a different mix might be involved at different stages of the wingbeat cycle, and even over different parts of the wing at a given stage of the cycle.[10] I am not ready for the challenge, other than to suggest that some insights might come from simple macrophotographs to confirm the geometry of veins and the geometric relation between orientations of veins and of shingling of scales, combined with strobe photos like Figure 2.3, especially with frontal lighting to emphasize the rumplings of the wings at all stages of the wingbeat cycle.

All of these speculative mechanisms could contribute substantially to the aerodynamic efficiency of slow flight with a shallow wingbeat: the slithering mode; the roll of scales in reducing drag and fostering laminar flow; and the roll of veins in managing the formation and size of trailing vortices. Efficient slow flight is especially important for patrollers, which spend much of their active time in flight, and could visually detect and discriminate habitats and objects of interest more readily when cruising slowly.

*2.2.4 Varieties of Flapping Flight in Open Air*

Figure 2.3 also shows that the fore- and hindwings sometimes overlap sufficiently that they can be considered a single unit, but sometimes they are separated to produce a slot that may allow the hindwing to "manage" vortex-shedding by the trailing edge of the forewing, and the forewing to "manage" vortex-formation on the leading edge of the hindwing. Not only does this variable slotting provide an additional and varying parameter that is not considered in traditional analyses, it also means that the wing as a whole has variable integrity, to say nothing of an ambiguous width and therefore an ambiguous aspect ratio.

Figure 2.5 shows a variety of patterns of flapping flight, and in particular contrasts the usual flight of a Small Cabbage White (Figures 2.2 and 2.5A) with its strong down-clap, versus that of the Wood Nymph (Figure 2.5B), which up-claps in such a way that a ring vortex is directed backward and slightly downward, and spends an appreciable proportion of the wingbeat cycle with its wings folded vertically, sliding through the air with an essentially vertical profile, an apparent adaptation to navigating the vertically structured, grassy habitat of most satyrines. The Pearly Eye (Figure 2.5D) shows this behavior as well,[11] as do most satyrines to greater or lesser degree. This mechanism, clapping at the end of the upstroke and expelling a ring-vortex backward and somewhat downward or upward depending on attitude, helps to explain the characteristic "indolent but dancing movement" of satyrines in flight (Scudder 1899, p. 289), and also their tendency to fold their wings ventrally when captured, the better to escape with a powerful upstroke.[12] Once I realized that satyrines tend to flee downward (e.g., Weed 1917, p. 218), it became much easier for me to capture them by aiming my net slightly below the butterfly, rather than centering it in the mouth of the net, as I do with other species.

The Pearly Eye (Figure 2.5D, also Figures 7.5, 7.6, 7.7, 7.9, and 7.11) uses all modes of wingbeat: down-clap, up-and-back-clap, peel, glide, and fan, in complex maneuvers to find and inspect potential mates as described in Chapter 7.

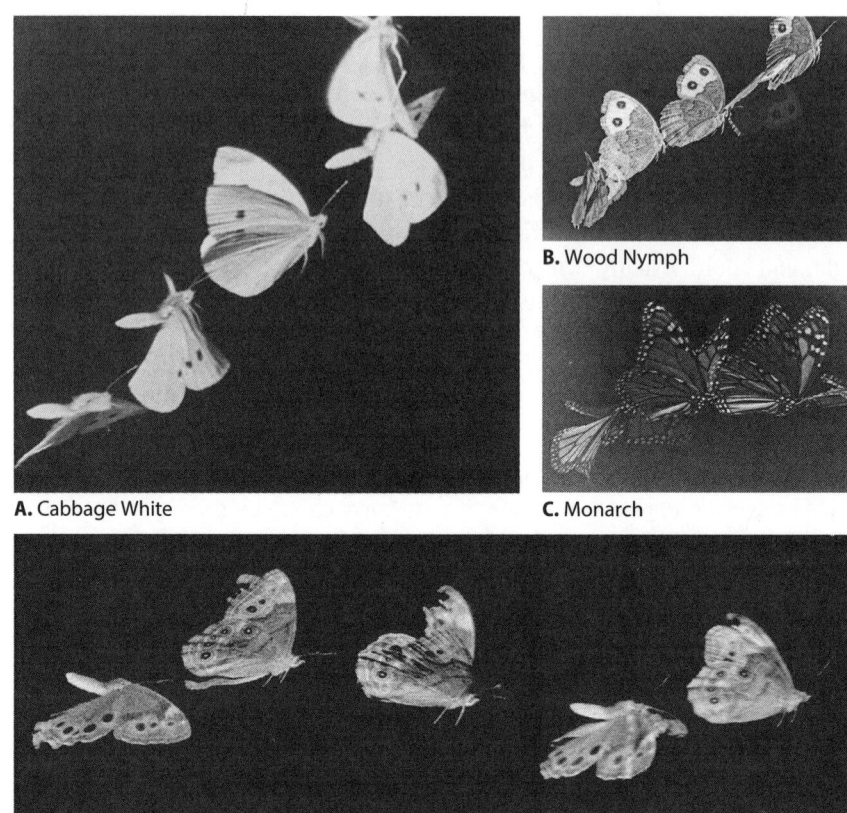

**A.** Cabbage White  **B.** Wood Nymph  **C.** Monarch  **D.** Pearly Eye

FIGURE 2.5. Varieties of flapping flight (strobe photos at 20-millisecond intervals). A. The Cabbage White is buoyed upward by reaction against an expelled ring vortex as the wings clap at the end of a downstroke (see also Figure 2.2C), and hauled upward by air entering from above as wings peel apart early in the downstroke (See also Figure 2.2A). In between, the wing acts as a flexible flipper, essentially fanning the air backward. The downstroke is a power stroke. B. In contrast, the Wood Nymph and other satyrines, e.g., the Ringlet, the Eyed Brown, and the Pearly Eye, may expel a ring vortex back and slightly downward as the wings clap at the end of the upstroke (see also Figure 2.2E). So the satyrines spend much of the wingbeat cycle with wings folded directly upward, which can facilitate passage between clumps of grass or other vegetation with strong vertical structure. C. The Monarch glides with wings in a stabilizing dihedral for much of the wingbeat cycle, maintaining forward progress and altitude by fanning with wings wide apart (see also Figure 2.2F). D. The Pearly Eye uses all of these modes: down-clap, up-and-back-clap, peel, glide, and fan (see also Figures 7.5, 7.6, 7.7, and 7.9).

**E.** Tiger Swallowtail

**F.** Fritillary

FIGURE 2.5. continued. More varieties of normal flight. E. The Tiger Swallowtail moves like the Monarch, but is buoyed by a clap at the end of the downstroke. F. The Great Spangled Fritillary uses all modes, but predominantly shallow flapping and slithering. It also shows slotting, in which the forewing and hindwing separate slightly and interact to manage airflow between them.

The large-winged species found characteristically in open habitats, Monarch (Figure 2.5C), Tiger Swallowtail (Figure 2.5E), and Fritillary (Figure 2.5F), fly mainly by shallow flapping and sailing, perhaps using the slithering mechanism described above (Section 2.2.2). The Tiger Swallowtail also episodically down-claps, resulting in the sudden change in attitude shown on the right of Figure 2.5E.

The variety of wingbeat cycles suggested by strobe photos under artificial conditions is confirmed by frame-by-frame analyses of videotapes of free-flying

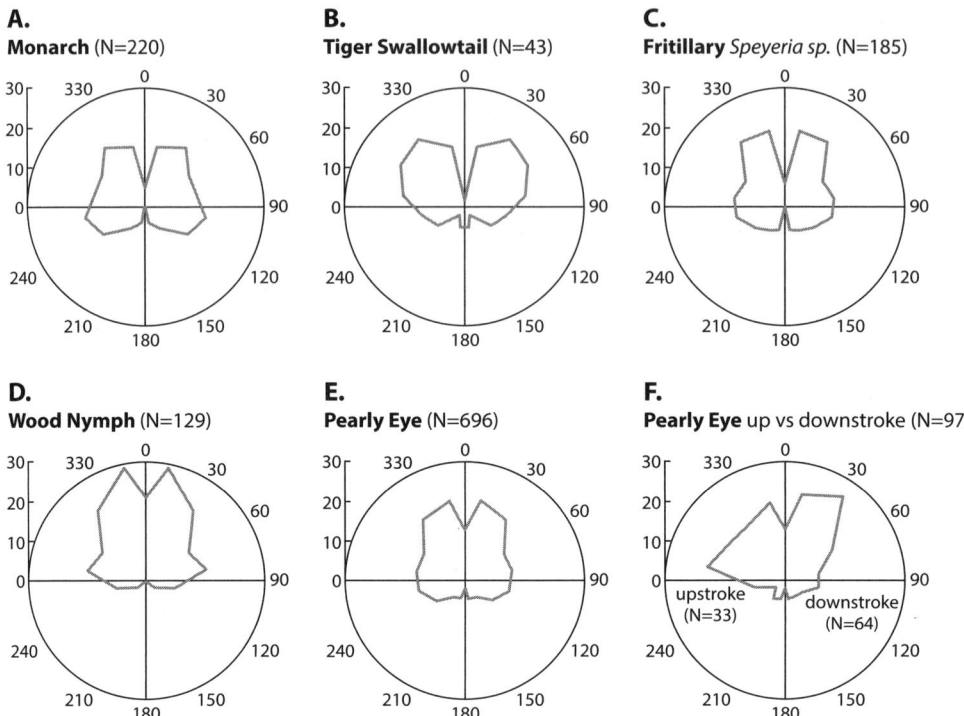

FIGURE 2.6. Phase diagrams of wing angles from frame-by-frame analysis of videotapes. Plotted in polar coordinates centered on the body axis is the percentage of frames in which the wings were at a given angle from vertical, during flapping flight. Frames were classified visually on a coarse scale (class width ca. 22.5°), and adjacent categories were averaged for the plot. A–C. The Monarch, Swallowtail, and Fritillary, are habitual sailers and, even in flapping flight, have their wings apart for much of the cycle. D–E. In contrast, the satyrines, the Wood Nymph and Pearly Eye, fly with wings nearly vertical for much of the cycle. F. Upstroke and downstroke are plotted separately for the Pearly Eye for those frames that could be classified as such unambiguously by reference to adjacent frames. See Section 2.2.4 for discussion of Figure 2.6F as evidence that the upstroke is the more powerful stroke for the Pearly Eye, and by analogy, for other satyrines.

butterflies in their appropriate habitats.[13] Figure 2.6 shows phase diagrams of flapping flight in open air for five species. The Monarch, Tiger Swallowtail, and Great Spangled Fritillary (Figures 2.6A–C), as befits their typically open habitat, separate the right and left wings for most of the cycle, with rare up-claps, and virtually no down-claps, resulting in little net movement up or downward. The satyrines, Wood Nymph and Pearly Eye (Figures 2.6D–E), spend most of the wingbeat cycle with wings above the back or in a narrow V. Among the 696 video frames analyzed for the Pearly Eye, there were 97 in which it was possible to distinguish unambiguously between upstroke and downstroke. When these are plotted separately in Figure 2.6F, it is clear that the approach to vertical in the upstroke (left side of Figure 2.6F) is more rapid than the departure from vertical in the downstroke (right side of Figure 2.6F). This confirms the previous inference that the upstroke is the faster and thus the more powerful stroke for the Pearly Eye, and by implication, for other satyrines. Few peels or down-claps were documented in the videos or the staged strobe photos of the Pearly Eye, but they are seen in strobe photos of Pearly Eyes outdoors (e.g., Figures 7.5, 7.6, 7.7, and 7.9A for peels and 7.5, 7.7, and 7.9B for down-claps).

*2.2.5 Speculations on Flight and Mimicry*

Both the Monarch and its well-known mimic the Viceroy often alternate between bursts of shallow flapping and longer sails with fixed wings and a slow rate of descent. However, their wings are typically held at different angles in the sail (Clark 1932). The Monarch sails with wings in a shallow V, generating the "dihedral effect" of aeronautics, in which side-slip automatically places the wings on the lower side at a greater angle of attack, increasing their lift and "correcting" both roll and yaw (e.g., Dalton 1977). Consequently, the Monarch can sail at high velocity, but its very stability limits maneuverability. Conversely, the Viceroy typically sails with the wings flat, or even slightly depressed, enhancing maneuverability but limiting stability at speed. However, when a Viceroy is startled, e.g., by a butterfly net, it tends to sail with wings in a shallow V, like the Monarch.[14] I have no doubt that this shift by the Viceroy is due to entering a different aerodynamic flight mode, but it has the effect of behaviorally enhancing the Viceroy's mimicry of the Monarch. Flight attitude and behavior have been shown to be elements of both Mullerian mimicry, i.e., mutual convergence within distasteful mimics (Srygley 1999), and Batesian mimicry, i.e., convergence of a tasty mimic on a distasteful model (Kitamura and Imafuku 2015).

The genus of the Viceroy, *Limenitis*, has several other mimics, including the Red-spotted Purple *L. arthemis astyanax* as a postulated mimic of the Pipevine Swallowtail *Papilio troilus* (Platt et al. 1971; Platt 1983). The White Admiral *L. arthemis arthemis* is traditionally considered to be nonmimetic, but I have been struck by its resemblance in slow flight to the much swifter White-winged Widow Skimmer Dragonfly *Libellula luctosa*. Figure 2.7 shows the two species. Indeed, I have often mistaken one for the other on a quick glance.[15] So it is conceivable that even the traditionally nonmimetic local *Limenitis* is a "racing stripe mimic" of a dragonfly.

**A.** White Admiral *Limenitis arthemis*   **B.** Male Widow Skimmer *Libellula luctuosa*

FIGURE 2.7. "Racing stripes" on White Admiral butterfly (Princeton, NJ, Aug. 2016) and male Widow Skimmer dragonfly (Washington Crossing, PA, June 2015). There is no mistaking the difference between these species at rest, but in flight they may look alike despite the slow sailing of the butterfly and the rapid darting of the dragonfly. See the text of Section 2.2.5 for speculation of possible mimicry by the White Admiral.

## *2.2.6 Impaired Flight, Failed Cues, Predation, and Wear and Tear*

For species that normally fly in open fields, the configuration of sky, vegetation, and horizon are crucial to orientation. When open-field species are given the task of flying from a darkened environment toward a visually narrow patch of light, their wingbeats may be normal, but they are so poorly aligned with gravity that the flight is distinctly maladaptive. Figure 2.8 shows typical misadventures: a Monarch flapping without progression, a Fritillary turning to fly sideways, and a Small Cabbage White power-diving to a crash.

Figure 2.9 shows damage that butterflies regularly incur from attempted predation or normal wear and tear with age. Such damage has surprisingly little effect on the flapping and sailing flight of habitual sailers, as long as the leading edge of the forewing remains intact. But the loss of wing surface changes the launching and maneuvering capabilities of all species, and in particular, loss of hindwing surface area has dramatic consequences for satyrines, which lose the effectiveness

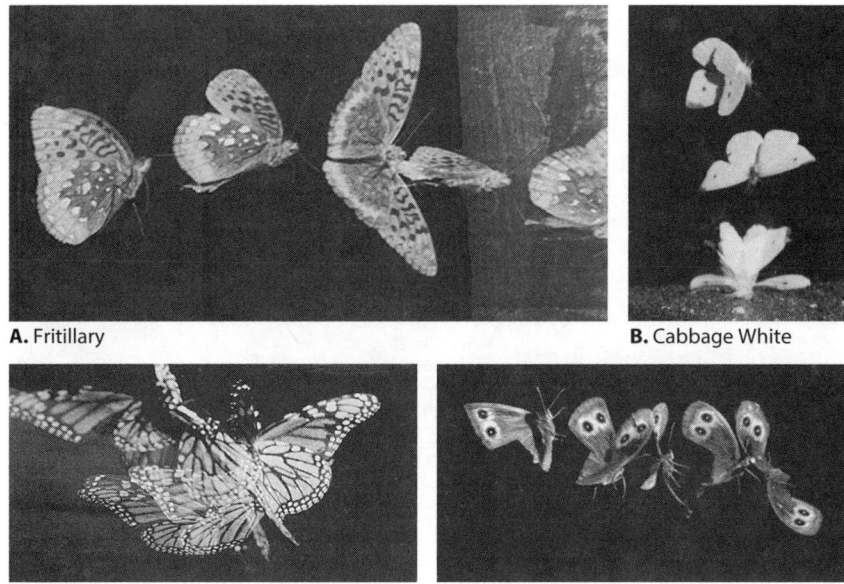

FIGURE 2.8. Flying misadventures: a sample of maladaptive flight patterns due to failures of orienting cues or wing morphology. The photos also show that viscous interactions between wing and air can dominate gravity and inertia under visual disorientation. A. The Fritillary rolls sideways. B. The Cabbage White crash-lands under full power. C. The Monarch has lost sight of the "light at the end of the tunnel," and without a horizon, it goes nowhere. D. The Wood Nymph is missing most of its hindwings and a bit of its forewings, probably after escaping from the beak of a bird. Without the full hindwing, it cannot shed a coherent ring vortex at the end of the upstroke, and so cannot fly in its normal way. Comparison with Figure 2.5B shows that the aberrant flight is far less effective than normal flight (for a contrast of species that can fly almost normally with heavily damaged wings, see Figure 2.9B-D).

of up-clapping, as well as important components of their linear flight (see Figure 2.8D, where a Wood Nymph with a damaged hindwing has lost all semblance of the normal flight shown in Figure 2.5B).

It is likely that normal wear and tear has a self-reinforcing negative effect on survival and mating success, though the evidence is indirect. Wear strips scales from the wings, increasing drag. Tear decreases wing area, changes wing profile, and destroys surfaces that are presumably adapted for the efficient generation, coalescence, and shedding of trailing edge vortices. The resulting abnormal flight is less efficient, less maneuverable, and likely to foster further wear and tear. Intact wings and normal scales are also important for thermoregulation, including both heating and radiative cooling (Kingsolver 1985b, Tsai et al. 2017). Maneuverable flight is crucial to predator avoidance, and maneuverable and/or sustained

**A.** Bird-clipped Pearly Eye

**B.** Bird-clipped Red Spotted Purple

**C.** Tattered Purple x White Admiral

**D.** Tattered Viceroy (Mercer County, NJ)

FIGURE 2.9. Butterflies with extensive wing damage. The Pearly Eye (A) was unable to fly in its normal way, but beat its wings rapidly at a likely great expense of energy (see Figures 2.8D versus 2.5B). The others were capable of seemingly normal flapping flight, and even of sailing at a low rate of descent, though they did not launch as readily as undamaged conspecifics.

efficient flight is required for finding and inspecting potential mates. So there are self-reinforcing physical reasons for wear and tear to decrease both survival and mating success.

### 2.2.7 Aerodynamic Parameters for Selected Species

I started with the hope that a more detailed technical aerodynamic analysis would more clearly separate perchers from patrollers. What follows does not do the job, but it does suggest speculative insights that may help in future attempts. In particular it highlights a potentially important difference between hindwing and forewing in their aerodynamic function, a difference that does not appear in traditional analyses, which treat the wings as a single coupled airfoil.

A first step is to calculate the Reynolds number, a theoretical measure of the ratio of inertial to viscous forces acting on the wing as it moves through the air with a resultant velocity that combines forward progression with the maximum velocity induced by flapping the wing itself. The frictional forces acting on the wing are generated by air moving with this resultant velocity over the wing over the "wing chord," the maximum width of the wing in the direction of that velocity. The Reynolds number, Re = (resultant velocity)(wing chord)/$(1.5 \times 10^{-5})$ (see chapter 5 and table 2.1 in Vogel 1994 for a terrific introduction to Reynolds number, with appropriate formulae and parameters).

Measurements from appropriately scaled versions of Figures 2.5 and 2.6 allow approximation of aerodynamic parameters for six species, and these parameters are presented in Table 2.2. Forward velocity was measured as the sum of distances between mid-thoraxes of successive strobe images in Figure 2.5, divided by accumulated 20 msec intervals. Wing-tip velocity was measured as (forewing length)($\pi$)(proportion of 180° swept by wing in one beat)/(time to take one beat); the proportion was estimated from Figure 2.6. Wing-tip velocity is in turn a rough estimate of the maximum velocity of airflow over the distal wing surface as displaced air flows around the wing edge in the middle of the up- or downstroke. Contrary movement of the butterfly relative to the adjacent air will make this an overestimate, but not by much. The forewing length is approximately orthogonal to the direction of flight. So the resultant maximum velocity over the distal surface of the wing was calculated as the hypotenuse of a triangle with right sides forward velocity and wing-tip velocity, and the angle of the resultant velocity to the body axis is its angle to the forward velocity in that right triangle (Figure 2.10 shows that this is not quite so, but the simplification is not ominous). Wing chord was measured as the maximum width of both wings with the average angle maintained in flapping flight, as shown in Figure 2.5. Wingbeats per second was calculated

TABLE 2.2. Aerodynamic Parameters for Six Species

| Parameter | Cabbage White | Pearly Eye | Wood Nymph | Fritillary | Tiger Swallowtail | Monarch |
|---|---|---|---|---|---|---|
| Forward velocity | 1.4 m/s | 1.9 m/s | 1.5 m/s | 1.6 m/s | 1.8 m/s | 2.0 m/s |
| Wing-tip velocity | 1.6 m/s | 0.9 m/s | 1.4 m/s | 1.6 m/s | 1.3 m/s | 1.4 m/s |
| Resultant velocity | 2.1 m/s | 2.1 m/s | 2.0 m/s | 2.2 m/s | 2.2 m/s | 2.4 m/s |
| Angle to body axis | 49° | 25° | 44° | 45° | 35° | 35° |
| Flying chord | 3.4 cm | 3.7 cm | 4.2 cm | 4.9 cm | 5.7 cm | 6.7 cm |
| Forewing length | 2.6 cm | 3.2 cm | 3.4 cm | 4.0 cm | 4.4 cm | 5.6 cm |
| Wingbeats per second | 20 bps | 17 bps | 17 bps | 12 bps ? | 17 bps | 13 bps |
| Reynolds number (Re) | 4,600 | 5,200 | 5,600 | 7,100 | 8,200 | 11,000 |

*Note:* See text (Section 2.2.7) for parameter formulas, details, and interpretations.

from cycle lengths measured in Figure 2.5. The uncertainty in this parameter for the Fritillary is due to the facts that Figure 2.5F does not clearly resolve down- versus upstroke, and Figure 2.8A is complicated by a quarter roll.

Given the dramatic differences in style of flight, especially as documented in Figures 2.5 and 2.6, I was surprised and mildly disappointed to find a narrow range among velocities and wingbeat frequencies (Table 2.2). Differences in Reynolds number, at least among the species I have measured, are due mainly to differences in wing and body size. The main import of Table 2.2 is that the larger species have large enough Reynolds numbers to suggest that they are near the transition from laminar to turbulent flow regimes (i.e., $10^3$ to $10^5$, Vogel 1994, Lian and Shyy 2007), and that adaptations—to reduce drag, to bias airflow toward laminar, and to shed separated vortices efficiently—could be very important, especially in rapid flight and when sailing with wings fixed (Sections 2.2.2 and 2.2.3).

Figure 2.10 gives speculative insights into the flight of a Monarch. Figure 2.10A shows postulated low-friction paths between veins and across scale rows. Figure 2.10B shows approximate airflow in mid up-flap; each arrow is the vector sum of forward velocity plus pivot-fugal airflow. On the hindwing, airflow aligns roughly with the low-friction paths, which should reduce the energy expended in shed vortices. On the forewing, however, IF the low-friction paths redirect airflow, they would have to do so over appreciable angles; this would tend to redirect

ADAPTATIONS TO HABITAT AND SOCIETY 51

A. Postulated Low-Friction Paths

B. Approximate Mid-flap Airflow

C. Wing Flexure at Peel of Downstroke

D. Slotted Wings at Mid-flap

FIGURE 2.10. Speculative insights into Monarch flight (see Section 2.2.7 for details and interpretation). A. Low-friction paths *if* drag is lower normal to scale rows and parallel to veins (Section 2.2.3). B. Very rough approximation of airflow in mid-up-flap. A conclusion, which is little affected by the unrealities of the approximation, is that expected airflow is along paths of low friction for the hindwing, but at an appreciable angle for the forewing. C–D. Wing flexure is appreciable during the downstroke. And fore- and hindwings vary during a flap from acting as a tightly coupled unit (C), to being slotted and loosely coupled (D).

shed vortices toward the outer edge of the forewing and away from the hindwing. Otherwise, the effect of the low-friction paths on the forewing would be to reduce the component of drag due to flapping per se, without reducing the drag due to forward progression. Potentially complicating factors are flexure of the wings and shift within a single wingbeat between fore and hindwings acting as a tightly coupled unit to being slotted and only loosely coupled (Figures 2.10C–D). So although the traditional aerodynamic parameters of wings show little adaptively interpretable variation among species, there might be interesting patterns of variation between forewing and hindwing in size, shape, flexure, orientation relative to venation, and coupling over the wingbeat cycle.

## 2.3 VISION

### 2.3.1 Compound Eye View: a Shifting Mosaic of Landscape, Movement, and Objects

The butterfly eye is a compound eye, a hexagonal array of lenses, wave-guides, and sensors, aimed in different directions to cover a large angular field of view. There is some overlap in the field of view of adjacent sensors. Just how and at what level of optical or neurological integration a composite image is formed is the subject of much technical research, and the answers seem to vary with species and setting. The further question of how, and indeed whether, the composite image is perceived as a whole by the insect, has yet to be answered definitively, but neurological detectors of colors, edges, lines, flicker, and optic flow have been widely documented (Cronin et al. 2014), starting for butterflies with Swihart's (1967) pioneering demonstration of neuronal electrical responses for color, movement, and their interaction.

The best analogy would be a coarsely digitized, but precisely oriented, motion picture taken through a slightly blurry fish-eye lens—a panoramic movie in needlepoint. This is very different from the "insect-eye view" of popular culture, exemplified in toy goggles with a large number of tiny, detailed pictures with only slightly different perspectives, viewed through a hexagonal array of lenses with nearly parallel axes. This misleading view originated in early photographs taken through what was represented as an "actual insect eye," but was only one element of the eye, the cornea, and that flattened onto a microscope slide.

Viewing life as a needlepoint movie does not necessarily show what a butterfly perceives. Instead it removes the kind of continuous detail that we are so used to interpreting, and opens our eyes to the huge amount of information about the environment that could be perceived as color, edge, line, flicker, and flow. Indeed, our saccadic eye movements, which help the fovea to scan the environment to build a detailed picture that is stabilized relative to our body movement, may obscure flicker and flow that give information about details of the environment in a very different way.

The angular acuity of the compound eye depends on the angle between optical axes of adjacent ommatidia, and this differs between species and even over the field of view of an individual. Typical parameters for the compound eye of a butterfly are as follows (Rutowski and Warrant 2002, Land and Nilsson 2012):

- number of ommatidia = 6,300–7,500 per eye
- total angular range = 160 solid degrees per eye
- binocular overlap between eyes = 10°
- angle of view of individual ommatidia = 1–2°
- angle between optical axes of adjacent ommatidia = 1.4–2°
- overlap in visual fields between adjacent ommatidia = 1°

Angular acuity varies among species and individuals with sex (generally higher in males, Rutowski 2000) and size (generally higher in larger eyes, Rutowski et al. 2009), and over areas of the visual field within individuals (Stavenga et al. 2001, Rutowski and Warrant 2002, Rutowski et al. 2009). Rutowski and his colleagues find that interommatidial angles of males are smaller, and hence vision is more acute, in the 10° or so of frontal binocular overlap, and that this is the area that is usually aimed at the likely places to spot a female when the male is on its perch (as it is for my Viceroys in Section 6.3.1). The equatorial region of the eye is also more acute than the rest of the field of view, and this may be the region in which optic flow (discussed below in Section 2.3.4) is potentially most informative.

Since my butterflies can respond to stimuli that subtend only 2° (Table 2.1 and Figure 2.1), the signal need only stimulate one ommatidium and its next neighbors.

### 2.3.2 Eyeshine Variations Suggest Adaptive Chromatic Resolution.

A tracheal tapetum behind the photosensitive surfaces of the eye forms an interference filter that reflects certain wavelengths of light back for a "second try" at the photosensor (Stavenga 1979). This means that the butterfly's eye has enhanced sensitivity at those wavelengths, and enhanced contrast between those and neighboring wavelengths (Bernard 1971, Goldsmith and Bernard 1974). Sexual variation in the most sensitive color has been shown to be adaptively related to differences in the importance of colored visual cues for behavior (Bernard and Remington 1991). Variations between species and variations within species over the angular segments of the visual field (Miller and Bernard 1968) are similarly related to the differential importance and orientation of visual cues, though the interpretation is not always straightforward. The most general pattern includes reflection of short wavelengths (blue-biased) dorsally (Stavenga et al. 2001), suggesting sensitivity and discrimination of patterns of skylight versus shading vegetation. Stavenga and colleagues have also shown many instances of local heterogeneity of eyeshine, in which neighboring ommatidia reflect different hues, often dramatically different hues. The function of this local heterogeneity has not been confirmed, but I speculate that it may enhance sensitivity to chromic patterns by generating apparent flicker as a butterfly turns its head or moves through its environment (expanded in Section 2.3.4).

Such local heterogeneity of chromatic sensitivity over the eye's field of view can only be observed clearly with narrowly axial illumination (Stavenga 1979), and the regional spectrum of sensitivity inferred by averaging it locally may badly misrepresent what a butterfly might perceive. Nevertheless, an ophthalmoscope, used with appropriate precautions and conservative interpretation, is a valuable tool to prospect in the field for species that might promise dividends in the lab. Accordingly, I examined dark-adapted eyes of several species with an ophthalmoscope, at the personal suggestions of Sibatani (1973) and Stark (et al. 1979). To bias my observations toward reflectance from the tapetum, and away from reflection from the cornea, I only recorded clearly colored reflectance that faded with light adaptation, and that did not change in quality as the ophthalmoscope approached the butterfly's eye. The following report is intentionally vague, given the cautions expressed above, the paucity of observations, and the variability of results among individuals. I examined males of all five main species and the Little Wood Satyr *Megisto cymela*, females of all but the Fritillary, and one female Wood Nymph *Cercyonis pegala*. Where such reflective colors were observed, they varied over regions of the field of view of the eye, and their patterns varied among species and among individuals. The dark-adapted eyes of female Ringlets and both sexes of Eyed Brown showed no evidence of tapetal reflectance. Among the individuals that showed clear color reflectance, green was most frequent and extended over the majority of the field of view, and the color shaded to blue dorsally in several instances. This suggests potential enhancement of chromic and contrast sensitivity to vegetation and skylight. The male Fritillary and female Wood Nymph showed strong red reflectance laterally. The Fritillary responds to orange lures (Table 2.1), and the Wood Nymph often congregates at magenta flowers (wild marjoram *Origanum vulgare* and starthistle *Centaurea maculosa*). Male and female Viceroys are variable, but a usual pattern is green overall, with dorsal blue and a lateral spot or equatorial region of orange, red, or yellow. Viceroys are predominantly orange, though the most distant responses of males were to backlit objects moving laterally across the field of view (Table 2.1 and Figures 2.1 and 6.4).

The most tantalizing observation comes from the male Little Wood Satyr, whose pattern of reflectance is green overall, shading to blue dorsally, but with a frontal yellow spot. The frontal spot seemed discrete, but I was unable to map it accurately because whenever I shifted the ophthalmoscope to the edge of the spot, the butterfly turned its head so as to keep the spot strongly lit. Little Wood Satyrs fly with the typical jerky path of satyrines, but they add frequent dives toward sunflecks on the forest floor. Turning so as to keep a sunfleck in the frontal field of view is a potential mechanism for diving toward it.

### 2.3.3 Views through Compound Eye-Glasses

To gain some insight into how butterflies may see the world, I designed and constructed an analog compound eye, a device described in detail in Appendix C.4. It looks at a hexagonal array of out-of-focus images, each with a narrow field of view overlapping its neighbors, with parameters chosen to mimic a typical butterfly's eye (after van Leeuwenhoek *fide* Bernhard 1965, Exner 1891, Eltringham 1919, Gould 1979). Of course this analog eye does not represent what a butterfly actually perceives, but it does remove details that the butterfly eye simply lacks the detail to resolve. This makes alternative details obvious—details that we might otherwise not notice. In particular, a portable viewer gives advantages over the traditional stationary pixelization of a photograph because of the flicker that Horridge (1986) argued to be a crucial aspect of visual perception in insects. The analog compound eye was used in the field to view scenes in which I thought that visual cues might be important, and several insights might not have been obtained otherwise.

Figure 2.11 constructs a static version of the simulated compound eye view. The scene of Figure 2.11A is pixilated in Adobe Photoshop to form Figure 2.11B. Details of the pixelization are given in Appendix C.4. Many specific details of the real scene are preserved in the digitized version. The Viceroy in the scene contrasts with its surroundings despite subtending the field of view of only a couple of simulated ommatidia, and being blurred with adjacent ommatidia. The orientation and configurations of distant and near horizons are faithfully represented, as is the artificial low spot in the horizon generated by giant mirrors. Movement of the simulated viewer produces flicker that could encode subtle gradations of texture in all vegetation, and could also make the butterfly even more obvious.

Looking at the world through compound eyeglasses emphasizes the importance of edges, gaps, and subtle configurations of the horizon. The shape of the horizon may help to orient hill-topping (Shields 1967), valley-bottoming (cf. Scott 1973), and traplining (Ehrlich and Gilbert 1973, Alcock et al. 1976, 1977). Butterflies perceive ultraviolet light (Wakakuwa et al. 2007), which is scattered even more by the atmosphere than the blue that makes the sky appear blue to us (viz. Cronin et al. 2014). We see nearby horizons more clearly than distant mountain ranges that blur into the blue light scattered by the intervening atmosphere. It is likely that butterfly vision can detect the relative configuration of vegetative "horizons" that are nearby at even smaller distances against a background of

**A.** Male Viceroy oriented to mirrors forming an artificial low spot in the horizon

**B.** The above scene digitized to the typical resolution of a butterfly's eye

FIGURE 2.11. Simulated compound eye view of a real scene. Many details that we notice in the real scene are obliterated in the simulation. But preserved are: orientation and configuration of the horizon (including the artificial one), location of small objects of contrasting color and luminance (e.g., the Viceroy, some goldenrod *Solidago* flowers, nearby sunlit leaves, shady patches). A moving simulation adds flashes and flickers that can discriminate fine differences in texture of light versus shade, color, and structure of vegetation. See Section 2.3 for more insights.

scattered ultraviolet. Configurations of distant and near horizons, as visual cues, will enter the interpretations of choice of habitat and movements through the habitat for all five of my species in Chapters 3–7.

A striking phenomenon appears when one uses the analog compound eye to find and approach an image of a butterfly (Horn 2013). Figure 2.12 is a static simulation of this process for a Great Spangled Fritillary. Figures 2.12A–B show that

**A.** Great Spangled Fritillary  **B.** Copy of A, digitized to about 6,300 pixels

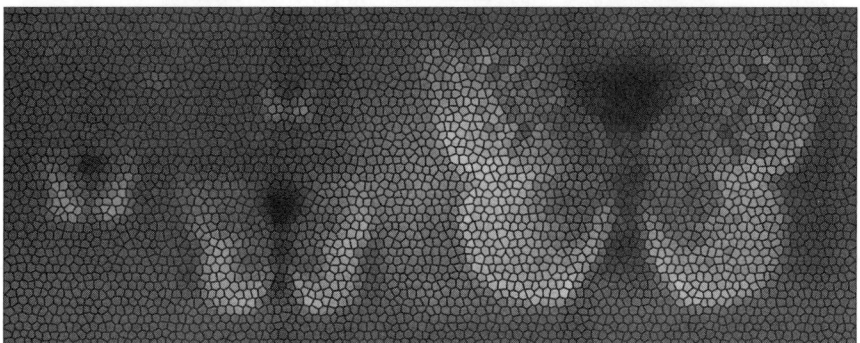

**C.** Simulated compound eye views being approached by Great Spangled Fritillary A.

FIGURE 2.12. 6,300 pixels is the order of magnitude of the number of ommatidia in the total visual field of a butterfly. So B is the minimum resolution that preserves anything like the pattern that we can easily see in A. Yet it is far beyond the capability of a butterfly, given that the simulation in B takes no account of the extreme curvature in the real field of view at the simulated object distance of 1.5 centimeters. C. Simulated compound eye views approaching, or being approached by, the Great Spangled Fritillary of A. Object distances in C: extreme upper left (and barely visible) 2 m, then in a row at 1 m and 50 cm, then in a second row at 25 cm and 12 cm, and on the right 6 cm. The barely visible image, at 2 m, is near the largest interaction distance observed (3 m, Table 2.1). With the analog compound eye, over simulated distances of 25 cm to 10 cm a striking phenomenon occurs; if the eye or the subject moves slightly, large portions of the field of view seem to flash between all orange and all dark brown. It may be more than coincidence that the typical courtship distance for this species is about 18 cm. So the regularity of the compound eye may act as a cross-correlation filter for the regularity of the spotted wing design. See Section 2.3.3 for details and cautions about this interpretation.

even a blurred representation of patterns that are obvious to the unaided human eye are beyond the resolution of the butterfly eye. Figure 2.12C shows simulated views from 2 m, about the farthest distance from which another Fritillary can induce an approach (Table 2.1), and then 1 m, 50 cm, 25 cm, 12 cm, and 6 cm. The analog compound eye shows a dynamic pattern that the static photos cannot convey: over the distances from 25 cm to 10 cm, if the eye or the subject moves slightly, large portions of the field of view seem to flicker between all orange and all dark brown. The typical aerial courtship distance for this species is about 18 cm. Unfortunately, the only courtships that I have observed in the field have been unsuccessful. Courtship, or rather rejection of a suitor, involves a hovering male and slow flicks of the female's wings. So additional signals are potentially available from physical buffeting by ring vortices and/or from entrained pheromones. Nevertheless, a potentially easily detectable visual signal is also available that encodes cross-correlation of the regularity of the spotted wing design.[16]

### *2.3.4 Optic Flow Encodes Relative Movement, Distance Parallax, and Moving Objects*

The movement of a butterfly in a patterned field of view generates a particular form of systematic flicker called "optic flow." Optic flow is the basis of classical experiments, in which a striped pattern is rotated around a loosely tethered insect, and in a so-called optomotor response, the insect tends to turn its head or body in a direction that slows or stops the flow of the image across its visual field. A neurosensory basis for detecting optic flow was shown in a butterfly by Swihart (1967), who recorded neurons that fired when a color pattern was moved in one direction and other neurons that responded to movement in the opposite direction, and behavioral responses to chromic contrasts in moving patterns have been shown by Stewart (et al. 2015). Many experiments in orientation, movement, and navigation of locusts, dragonflies, bees, wasps, ants, and flies have been interpreted in terms of optic flow (Horridge 1986, 1992, Horridge and Marcelja 1992, Srinivasan et al. 2000, Cronin et al. 2014). Recent workers with flies (Fenk et al. 2014) and dragonflies (Mischiati et al. 2015) have shown that complex behavior, like orienting and moving toward small targets, can be interpreted by simple rules of behavior in optic flow fields, and they have also proposed speculative models of very simple neural networks with compound-eye-like input that could be sufficient to coordinate the required movements. Such visual input and biomimetic circuitry has been designed for robots, which have enacted the predicted, seemingly complex behavior (Webb 2002, Floreano et al. 2013, Pericet-Camara et al. 2015).

Figures 2.13 and 2.14[17] give a heuristic demonstration of features of the visual field that may stand out as a result of optic flow (Figure 2.13) or that could be detected against a background of optic flow (Figure 2.14).

Flying straight ahead produces symmetrical flow backward in both eyes, radiating from the de facto goal (Figure 2.13A). This means that a butterfly can fly toward (and follow) a visual stimulus by keeping that stimulus in the frontal region of binocular overlap in its visual field, as was suggested in Section 2.3.2 for the Little Wood Satyr as it dives toward sunflecks in the forest.

Turning the head or turning the body in flight produces front to back flow in one eye and back to front in the other (Figure 2.13B). The difference between forward motion (Figure 2.13A) and turning (Figure 2.13B) produces a dramatic difference in optic flow, and this provides a plausible signal to coordinate a pattern of movement that alternates forward movements with turns. A forward-biased random walk will be elaborated as the Tactical Forward Vagrancy Model for search behavior in Chapter 3. That model realistically mimics the movements of Ringlets (Chapter 3), and is the basis of a simple model that mimics seemingly complex interactive behavior of Pearly Eyes (Chapter 7). When a perched butterfly turns its head, any fixed pattern of contrasts in the field of view should be enhanced by flicker, though there will be little or no parallax shift for objects at varied distances.[18]

Keeping a fixed object in a lateral region of either eye while otherwise flying forward would cause the butterfly to circle the object while producing overall flow like that of turning. Such circling behavior can be seen in male-male aerial encounters of many butterfly species, and has been picturesquely named "spinning wheel flight" by Wickman and Wicklund (1983). This may also be the mechanism that produces the helical flights of male Pearly Eyes around tree trunks as they return to perches after breaking from encounters with another butterfly (described above in Section 2.1.2 and in Chapter 7).[19]

Figures 2.14A–C show the flow field generated by a simple scene in Figure 2.14B. Near elements flow faster and farther than distant elements, providing the information needed for depth perception, and early experiments confirmed such depth perception in honeybees (Horridge et al. 1992, Srinivasan et al. 1990). Figures 2.14C–D contrast a moving versus a stationary butterfly's view of a moving and a stationary object. It could be difficult for a moving butterfly (Figure 2.14C) to perceive a stationary object unless it contrasted sharply in hue or luminance from its surroundings, whereas a moving object might be easier to detect because of the added contrast of flow relative to the flow of the rest of the visual field. A stationary butterfly (Figure 2.14D) should be able to detect either a moving or a stationary object of any appreciable degree of contrast. The contrast of an object that is moving relative to a stationary butterfly will be enhanced by flicker, but such flicker could also be generated by the butterfly's moving its head.

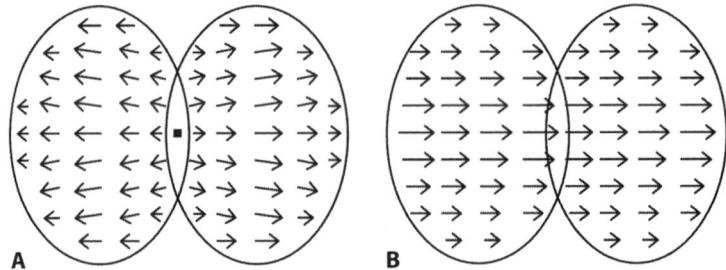

FIGURE 2.13. Optic flow of scenes past the eyes of a butterfly (after Cronin et al. 2014, p. 234): A. Flying straight ahead toward an object represented by the central square. Forward motion produces symmetrical flow in both eyes, radiating from the de facto goal. B. Turning continuously toward the left. Rotation produces asymmetric parallel flow in both eyes. The difference between forward motion (A) and rotation (B) should produce a very clear signal for a variety of backgrounds.

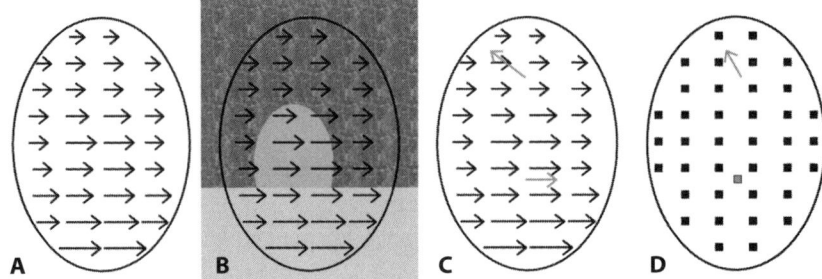

FIGURE 2.14. Optic flow in the right eye of a butterfly: A–C. Flying forward past the scene in B, forest in the background, a shrub in the middle distance, and a field extending to the near foreground. Near elements of the scene flow faster and farther than distant elements, providing potential depth perception. C. The upper colored arrow represents a flying object, perhaps another butterfly, and the lower arrow represents a stationary object. D. The same objects are represented in the field of view of a stationary butterfly. C versus D. It would be difficult for a flying butterfly to perceive a stationary object unless it contrasted sharply with its surroundings, though a moving object could have a strongly contrasting flow compared with the overall field of view (C). It could be easier for a stationary butterfly to perceive either moving or stationary objects (D versus C), though contrast of color or luminance is still helpful.

Fortuitous observations of head movements came from my first attempts to photograph the head of a male Pearly Eye for Figures 2.17A–B, using the photographic "studio" of Appendix C.5. The initial photos were blurred because the butterfly rapidly turned its head back and forth in apparent response to vibrations transmitted through the apparatus as the mirror in the camera shifted just prior to exposure. Such movements would have the effect of causing elements of pattern in the field of view to flicker, enhancing detection, perhaps in adaptive response to vibration stimulating increased alertness.[20]

*2.3.5 Visual Orientation and Flight within Habitat*

For a butterfly flying above the vegetation in an open habitat the visual field below will be dominated by the flicker generated by moving past dappled vegetation, but the visual field above will be dominated by sky, which will flicker less. So a butterfly can stably fly above the vegetation by moving so as to generate more flicker below than above. For a butterfly that routinely dips down into the vegetation it would be helpful to fly with an up-and-down motion that generates additional flicker, added to that of forward motion, as it crosses from exclusively open sky above and laterally to a mixture of sky and vegetation. This increased vertical resolution of habitat structure may provide an added benefit to the "dancing" flight of satyrines (Sections 2.2.2 and 2.2.4).

*2.3.6 Pseudopupil Variations Suggest Adaptive Spatial Resolution*

The primary pseudopupil is a dark spot that appears when you are looking down the axis of an ommatidium, because axial or nearly axial light is absorbed by proximal pigments. Accordingly, approximate inferences can be made about the angle between axes of adjacent ommatidia, and hence variations in acuity across the field of vision, by counting ommatidia in the darkest part of the primary pseudopupil. My macrophotographs (in Figures 2.16–2.17) are not sufficient for this purpose, however, as it requires a microscope of narrow aperture with matching aperture in the light source (Stavenga 1979). However, even with macrophotographs, the pattern of light and dark around the primary pseudopupil can be used to infer patterns of pigmentation between adjacent ommatidia (Yagi and Koyama 1963, Sibatani 1973, Stavenga 1979). Figure 2.15 caricatures dramatically different patterns of primary and secondary pseudopupils as well as the patterns of pigmentation around individual ommatidia that can produce those pseudopupil patterns (after Yagi and Koyama). Figure 2.16 shows photographs of species with interestingly different pseudopupil patterns and hence different patterns of inferred interommatidial pigmentation that should be adaptive to their habitats.

No pseudopupil is visible at all in some species in which the pigmentation is dark and distally extensive so that little incident light is returned in any direction, and nearly all illumination along the axis of a given ommatidium reaches

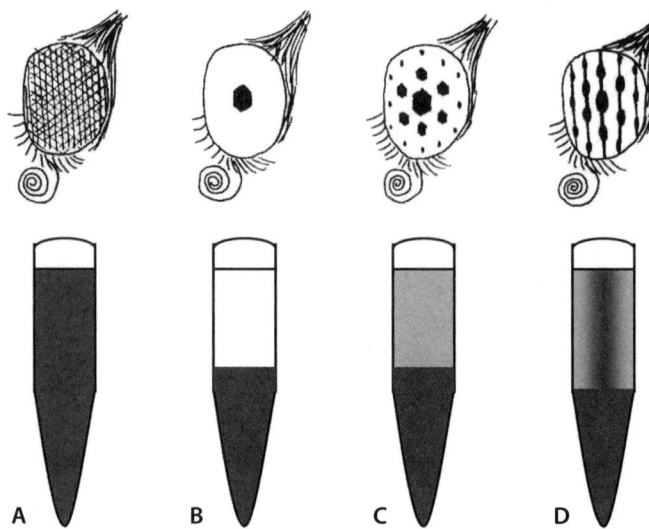

FIGURE 2.15. Pseudopupil types, with schematic representation of pigmentation surrounding the optically clear portion of an ommatidium (after Yagi and Koyama 1963). A. No apparent pseudopupil is seen in many species that are active mainly in the open (e.g., Viceroy). Widespread pigments absorb light, isolating adjacent ommatidia, increasing spatial resolution while decreasing sensitivity, which is compensated by the high ambient light intensity of open habitats. B. A single pseudopupil is seen in many species of open habitats, perhaps implying slightly less spatial resolution and more sensitivity. C. A primary pseudopupil and six secondaries, sometimes with higher orders in hexagonal array, are seen in species that move between open and semi-closed habitats (e.g., Fritillary and Eyed Brown). Pigments in the distal surrounding cells may migrate distally to favor spatial resolution in high light or proximally to favor sensitivity in low light. Some species (e.g., *Vanessa*, *Nymphalis*, and *Polygonia*) have reflective pigments that migrate oppositely to like effect. Such reflective pigments are suggested by bright spots between the primary and secondary pseudopupils. D. Some hexagonal arrays of pseudopupils are connected by dark striations (e.g. satyrines, including the Ringlet and Pearly Eye). According to Yagi and Koyama, this is due to distal cells and interstices that are more heavily pigmented above and below than fore and aft. See Section 2.3.6 and Figure 2.17 for my interpretation of a possible adaptation to habitat for pattern D.

its photoresponsive cells (Figure 2.15A). The Viceroy shows this pattern (Figure 2.16A), as befits its open habitat, where high light intensities allow the low sensitivity that extensive pigmentation entails. Optically isolated ommatidia produce higher resolution over the visual field, and the Viceroy is indeed the species with the longest distance of visual response to a potential mate, and the smallest angle that the potential mate need subtend to elicit a response. (Table 2.1 and Figures 2.1 and 6.4).

A hexagonal array of secondary and tertiary pseudopupils is produced by lighter distal pigments surrounding each ommatidium, partially isolating them optically from one another (Figure 2.15C). This pattern is found in the Clouded Sulphur (Figure 2.16B) and the Great Spangled Fritillary (Figure 2.16C), which fly mainly in open habitats. The slight loss in acuity in the Sulphur compared with the Viceroy may be compensated by the fact that the Sulphur's wings reflect ultraviolet and hence contrast strongly against background vegetation (Papke et al. 2006). The Fritillary episodically dips down into openings in dense vegetation, and Figure 2.16C suggests that its visual acuity may be stronger dorsally, in the direction of open air above.[21]

The hexagonal array may show bright spots between the pseudopupils, representing distal reflective pigments that migrate toward the ommatidium to

FIGURE 2.16. Pseudopupil types. A. Viceroy male with no apparent pseudopupil. B. Clouded Sulphur female *Colias philodice* with a single primary pseudopupil with weak secondaries and tertiaries. C. Great Spangled Fritillary female with a hexagonal array, stronger dorsally than ventrally. D. Red Admiral female *Vanessa atalanta* with a hexagonal array with evidence of reflecting pigmentation. E. Ringlet male with a hexagonal array with weak vertical striations. F. Pearly Eye male with a hexagonal array with strong vertical striations. A–B. The Viceroy and Clouded Sulphur fly mainly in open habitats. C. The Fritillary flies mainly in the open, with episodic excursions into the shade of grasses and herbs. D. The Red Admiral divides its time between open and lightly wooded habitats. E. The Ringlet flies mainly in the open, but episodically dips down into dense grass. F. The Pearly Eye spends most of its time in open woodland, glades, and woodland edges. These habitat prevalences correspond to the adaptive speculations of Figures 2.15 and 2.17 and Section 2.3.6.

FIGURE 2.17. Speculative relation between satyrine vision and habitat. A. Vertical stripes in a male Pearly Eye's pseudopupil suggest ommatidia with darker distal pigments above and below. B. Frontal view suggests a forward-directed binocular "sweet spot." Interommatidial bristles are widely distributed over the eye. C. Model portraying the postulated isolation of ommatidia vertically, but their connectedness laterally (cf. Figure 2.15.D). D. Habitat at a woodland edge typical of those favored by Pearly Eyes. The photos were taken in August 2015 near Morristown, NJ. The subject was held in a marking clamp (Appendix C.1 and C.5), and was literally alive and kicking. See Section 2.3.4 for enlightening details of its behavior. Make a copy of Figure 2.17C on transparent acetate. Then "fly" it in front of Figure 2.17D. When your acetate butterfly is flying at an appreciable angle to the vertical structure of the vegetation, a strong Moiré flicker is generated. The flicker disappears when the butterfly is oriented so as to pass most easily through the habitat with wings over its back (cf. Figures 2.5B, 2.5D, and 2.6E). See Section 2.3.6 for further details.

ADAPTATIONS TO HABITAT AND SOCIETY

increase sensitivity in low light (Yagi and Koyama 1963). The Red Admiral (Figure 2.16D) shows this pattern, and indeed it spends continuous time alternately in sun and in shade.

A peculiar pattern of pseudopupils is shown by many nymphs and satyrs (Nymphalidae: Satyrinae; the "Browns" of Europe). They typically show a hexagonal array of dark spots, but the spots are connected vertically by roughly parallel dark lines (see Figures 2.15D–F, and 2.17A–B). Yagi and Koyama (1963) found that in the satyrs the distal absorbing pigments are concentrated on the dorsal and ventral sides of the ommatidium. My highly speculative interpretation of this peculiarity is that it tends to isolate ommatidia optically from their dorsal and ventral neighbors, but that it may leave them "leaky" to light laterally. Under this interpretation, the eyes of nymphs and satyrs could act as moiré analyzers of vertical and horizontal structure in their visual fields. This mechanism is better explained by playing with Figure 2.17 as suggested in its legend, rather than by puzzling through a verbal description.

The inferred distribution of interommatidial pigments shows adaptive relation to the light environment, from isolating pigments in the open, to adjustable reflective pigments in varied dappled shade, to few pigments in constant shade. The striped pseudopupils of some satyrines suggest a potential moiré analyzer of environmental structure, inviting optomotor experiments to look for predictable variation in response to changing the angle of the axis of rotation of the visual field.

## 2.4 OLFACTION

Most of the behaviors that I have observed can be parsimoniously interpreted as responses to long-distance visual stimuli, but there are some olfactory cues that are so obvious that even I can perceive them. Accordingly, I have looked for behaviors that might be hard to explain from visual cues alone, and found them, particularly in the Fritillary and the Pearly Eye.

### 2.4.1 Fritillaries Produce Distinctive Odors.

Characteristics odors have been noted for both sexes of the Great Spangled Fritillary, with that of the male being "strong and pleasant," and that of the female, produced from abdominal processes everted when the female is gently squeezed, being "strong and nauseating" (Clark 1926).[22] Clark interpreted the female odor as a defense against vertebrate predation, but noted that the odor and the organs producing it were similar to those of female Neotropical Heliconians, near relatives of Fritillaries, whose odors are now known to be pheromones by which males find and preemptively guard female pupae prior to emergence of the adult and subsequent mating (Estrada et al. 2010). Odorous organs have been described for females of a European Fritillary, *Argynnis paphia*, and shown to elicit enthusiastic behavior in males (Treusch 1967). Male scents have been detected in the same species, and late stages of courtship place the scent-producing scales of the males next to the antennae of the females (Magnus 1950). Olfactory discrimination is traditionally accepted as the means by which Fritillaries, both European *Argynnis* and North American *Speyeria*, prevent hybridization between recognized species that differ little in visual morphology (Scott 1972–3, Hammond et al. 2013). Courtship distance and behavior of my large Fritillaries is consistent with visual discrimination by the wildly speculative mechanism above (Section 2.3.3 and Figure 2.12), OR olfactory discrimination OR the one reinforcing the other.

I am able to detect the odor of male Fritillaries that have been enclosed in an otherwise odorless box for a few minutes. I cannot detect the odor of females with similar treatment, but I have not tried squeezing them first. Male Fritillaries pre-emerge the females by nearly a month, and early in their flying life they behave very like females in the exploratory flights prior to oviposition (Section 2.4.3 below). It is conceivable that once the females have emerged, the males are orienting to a female pheromone, but I have no evidence to confirm or to deny such a long-distance response to a pheromone.

## 2.4.2 Pearly Eyes Respond to Odors

Both male and female Pearly Eyes clearly use olfaction initially to locate slime fluxes on trees. Their daily schedule of behavior is adjusted to times that are optimal for tacking upwind to find the source of an appropriate odor. Anemotaxic searching flights, tacking upwind across odor plumes, are especially common and apparently efficient in the early afternoon of otherwise windless days, when a light breeze (ca. 0.5–1 m/sec) flows laminarly from woodland to field. Those conditions are optimal for long distance olfactory orientation (Cardé and Charlton 1984, Murlis et al. 1992, Vickers and Baker 1997). After discovering that I myself could detect and navigate to artificial ferments under those conditions, and that the ferments were most effective in attracting Pearly Eyes when fermentation had reached the vinegar stage, I tried a cellulose sponge with distilled vinegar alone as an attractant. It brought male Pearly Eyes from many meters, but they would not settle unless I added crushed fruit or fruit-flavoring extract. I tentatively suggest that acetic acid suffices as a long-distance attractant, but some other components

ADAPTATIONS TO HABITAT AND SOCIETY  67

of the ferment, speculatively esters, are required to induce settling.[23] Olfaction may also be involved in the attraction of male Viceroys to fermenting slime fluxes.

The male Pearly Eye is known to produce an odor from androconial scales on its wings. Weed (1917, pp. 219–220) puts it charmingly, "The males possess, perhaps to a greater degree than any other of our native butterflies, the ability to give off a peculiar, pleasant aroma which is noticeable whenever the insects are collected and which at least one careful observer has been able to detect in the open air as the butterfly flew near." I can detect such a male-specific aroma, but only after the butterfly has been confined for a few minutes in an otherwise odor-free box. Courtship ultimately involves the male approaching and butting a stationary female from the side with occasional wing-flicks. So there is ample opportunity for the pheromones to be locally broadcast and received. Given the observation of Weed's "careful observer," entrainment of pheromone in the vortex wake of a flying male is possible, as suggested above in Section 2.2.2. However, in flight the male typically follows the female, rather than vice versa, and so wake-following of male pheromone is not likely important to long-distance courtship. Nevertheless, it is possible that local concentrations of pheromone, especially in the vicinity of slime fluxes attended by many males, could give females an integrated clue of male numbers and intensities of activity.

### 2.4.3 Olfaction and Oviposition

Discrimination for oviposition is most likely preliminarily by vision at a distance and ultimately by taste, which is essentially olfaction with direct contact, though there is a possibility for aerial olfaction at intermediate distance. Direct observations of oviposition are documented in Appendix B.3.

The Ringlet lays eggs on thin bladed grass, green or dried, and, in confinement, on cotton threads of about the same diameter as the preferred grasses. Accordingly, I suspect that once the female is in the appropriate open habitat, something of about the right size visually and physically is likely to be either a host plant or something near a host plant. Larvae from eggs that are not very near a host plant are unlikely to survive, because the unfed larva is mostly head capsule, with a tiny body and miniscule legs that can barely grip the host grass. The large head capsule is needed to house mandibles large enough and muscles powerful enough to take the first few bites of the tough host plant.

The Pearly Eye lays eggs on wide-bladed species of grass or sedge, on the underside of a leaf at a peak where the leaf bends over sharply. The female approaches the leaf as though perceiving it visually from a distance, and alights on the horizontal landing pad at the peak where the leaf flops over of its own weight. The female then bends the tip of her abdomen to tap the underside of the leaf, and lays a small cluster of eggs if the leaf is appropriate. The signal to oviposit may be either physical or chemical; my guess is chemical because I see an overlapping range of colors, textures, hairiness, and indeed, species, when comparing instances of acceptable versus unacceptable hosts for oviposition. The Eyed Brown displayed similar behavior in the two instances of egg-laying that I observed.

Female Fritillaries do not lay eggs until a month or more after emergence from the chrysalis (e.g., Weed 1917). During that time their typical flight pattern is direct cruising over many meters, with episodes of dropping into a patch of deep vegetation, often with the food plant, violets (*Viola*) in the understory, and fluttering through that vegetation before rising to cruise to the next patch. The male displays similar behavior soon after emergence, but gradually shifts to direct flights between episodes of visiting and nectaring at flowers in the overstory, near patches of vegetation where violets are in the understory. The behavior of both sexes is described in more detail and interpreted in Chapter 5, but it is hard to imagine a visual clue that will signify spring-flowering violets in the understory of the much more complex canopy of summer vegetation above. Accordingly, I suspect that violets may produce a characteristic odor that accumulates in the vegetation above them, and that Fritillaries of both sexes can detect.

Chemical cues are more strongly suggested by the fine-scale movements that I observed by a female Fritillary (in this case *Speyeria aphrodite*) within a patch of vegetation that did not include violets (Appendix B.3). She approached several species with dark green leaves, landed on those with small rounded leaves, immediately left one with a hairy leaf (*Hieracium*), and walked about on the others while tapping them repeatedly with the tip of her abdomen before departing. During most of this episode, she walked from place to place with her wings tightly folded.

### 2.4.4 Vision versus Olfaction

Vision is a sufficient modality for the interactions that I observe at long distances (meters to centimeters), and olfaction is likely involved in closer interactions (centimeters and less) of butterflies with plants and with each other.

Initial responses of males to other butterflies are initiated at distances of one to several meters. They are probably responses to visual cues that can be small and subtle. Subsequent interactions are at distances of the same order of magnitude as both the size of a butterfly and the size of ring vortices left in a flying butterfly's wake. They could involve visual, mechanical, and olfactory cues in ways that cannot be disentangled by solely visual observations, no matter how sophisticated.

## 2.5 HEARING

Male Pearly Eyes are initially very skittish, as I discovered when trying to move close to capture and mark those which had come to my fermenting lures. I soon noticed that even when I approached very slowly and from the woodsy side so as to present as little visual contrast and movement as possible, an aggregation of males would often disperse if I stepped on a twig and snapped it audibly. I could hardly believe that it was the sound of the twig snapping that spooked them, but when I tried the experiment of standing still before they dispersed, holding a twig with my hands behind my back—when I snapped the twig, they took flight. Satyrines, the subfamily of the Pearly Eye, are known to have potential organs for hearing in the enlarged base of the leading vein of the forewing (e.g., Scott 1986). So perception of airborne vibration is possible. Recall that at the end of Section 2.3.4, I described a behavioral response of a male Pearly Eye to substrate-borne vibration. Responses to vibrations, both substrate-borne and airborne, would be particularly adaptive for the Pearly Eye because it faces potential predators that are attracted to the ferments on which it feeds, namely hornets *Vespulidae*, and to the aggregations of the butterflies themselves, namely birds (Section 9.3.4).

## 2.6 SUMMARY OF RESULTS *AND SPECULATIONS*

The following summarizes results and definitive inferences *with speculations in italics* or in permissive construction. The speculations are *not* to be interpreted as new findings; they are merely suggested as potential mechanisms worth further study. This summary is ordered by level of organization of behavior, whereas the body of this chapter is more by the physical mode of behavior. So for each statement or paragraph below there is a terminal reference to the appropriate section in the body of the chapter. Where appropriate, this reference is preceded by key words from the subtitle of the book.

### 2.6.1 Finding and Navigating Habitat

Patterns of pigmentation around the distal part of each ommatidium in the compound eye can be inferred from the external appearance of the pseudopupil, a pattern of dark spots over the surface of the eye. *Species of open high-light habitats have dark shielding pigments, optically isolating neighboring ommatidia and allowing maximal spatial resolution. Species that inhabit shaded habitats have less such pigmentation, decreasing resolution, but enhancing sensitivity. Species that move between light and dark habitats have laterally migrating reflective pigments to adjust the trade-off between resolution and sensitivity. As a butterfly moves through its environment, flickers in different regions of the visual field can provide detailed information about the configuration of the horizon, and about patterns and textures of light, shade, and color.* (Cues: Sections 2.3.3 & 2.3.4).

Ringlets favor open habitats by moving toward regions of high light. Pearly Eyes prefer open woodland by moving toward regions of patchy shade. *Male Eyed Browns may organize their movements in part by moving toward V-notches in the near vegetative horizon.* Fritillaries organize their flight paths by moving toward V-notches in the far sky-horizon, as do female Viceroys; accordingly, male Viceroys choose perches overlooking such V-notches (Cues & Rules: Section 2.1.2).

The Pearly Eye and Wood Nymph *and other satyrines like the Ringlet and Eyed Brown* incorporate a powerful upstroke in their flight, and spend an appreciable proportion of the wingbeat cycle with the wings folded vertically. *This is adaptive for the vertical structure of gaps in their grassy habitats. Furthermore, the ups and downs of the flight path in each wingbeat cycle may generate a pattern of flicker in the optic field that increases visual resolution of habitat structure.* (Behavior & Cue: Sections 2.2.4 & 2.3.5).

Another satyrine characteristic is an eye that shows a vertically striated pseudopupil. This is consistent with pigments that shield ommatidia from dorsal and ventral neighbors *and that may provide a moiré analyzer of orientation in a vertically structured habitat* (Cue: Section 2.3.6 and Figure 2.17).

Local patterns and regional differences in chromatic sensitivity over the field of view of the compound eye, and their variation between sexes and among species are worth exploring for potential adaptive patterns (Cue: Section,2.3.2).

Patterns of flight and vision are adapted to geometrical constraints of habitat, and carry implications for escape from predators, and for the rapid maneuvers that the Viceroy (Chapter 6) and the Pearly Eye (Chapter 7) make during their interactions with each other. In particular the optical properties of the eye preserve detailed information about edges, textures and regularities of contrast, and bearings of contrasting objects—information that *may be used* to choose and to orient with respect to habitats, perches, food plants, and potential mates. Accordingly, I find that many aspects of behavior in the field are correlated with relatively simple environmental cues, emphasizing visual edges, horizons, and patchy texture (Cues & Behavior).

*2.6.2 Orientations of Scales and Veins Affect Mode and Efficiency of Flight.*

Scales on the wing are shingled in rows at right angle to the major veins; this lowers aerodynamic drag parallel to the veins. *This in turn may direct airflow from the fulcrum toward the edge of the wings and manage the formation, coalescence, and shedding of trailing edge vortices; it may also increase the airspeed at which turbulent drag becomes important, favoring efficient gliding and "slithering" through the air with a slow and shallow wingbeat, a potential adaptation for patrolling.* These speculative mechanisms deserve empirical research, as do the effects produced by observed episodic separation between forewing and hindwing in several species (Behavior: Sections 2.2.2–4).

Interpreting the mechanisms of flight is difficult because butterflies often operate near the transition from laminar or turbulent flow. So lift and propulsion *may be due* to any of several effects or a combination of them—Bernouli effect (generated by wing profile or by separated vortices); reaction against inertia of air or against momentum of a shed vortex; the wing functioning as a paddle, a flipper, or an airfoil—or even different parts of the wing playing different roles in different parts of the wingbeat cycle. The cycle itself can be continuous and repetitive, or interrupted by claps and flings. Camber *may be* a passive consequence of inertial following of air, and/or it can combine one or more functions: an airfoil for lift-generation—stiffening normal to the camber—or vortex-management (Behavior).

### *2.6.3 Consequences of Flight Behavior for Survival*

Wing damage from attempted predation or from ordinary wear and tear can potentially shorten life and decrease reproductive success (Life History: Section 2.2.6).

A startled Viceroy often assumes the sailing profile of its model, the Monarch, though *whether this is active behavioral mimicry* or just a consequence of adopting more rapid flight is not clear. The White Admiral, a traditionally nonmimetic congener of the Viceroy, can be mistaken in its slow flat-profile flight for the much swifter White-winged Widow Skimmer Dragonfly. (Behavior & Life History: Section 2.2.5).

### *2.6.4 Searching For and Detecting a Potential Mate*

Among all my local species, patrollers, which spend most of their active time on the wing, tend to have broad, rounded wings, adapted to slow, flapping flight; perchers, which dart out from rest to approach and maneuver around potential mates, have narrower forewings with more nearly straight or even concave outer edges, adaptive to acceleration and rapid flight. However, avoidance of aerial predators requires rapid maneuvers and potentially limits the extent to which patrollers can achieve most efficient sustained flight. Patrollers of open habitat tend to fly with a slow and shallow wingbeat compared with perchers (Behavior:

Sections 2.2.1 & 2.2.4), and they become badly disoriented when a distant horizon is not in view (Cue: Section 2.2.6).

Males of all five species respond to visual cues of a potential mate, even when the signal subtends only the field of view of a single ommatidium and its near neighbors. The cue itself varies with the species: reflectance, color, flutter, and/or lateral movement (Cue: Sections 2.3.1 and 2.3.3). *Experiments with a simulated compound eye view suggest that this is a plausible ability* (Figure 2.12). Interspecific differences in interaction distance are consistent with differences in likely visual acuity and in contrast of potential mates with their backgrounds (Cue & Behavior: Section 2.1.2).

For the Ringlet, the only three courtships that were observed from encounter to copulation involved a sitting female that rose to a patrolling male. So the male is not so much "searching for" a female, as putting himself in a position to be "found by" a female (Behavior: Section 2.1.1). Mechanisms of object recognition in the presence of optic flow suggest that a stationary subject finding a moving object contrasting with a simple background is far easier than a moving subject finding a stationary object of low contrast with a complex background (Cue & Behavior: Section 2.3.4).

Tantalizing experiments with a simulated compound eye view of another butterfly show a dramatic full-field flicker for a regularly repeated pattern of spotting on the wing at a particular distance (Cue: Figure 2.12). *This might generate a signal of species-specific patterns in close courtship.*

### 2.6.5 Optic Flow Unites Movement and Vision.

Optic flow and its variation over the field of view of the compound eye can potentially encode relative movement, distance parallax, and the location of moving objects. Accordingly, such features will be cited as potential cues in descriptions of behavior in the following chapters (especially Chapters 3–7). For a butterfly moving forward, optic flow will be fore to aft in both eyes; for a butterfly turning left, optic flow will be forward in the left eye and backward in the right, and for a right turn, vice versa. Accordingly, information is available to detect and organize changes in flight paths by changes in optic flow. This will become important especially in models of movement for Ringlets (Chapter 3), and in models and interpretations of interactions between Pearly Eyes (Chapter 7). *Moving so that a fixed object is kept fixed in the lateral field of view results in circling that object*, a behavior seen especially in the Pearly Eye (Chapter 7). When a perched butterfly turns its head, *contrasting patterns in the field of view should be enhanced by flicker*. (Cue & Rule & Behavior: Sections 2.1.2 and 2.3.4)

Flying just above the prevailing vegetation produces an optic flow pattern that flickers more in the ventral field of view than in the dorsal. *So moving so as to maintain such a pattern of flicker may provide an autonomous mechanism for regulating flight just above the vegetation.* (Cue & Rule: Section 2.3.5)

*2.6.6 Physical Modes of Behavioral Interaction*

The flight of butterflies leaves a trail of potentially coherent ring vortices. These vortices may affect observed behavior in several ways: . . . entraining following butterflies, . . . leaving a wake of concentrated pheromones, . . . and buffeting interacting butterflies from an appreciable distance. This makes it difficult to disentangle physical versus chemical versus strategic modes of interaction from visual observation. (Cue & Behavior: Section 2.2.2).

Most of my observations are consistent with butterflies using vision to initiate and mediate long-distance interactions with their environment and with each other, and using olfaction only for close interactions (Section 2.4.4). However, Pearly Eyes appear to follow odor plumes upwind, at least initially, to locate the fermenting slimes fluxes on which they congregate (Section 2.4.2). *And Fritillaries may use odors of their larval food plant and of each other to approach from an intermediate distance of one to several meters* (Section 2.4.1). *Conversely, female Fritillaries, Eyed Browns, and Pearly Eyes may use vision more than olfaction at the close range of centimeters, to approach favorable places to lay their eggs* (Cues: Section 2.4.3).

Male Pearly Eyes launch from perches in response to substrate-borne vibration *and to airborne sound, perhaps as a way of avoiding predators* (Cue & Rule: Section 2.5).

The fact that butterflies are intimately coupled to the air around them means that they can potentially communicate either mechanically or chemically through puffs of air that they throw at each other or leave in their own wakes. In turn they can be buffeted by these puffs of air, making it difficult to determine whether close apparent reactions are more than passive. This insight will appear especially in interpretation of the complex behavior of interacting Pearly Eyes (Behavior: Chapter 7).

CHAPTER THREE

# Tactical Forward Vagrancy

*The Ringlet* Coenonympha tullia inornata

> Two butterflies went out at noon—
> And waltzed upon a Farm—
> —Emily Dickinson, 1862

Matchett (1962) reviews Dickinson's own ambiguities about the fate of her waltzing butterflies. My job is to describe the pattern of the waltz, and to speculate about the fate of the butterflies, including the question of who is leading.

The Ringlet inhabits open fields and parkland, and grassy edges of marshes (Shapiro 1974). The males patrol all day long with a bouncy flight, right above the grass-tops (Opler and Krizek 1984), but they remain perched when the weather is cold (Heinrich 1986a). The males may stop to sip nectar during their adult lifespan of only 2–3 days (Turner 1963 in Wales, UK). In copula, the female carries the male (Opler and Krizek 1984). Females lay eggs, and larvae can be raised, on a variety of grasses (Scott 1986), including, from my own work, poverty grass *Danthonia spicata*.

The behavior of the Ringlet at my study site is very like that of the same species in Wales, where it is called the Large Heath (Turner 1963). Male Ringlets wander about their natal habitat, "reflecting" from the boundaries of adjacent habitats. This behavior is compared with simple quantitative models of a bounded forward-biased random walk, ultimately derived from the classic equations of Skellam (1951; see also Root and Kareiva 1984, Levin 1986), and rechristened the Forward Vagrancy Model. The results are not so much a test of the models as a use of the models as standards against which to gain insights into the behavior of the butterfly.

## 3.1 RATIONALE FOR MODELING MALE PATROLLING

Within their preferred habitat, the flights of male Ringlets are close enough to a forward-biased random walk to allow modeling of their searching behavior. That, combined with local demography, will allow an estimate of how many eligible females a male might encounter within its detection distance, as a function of how much time it spends on the wing (cf. Bond 1980 for the Green Lacewing *Chrysopa carnea*). Heinrich (1986a) has explored the thermal ecology of the Ringlet. Even if a male spends all of its thermally permissible lifetime in flight, it will come within the detection distance of only a very small number of virgin females—and females indulge only a single mating (Section 3.4.2).

The random walk models will show that mate-finding is a less consuming activity for females than for males. Nevertheless, it is sufficiently inefficient that wandering virgin females are sometimes found, and two of the three courtships watched from start to finish by me or Diane Wiernasz were initiated by the flight of a female from its perch toward a male that had passed within detection distance (cf. Rutowski et al. 1981 for sulphurs *Colias*, and Wickman 1986 and Wickman et al. 1995 for the Small Heath *Coenonympha pamphilus*).

In the search process, there are several asymmetries between males and females. Competition among males for females is more intense than that among females for males. Models show that if males wander and females remain still, mate-finding is nearly as efficient as if the females were to wander as well. Furthermore, visual physiology suggests that detecting a flying potential mate by a sitting one is much easier than the reverse (Section 2.3.4). Consequently, the wanderings of males may not be so much a "search" for females as an activity that puts males in a favorable position to be found by stationary searching females.

## 3.2 NATURAL HISTORY AND BEHAVIOR

### 3.2.1 Habitat and General Movement

Figure 3.1 is a 1973 aerial view that situates the main field in which my study of Ringlets was done from 1974 to 1981. The field comprises 0.76 hectares, roughly a 56 × 150 m rectangle. Figure 3.2 has two ground views of that field in 1974 and 1975.[1] The Ringlet wanders about in dry upland fields, without any discernible pattern. Males appear to encounter females at random, and females may encounter at random the preferred sites for oviposition on thin-bladed grasses.[2]

Ringlets tend not to transgress wooded boundaries, and they, particularly the males, even avoid shaded regions near woods and hedgerows. The few times that I have seen males move through gaps in a hedgerow, when I placed myself in the butterfly's original position, I could see either the horizon or an extensive patch of brightly lit open field on the other side (Figure 3.2B). Males will sometimes orient toward and briefly follow light patches of soil or trampled grass, and at the edge of a field they will sometimes follow along the lit side of the edge of a shadow for several meters before turning away from that shadow. Active females are slightly

FIGURE 3.1. Aerial photo of study site for the Ringlet (1973, courtesy of the USDA Agricultural Stabilization and Conservation Service). North is to the left. Colored in green are trees and woodlands that are sufficiently dense to impede movement of Plain Ringlets. The main study field is the confined area in the lower left, where Ringlets were common. Elsewhere they were found more rarely on the drier (lighter) hillsides. The buildings in the lower right of the study field are those of Figure 0.2.

more likely to penetrate boundaries, flying rapidly through the intervening matrix and slowly through patches of the preferred habitat, laying eggs as they go. Both sexes seem to obey the memorable dictum of Vince Lombardi (Lombardi and Heinz 1963), "Run to daylight!"

In homogeneous but nonpreferred habitat, e.g., lush grassland with wide-bladed species, wet meadow, marsh, sparse parkland, and pond, both sexes tend to fly rapidly in paths that are straight over several tens of meters. These flights over nonpreferred habitat are usually above the highest surface by about 40 cm, very seldom more than 1 m. The difference between rapid straight paths in non-preferred habitat and slow twisty paths (at least for the males) in preferred habitats is sufficient to ensure a bias toward preferred habitat in a heterogeneous landscape (Johnson et al. 1992 present relevant theory, but see Section 8.2.4 for critical

FIGURE 3.2. The preferred habitat of the Ringlet is dry upland field. A. Main field for study of Ringlets in July 1974, looking south from near northwest corner. Males would sometimes orient toward and briefly follow light areas like the tire tracks above. B. Episodically permeable gaps in hedgerow on east side of main study field in August 1975. Inset: Female Plain Ringlet #74.19 from the first brood, on timothy grass *Phleum pretense*, 26 June 1974. Ringlets will not transgress wooded boundaries like those in (A) above. They tend to avoid even areas of shadow like those along the hedgerow in (B). The very few times that I saw them move through gaps such as those in (B), if I positioned myself where the butterfly initially was, I could invariably see either the horizon or an extensive patch of brightly lit open field.

commentary). And avoiding or reflecting from boundaries helps to confine the Ringlets to the preferred habitat once they are there.

### 3.2.2 Local Population Demography

Given the general confinement of Ringlets in fields bounded by woodland, I have used the actual number of individuals known to be alive as a close approximation of relative numbers, for both survivorship (Figure 3.3) and phenology (Figure 3.4). Since the losses in Figure 3.3 combine both mortality and emigration, the figure is local residency, rather than true population survivorship, but the language of "survival" is retained for rhetorical purposes. For the Ringlet, most butterflies were recently eclosed when newly marked (74% for males and 67% for females). Of the remaining relatively worn individuals (26% for males; 33% for females), some were probably immigrants from beyond the main study field, and a few may have been present since eclosion, but undetected prior to marking. So marked individuals that were later not observed were likely dead or departed, with a very

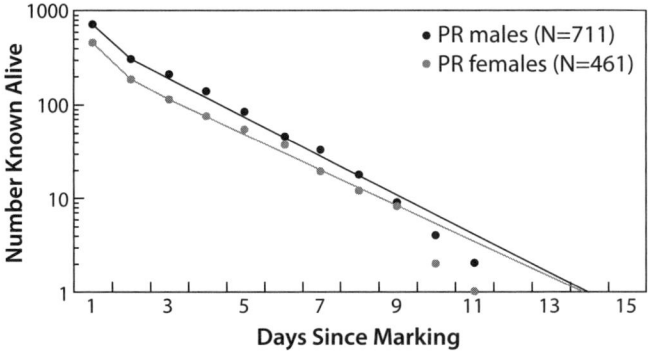

FIGURE 3.3. Ringlet local residency ("survivorship"). The number of individuals known to be alive includes all marked butterflies that were caught on that day or some later day. This is lower bound on the number of residents in or near the study area, and is amenable to demographic analysis of local "survival." Data are pooled from all years and both broods. Blue dots are males; pink are females. "Days after marking" approximates adult age, especially for males, because the proportions freshly eclosed when marked are 0.74 for males and 0.67 for females. The scale is semilogarithmic; so a straight line represents a constant loss rate per day, due to deaths and emigrations. The lines are maximum likelihood bilinear fits with constant survival after Day 2. First day survivals are 0.43 for males and 0.40 for females. Subsequent daily survivals are best estimated as 0.61 for males and 0.65 for females. Survivals beyond Day 9 are likely less, but not worth estimating given the low sample sizes, and the low probability of reaching that age. After Day 1, the remaining mean "lifetime" (literally local residency) is constant for both sexes, 1.6 days for males and 1.9 days for females.

few being locally hidden. In any case, dead individuals or permanent emigrants are lost to the local population, with all the calculable effects on local population dynamics (discussed more thoroughly in Section 9.3.4). With that proviso, the data of Figure 3.3 show a local survival rate after Day 1 of 0.61 for males and 0.65 for females. If those local survival rates are constant, as they are in Figure 3.3 until the data become few enough to be problematic, then the mean remaining local lifespan is also constant, independent of age. If daily survival is s, then the mean remaining lifespan is $\sum_{x=1,\infty} (s^x) = s/(1-s)$. So the mean remaining local lifespan is 1.6 days for males, and 1.9 days for females. Both sexes of Ringlet are short-lived as adults.

Dissection of a female can reveal the number of times that she has mated from remains of spermatophores, small elongate capsules, left by each mate in her bursa copulatrix, a small sac that is part of her reproductive system. Her oviducts contain developing eggs, the size of which can show how ready they are to be fertilized and laid. Table 3.1 shows the results of dissection of ten females. The data are consistent with the following generalizations. Females tend to mate only once (data are in Table 1.3). That mating is soon after eclosion, yet active virgin females are sometimes found by me, let alone by male Ringlets. Completed eggs are available about a day after eclosion.

TABLE 3.1. Female Ringlet Fecundity

| Date of dissection | Estimated days since eclosion | Spermato- phores upon dissection | Full-sized eggs upon dissection | Eggs laid in (days of) captivity | Notes |
| --- | --- | --- | --- | --- | --- |
| 10 Aug.75 | 0 | 0 | 0 | – | |
| 10 Aug.75 | 0–1 | 1 | + | – | |
| 10 Aug.75 | 1–2 | 1 | 0 | – | |
| 20 Aug.75 | 3+ | 1 | 2 | 15 (1) | |
| 21 Aug.75 | 3+ | 1 | 1 | 17 (2) | |
| 21 Aug.75 | 3+ | 1 | 0 | 9 (2) | larval parasitic wasp inside |
| 21 Aug.75 | 5+ | 1 | 6 | 10 (4) | |
| 21 Aug.75 | 6+ | 1 | 2 | 31 (5) | |
| 10 Aug.75 | 7 | 1 | + | – | |
| 10 Aug.75 | 9 | 1 | + | – | |

*Note:* Estimated days since eclosion is the number of days since the female was observed to have no wear whatsoever on the scales of wings or body. Spermatophores were easy to identify in the copulatory bursa of this species, and there was never any suspicious tissue that could have been a shriveled spermatophore. The data are consistent with the following: (a) females mate only once, (b) likely on the day after eclosion or the next, (c) completed eggs are available about the day after eclosion, and (d) eggs laid per day can vary from about 2 to about 6.

### 3.2.3 Male and Female Phenology

Figure 3.4 shows, by date in the second brood of 1975, the number of adults known to be alive for both males and females. The median date for male numbers is 11 Aug; that for females is 13 Aug. This suggests that males tend to pre-eclose females by about two days. Such "protandry" (Greek for "male first") is expected when there is intense competition among males for females that accept only a single mating (Fagerström and Wiklund 1982, Odendaal et al. 1985). Two days may not seem like much, but it exceeds the mean expected local lifespan of males. Also plotted in Figure 3.4 is the daily number of newly marked, mostly freshly eclosed, females, which are likely to be the only candidates for mating on that day. The daily ratio of males known alive to females newly marked is an index of the number of males competing for a virgin female; that number averages $3 \pm 0.4$ males/female, with a range of 1 to 7. This confirms that competition among males is indeed intense.

Heinrich (1986a) studied thermal regulation and flight activity of Ringlets near Mount Blue State Park in western Maine, from mid-June to early July, the period corresponding to the first brood of my population. He found a sharp drop in activity at ambient temperatures below about 20°C. My observations are in concordance, though being both further south and later in the summer, my weather was uniformly more salubrious. Figure 3.5 shows that for only 2.5 days of the 22 days of observation was the temperature low and the weather poor; on all other days Ringlets were active for much of the day, as evidenced by the observations of Figure 3.4.

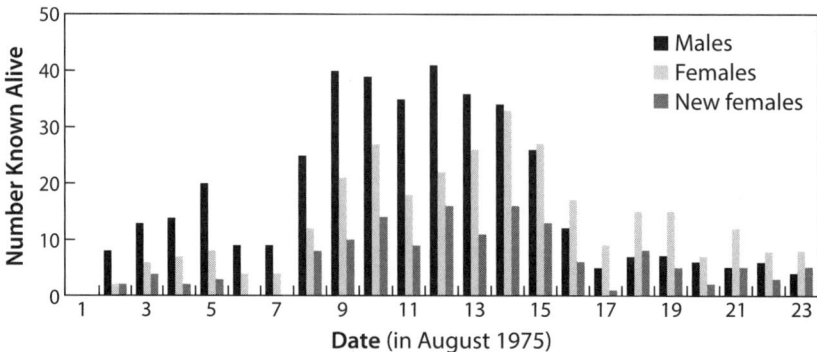

FIGURE 3.4. Number of individuals known to be alive during August 1975 in the 0.76 hectare main study field. Males are light blue and females are pink. Also shown in red is the number of females initially marked, hence freshly eclosed and potentially available for mating, on each day. Males tend to pre-eclose females by about two days.

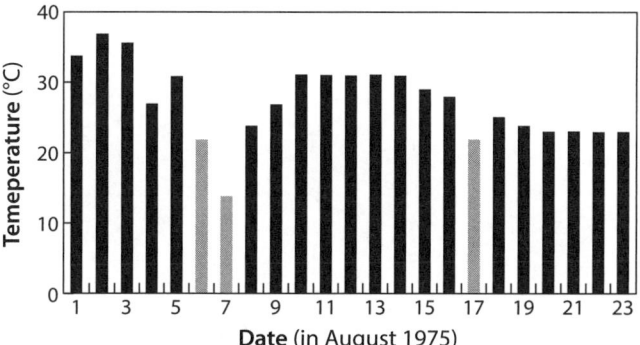

FIGURE 3.5. Maximum shade temperature versus date. On warm days, recorded in orange, the temperature was generally above 20°C (68°F) from 09:00 EDST to 18:00 EDST. The blue days were cool and rainy, 6 and 7 Aug. all day, and 17 Aug. all morning, . . . and Ringlets were not flying (cf., Heinrich 1986a). Hence the deficient numbers on those dates in Figure 3.4.

## 3.2.4 Meeting and Mating

In Chapter 2 I argue that a sitting butterfly can find a flying one more easily than a flying butterfly can find a sitting one (Section 2.3.4). Flying male Ringlets turn to investigate other flying Ringlets (usually other males) at a distance of about 26 ± 1 cm. They also investigate brown Ringlet-sized objects like the dried floral heads of star thistle (*Centaurea maculata*) at about the same distance. However, two of the three complete encounter-to-mating sequences that Diane Wiernasz and I have seen in the field were initiated when a perched female rose and flew about 1 meter directly toward a flying male. The pair then landed, and, at least for the pair that I witnessed, they were in copula within 1.5 seconds. These observations suggest that a male flight may be less the male's actively searching for a female than the male's putting himself in a favorable position to be found by a female.

## 3.2.5 Movements of an Exemplary Male

Figure 3.6 maps the movements of an exemplary male over a distance of 535 m in half an hour. This is the longest continuous path I have for a single individual. There is no obvious systematic pattern, and the wanderings overlap with those of other Ringlets, both male and female. This suggests that an initial model of Brownian motion, a blindman's buff in which the butterfly flies a short distance in a random direction and then chooses a new direction at random, flies a short distance, and so on. The classic paper by Skellam (1951) derives a relationship for

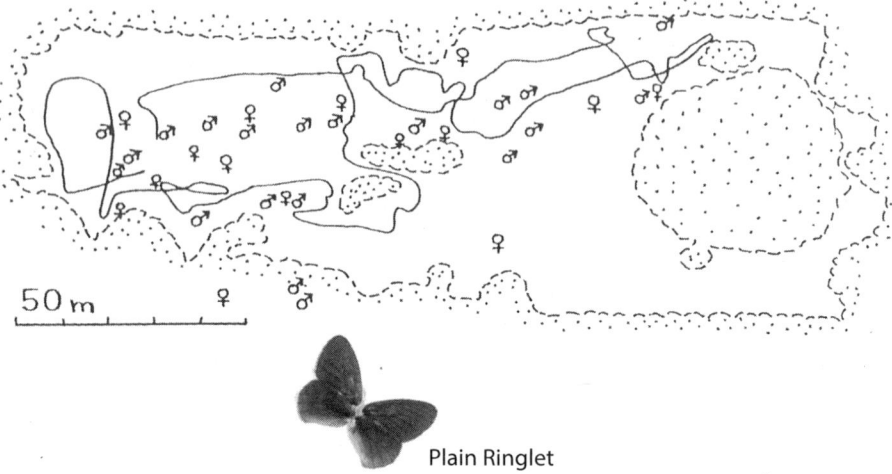

FIGURE 3.6. Track of a patrolling male Ringlet over a distance of 535 m in 30 minutes on 14 Aug. 1981 in the main study field. Stippled areas are wooded at sufficient density that Ringlets tended to avoid them. The gender symbols are the centroids of positions of 21 other males and 15 females that I located either while recording this track or in systematic searches of the whole field immediately before or afterward. This figure supports two inferences: (1) A male tends to wander about the preferred habitat rather than covering it systematically, and (2) the wanderings of one male overlap those of other males and of females.

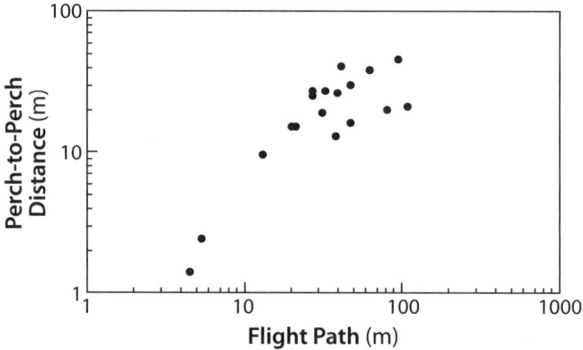

FIGURE 3.7. Male Ringlet displacement versus flight distance. Data are from segments of flights like the one in Figures 3.6 and 3.9. Both axes are scaled logarithmically; so linear movements would propagate as a line of slope 1, and a Brownian random walk would have a slope of ½ because displacement would be proportional to the square root of the path length. A very rough caricature of the data above could be linear movements of a few meters, Brownian motion over tens of meters, and a tendency to level off as movements begin to be limited by the geometry of the field over a range between its width (56 m) and its length (150 m).

Brownian motion in which the net distance moved is proportional to the square root of the total path length. A heuristic defense of this relation is that every move is equally likely to take the butterfly toward or away from where it eventually ends up. Figure 3.7 tests Brownian motion as a model for Ringlet movement with a log-log plot of perch-to-perch distance against flight path, with an expectation of a slope of ½ for true Brownian motion (following Kareiva and Shigesada 1983, and Levin 1986). The data of Figure 3.7 are scattered, with only a bare suggestion of straight-line movements over distances of a few meters, Brownian motion over tens of meters, and a tendency to level off as movements begin to be limited by the geometry of the field (peripheral dimensions: 56 × 150 m, Figure 3.6). Nevertheless, the upper bound of the data is more direct, i.e., straighter or more forward-biased, than Brownian motion.

## 3.3 RANDOM WALK MODELS

There have been two main approaches to modeling random walks that are more direct than Brownian. Currently fashionable is the so-called Lévy flight, which extends the straight-line movement between turns, specifically choosing the length of that line from a "fat-tailed" distribution, so that long moves are interspersed among smaller ones (Bartumeus and Levin 2008). Turns are still to a random heading, uniform over 0°–360°. The more strict definition of a Lévy flight requires that the distribution of moves be from a scale-free distribution, such as a power function, though Bartumeus and Levin sensibly argue that "Lévy-like" behavior can be obtained from a wider range of models.

A traditional alternative to modeling directed flight is to keep short moves but to bias the heading forward for each new move (Cody 1971). The new heading may be chosen from a number of potential distributions. I shall use the normal distribution, $N(\mu, \sigma)$,[3] which is nicely behaved for small standard deviation (i.e., small $\sigma$), and although circular distributions are more appropriate for models with wider variation in turning angle, such models generally converge on Brownian motion as the turning angle diverges.

### 3.3.1 Prelude to Modeling Movement: Behavior at a Boundary

A crucial component of any model of movement in a bounded arena is the representation of behavior at the boundary of that arena.

The first question to be settled is whether the boundary is semi-permeable or not. The Ringlet moves readily in open fields and meadows, but avoids dense shrubbery or woodland.[4] The whole study area is a mosaic of abandoned agricultural fields in a matrix of hedgerows, partially shrubby meadows, and woodlands (Figures 3.1 and 8.1). The main field in which Ringlets were studied (Figures 3.1 and 3.2) is surrounded by woodland and hedgerow that inhibit movement sufficiently for its boundaries to be impermeable in the models of Ringlet movement that follow in Section 3.3.2.

Figure 3.8A shows qualitatively different behaviors at a boundary and the quantitative changes in heading that are needed to realize them in simulations. Ignoring the boundary may seem silly as "boundary" behavior, but it is often used in models of movement well away from the boundary, under the technically correct, but misleading phrase "periodic boundary conditions." An Ignore Model is realized by having the butterfly reappear with the same heading at the opposite side of the arena, essentially mapping movement on a toroidal lattice. It is a mistake to consider such a model as unbounded; indeed, once an X by Y

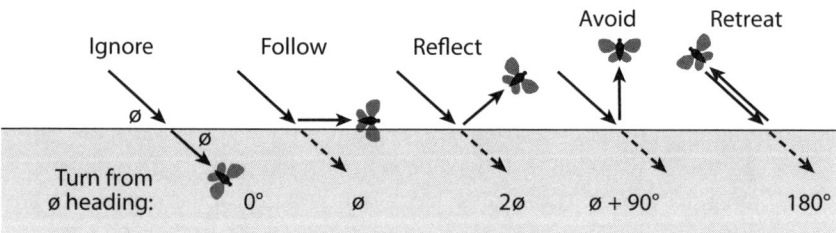

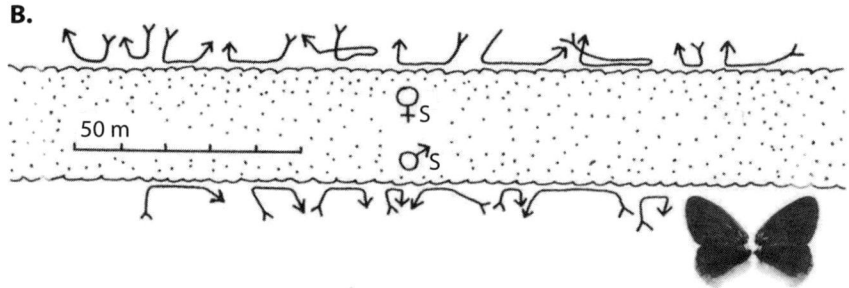

FIGURE 3.8. Behaviors of model butterflies and real Ringlets at a habitat boundary. A. Plausible behaviors of a model butterfly at a boundary, and the angles that it would have to turn from an initial angle of incidence of ø to the boundary. B. Real Ringlet tracks at the edge of a field. The behavior of real Ringlets combines the Follow Model for a variable but short distance with Avoid, such that Reflect is not a gross misrepresentation of the behavior. See Table 3.2 for further support and Section 3.3.1 for discussion.

lattice is wrapped into a torus, the farthest apart that any two points can be is the same as that on a bounded X/2 by Y/2 lattice. Accordingly, an Ignore Model on a toroidal lattice is inappropriate for simulating encounters between two butterflies. However, it is potentially useful for comparisons among models with different structure, or with different parameters for movement of a single butterfly in an open habitat, because it imposes no additional turn on the model behavior.

Models that Follow the boundary, Reflect from it, Avoid it, or Retreat, are realized by turns that range respectively from the angle of incidence to 180° (Figure 3.8A). Plausible cues for sensing such turns are available from optic flow of the visual field, which of course includes the boundary (see Chapter 2, Section 2.3.4). The mean and standard deviation of turns that the different behaviors impose can be calculated by assuming that the angle of incidence $\phi$ varies uniformly from 0° to 90°.[5] This is done in Table 3.2.

Figure 3.8B shows the behavior of real Ringlets at the boundary between field and woodland.[6] The behavior is variable, but combines a tendency to Follow the boundary for 2–10 m, then a turn to Avoid it; the sum of these two behaviors would have a theoretical standard deviation of ±37° (calculated from Table 3.2). The random walk model that best fits male Ringlet paths in the open is forward-biased with standard deviation of ±55°. Accordingly, to Reflect at the boundary (standard deviation = ±53°) is a better approximation of reality than either to Follow + Avoid (standard deviation = ±37°) or to Avoid alone (standard deviation = ±26°). This fortuitous accident favors modeling the ranging of a male Ringlet with a forward-biased random walk that Reflects from the boundaries of its arena.

To Ignore the boundary is the favored behavior for making comparisons among different models. It is easy to program for many kinds of ranging behavior, including the Lévy Walk. It avoids the need to correct for accumulations of model butterflies at the boundary seen with the Follow behavior, or deficiencies near the boundary seen with Avoid or Reverse. And it allows the next move of a Brownian Walk to have the same distribution of turns as the previous one (mean ± standard deviation = 180° ± 104°).

TABLE 3.2 Expected Turning Statistics for Model Behaviors at Boundary

| *Behavior relative to boundary:* | *Ignore* | *Follow* | *Reflect* | *Avoid* | *Retreat* |
|---|---|---|---|---|---|
| Turn from prior path at angle ø: | 0° | ø | 2ø | ø + 90° | 180° |
| Mean turn ± standard deviation* | 0° ± 0° | 45° ± 26° | 90° ± 53° | 135° ± 26° | 180° ± 0° |

* for ø varying uniformly between 0° and 90°

## 3.3.2 Random Walk Models Tuned to Real Ringlets

The parameters of either a Lévy flight or a forward-biased random walk can be visually "tuned" to match the appearance of the flight paths of male Ringlets. Figure 3.9A shows 20 flights of male Ringlets in the main study field. In Figure 3.9B is a model Lévy flight with step lengths chosen from a power law with exponent 2, which, in theory and in simulations, compensates for the exponent of ½ in a Brownian walk. The Lévy steps in Figure 3.9B are scaled by a factor 6, which gives the best visual approximation to the real paths of Figure 3.9A.

Figure 3.9C shows a model forward-biased random walk with a constant step size and a new heading at each step that departs from the previous heading by a

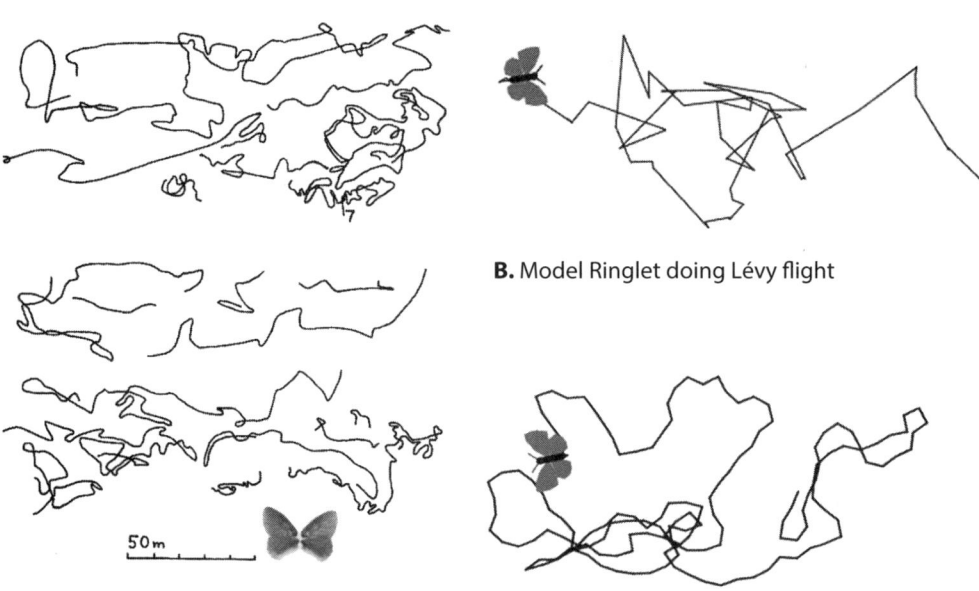

**A.** Paths of real Ringlets  **B.** Model Ringlet doing Lévy flight  **C.** Model Ringlet doing directed random walk

FIGURE 3.9. Paths of real male and model male Ringlets. A. The real paths are separated by individual, but oriented as they were in the main study field, whose rectangular geometry obviously influences their aspect ratios. B. The Lévy flight has step lengths chosen from a power law distribution with exponent 2, which, in theory and simulations, compensates for the exponent of ½ in a Brownian walk. The Lévy steps are scaled by a factor 6, which gives the best visual approximation to the real paths. The forward-biased random walk (C) has a constant step size and a new heading at each step that departs from the previous heading by a random angle chosen from a normal distribution with mean 0 and standard deviation 55° (hereafter abbreviated N[0°, 55°]). That 55° standard deviation gives a visual approximation to the real paths. Even 60° looked too twisty, and 50° too direct. See Section 3.3.2 for details, interpretation, and conclusions.

random angle chosen from N(0°, 55°). The standard deviation of 55° gives a good visual fit to the real paths in Figure 3.9A; even 60° looked too twisty, and 50° too direct; and the more extreme 45° and 65° gave poor fits. The forward-biased random walk of Figure 3.9C gives a better visual representation of the real flights of Figure 3.9A than does the Lévy flight of Figure 3.9B. Nevertheless, both models are close enough approximations to reality that their parameters are used to simulate comparisons among models in Section 3.3.3.

The same procedure can be applied to female Ringlet flights. The recorded female flights are more direct than those of males. The best visual fit for a female model is with a forward bias of N(0°, 40°); it fits the real turns but underestimates the length of straight segments; N(0°, 35°) fits the straight segments but not the turns; and N(0°, 45°) is too twisty. Furthermore, after landing, females often launch to continue in their landing trajectory. In most cases I could not get close enough to stationary females to determine whether they were laying eggs without potentially disturbing their flight rhythms or paths. An additional constraint is the size of the field itself (Figure 3.6), whose maximum width (56 m) or length (150 m), limited long, straight flights by female Ringlets. Accordingly, I have not illustrated the tracks of real females, and I do not take the model of their movement with a forward bias of N(0°, 40°) seriously, other than as a clear contrast with the movements of males.

Figure 3.10 plots displacement versus flight distance movements conforming to the model of Figure 3.9C, a random walk that is forward-biased by an angle N(0°, 55°) for males and N(0°, 40°) for females. The model output is plotted along with real flights of male and female male Ringlets, respectively, in Figures 3.10A and 3.10B. The dearth of short flights is due partly to facts that 1 meter was about the resolution of my mapping, and that 3 meters was about the lower limit of flights worth recording. The longer perch-to-perch distances were limited by the dimensions of the field (about 56 × 150 m; Figure 3.6), and the flights within those distances may have been further confined and straightened by channels between the wooded boundaries and internal vegetation. Model movements were simulated on a 31 × 31 m grid, wrapped on a torus; the maximum length of model flights is therefore the half-diagonal of the unwrapped grid, i.e., $(31/2)(\sqrt{2}) = 22$. These deficiencies do not affect the following conclusions.

The movements of males (Figure 3.10B) are within the cloud of points generated by simulation of a tuned forward-biased, at N(0°, 55°), random walk, but the intrinsic variation in both data and simulation make critical tests problematic. The variation also implies huge potential variation in the reproductive success of individual males that are all employing the same strategy. Following the rhetoric of Railsback and Grimm (2012), I do not propose that male Ringlets literally engineer a random walk for their search, but rather I suggest that a model of a

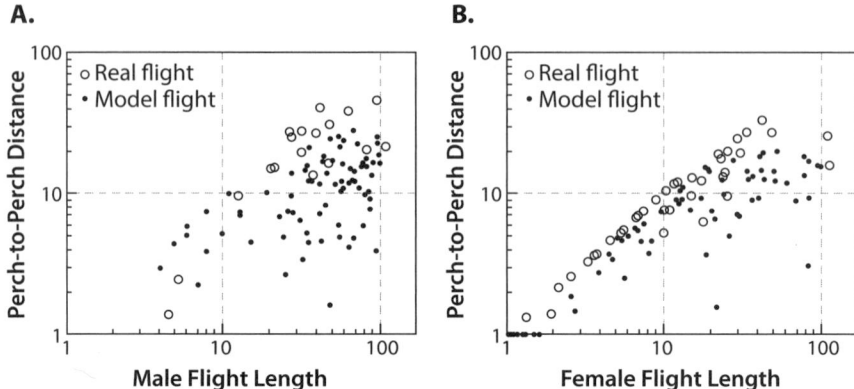

FIGURE 3.10. A. Male displacements vs flight distances are plotted on log-log scales for simulated flights like those of Figure 3.11C, and compared with the real flights of Figures 3.7 and 3.9A. The real male flights are well represented by the model flights, which are forward-biased with N(0°, 55°), though real flights of intermediate length have longer displacement (i.e., are more direct) at the upper bound. B. The same for females, the model flights being forward biased with N(0°, 40°). The real female flights are also well represented by the model flights. For both sexes, the rectangular aspect ratio of the field and the interior woody vegetation (Figure 3.6) may enforce narrow and straight corridors on medium and long flights. And both simulations, being on a 31 × 31 m square wrapped into a torus, are limited to a maximum displacement of about 22 m. The following conclusions are not affected by these biases of both reality and model. Female movements are more nearly linear over all distances than males. Appropriately Forward-biased Vagrancy is a plausible model for real movement of Ringlets, though the intrinsic variance shown by the model makes a critical test of the match problematic.

forward-biased random walk is an adequate description of a real Ringlet path to estimate his rate of encounters with females and to illustrate his challenge in finding a mate (Section 3.4.2). Accordingly, except where otherwise noted, the model for male Ringlet behavior will be a forward-biased random walk, with short movements at a heading that changes with a normal distribution N(0°, 55°) for each move, and reflection from boundaries at the angle of incidence.

Female movements are more nearly linear over all distances than male, both in reality and in the forward-biased random walk. Forward-biased Vagrancy is a plausible model for real movements of male Ringlets.

### 3.3.3 Brownian Motion, Lévy Flight, and Forward-Biased Random Walk

Figure 3.11 compares Brownian motion, Lévy flight, and two forward-biased random walks with different biases. Each simulation takes 960 steps on a 31 × 31 (= 961

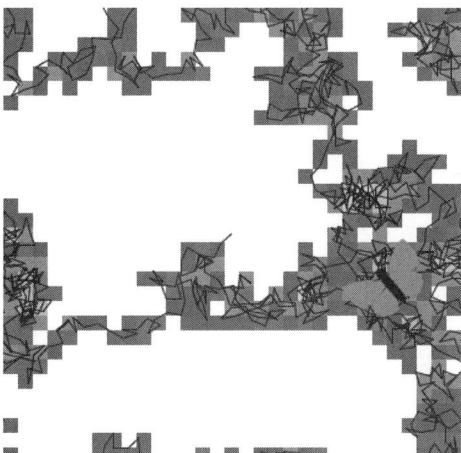

**A.** Brownian Random Walk:
Unit Steps at Rnd(0-359°)
Coverage = 37%

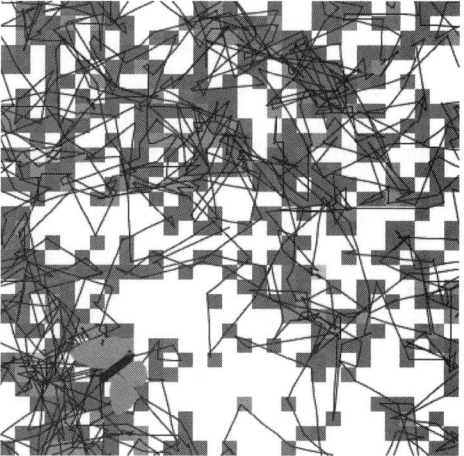

**B.** Random Lévy Flight:
Steps = 6*Rnd(1)² at Rnd(0-359°)
Coverage = 51%

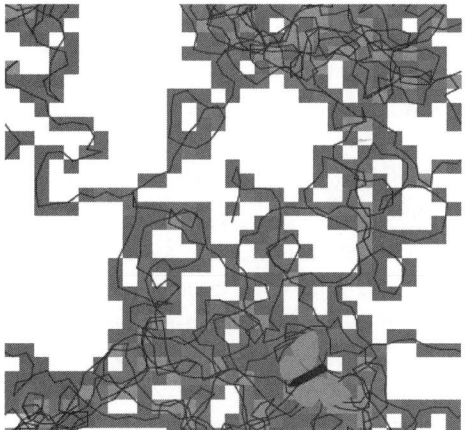

**C.** Forward-biased Random Walk:
Unit Steps at Heading + N(0, 55°)
Coverage = 54%

**D.** Nearly Straight Random Walk:
Unit Steps at Heading + N(0, 5°)
Coverage = 63%

FIGURE 3.11. Various forms of random walks are compared on a 31 × 31 (= 961 patch) toroidal lattice. Each walk takes 960 steps. Parameters of the random Lévy flight (B) and the forward-biased random walk = Forward Vagrancy Model (C) are adjusted to give a visual approximation of empirical short flights by Ringlets (see Figure 3.9). Unvisited patches are white; visited patches are colored by number of visits in a 10-grade warming sequence from blue (1 visit) through red (10 visits). "Coverage" is the percent of patches visited at least once. See text for details and interpretations.

patch) lattice. The lattice is wrapped on a torus, so that a creature flying off the top edge returns on the bottom; off the right, on the left; and both vice versa. Unvisited patches are white; visited patches are colored by number of visits in a warming scale of blue for one visit to red for 10 or more visits. The complete track of the wandering male is recorded as a black line. And the percentage of patches visited is given for each model.

Figure 3.11A records a Brownian random walk. The step length is a unit from one patch to its next neighbor. The heading for each step is a uniform random direction, equiprobably any number of degrees from 0° to 359°, or in shorthand, Rnd(0–359°). Given that 960 steps are taken to search 961 patches, the coverage of 37% is poor, and clumping of multiple revisits is dramatic. Multiple visits, suitably timed, would be adaptive for a renewing resource, but the clumping of Brownian motion is short term, and the resource of the male Ringlet, namely virgin females, does not renew within a day. Therefore, the Brownian random walk is both ineffective and inefficient.

Figure 3.11B records a Lévy flight with parameters tuned to match the short flights of a real Ringlet (step length = $6*[Rnd(1)]^2$; heading = Rnd[0–359°]; cf. Figure 3.9). The coverage of 51% is much better than that of the Brownian random walk, and the clumping and multiplicity of revisits are much reduced. However, the track of the male passes over many patches without searching them. So the Lévy flight is effective, but inefficient.[7]

Figure 3.11C records a forward-directed random walk with parameters tuned to match the short flights of a real Ringlet (unit steps at new heading = old heading + N[0°, 55°]). With a coverage of 54%, this forward-directed random walk is as effective as the Lévy flight of Figure 3.11B, and far more effective than the Brownian motion of Figure 3.11A. There are multiple revisits to some patches, but they are not severely clumped. And every patch flown over is visited. So a forward-biased random walk can be both effective and efficient.

Figure 3.11D records a forward-biased random walk with a heading so biased (N[0°, 5°]) that it is very nearly straight. This model is intended to imitate a systematic "raster" search with just enough random curvature to prevent its getting caught in periodic repeats that look like the bars that plagued ancient television screens. Accordingly, coverage is very good at 63%. Revisits are rare, scattered, and generally of low multiplicity. Thus the semi-systematic search of Figure 3.11D is somewhat more effective and efficient than the forward-biased random walk tuned to the real Ringlet (Figure 3.11C).

The overall result of Figure 3.11 is that the forward-biased random walk, tuned to visual resemblance to the real paths of Ringlets, is an efficient and effective searching algorithm, and it is therefore the basis for further explorations.[8]

## 3.4 INSIGHTS FOR RINGLETS FROM FORWARD VAGRANCY MODELS

### *3.4.1 Waiting Time and Female Movement*

So far, the models have implied that the male moves through the appropriate habitat searching for a stationary virgin female. Empirically, Diane Wiernasz and I encountered virgin females very rarely in the flying population, and two of the three complete courtships that we observed comprised a sitting female rising to an encounter with a passing male. So a natural question to pose is: Would mate-encounter be faster if virgin females actively searched, just as the males do?

It is difficult to intuit the answer. One could argue on the one hand that with both sexes moving, the velocity-dependent rate of potential encounter is effectively about doubled over that of one sex moving and the other remaining stationary. On the other hand, if one concentrates on any particular movement interval when the sexes are distant from one another, each party is equally likely to move toward or away from the initial position of the other. One quarter of the moves will then have both moving toward each other, one quarter away, and the remaining half will have them moving in the same direction, to a first approximation not changing their separation. So the velocity-dependent rate of encounter would be cut in half. The net effect would be that having the female move would have minimal effect on the encounter rate.

I simulated this problem with a model in which the male flies with a forward-biased random walk, with heading changing at each step by $N(0°, 55°)$ as in Figure 3.9, to encounter a female; and I accumulate the number of steps from start to encounter. After each encounter, a new female appears at a random location, and she either stays there or moves with the same rules as the male until encountered. In Figure 3.12, the resulting spectrum of times to encounter is subjected to the same kind of survival analysis as are the local persistences in Figure 3.3. The distribution of waiting times to encounter is very close to exponential with fixed exponent, and the maximum likelihood exponents are the same whether the female moves or remains stationary. In short, as long as the male moves, whether the female moves makes little difference to the rate of male-female encounter for empirically realistic parameters.

### *3.4.2 A Model Male in a Real Field*

How many virgin females is a male likely to encounter in his travels? Figure 3.13 simulates the wanderings of a model male Ringlet in a single day in the main

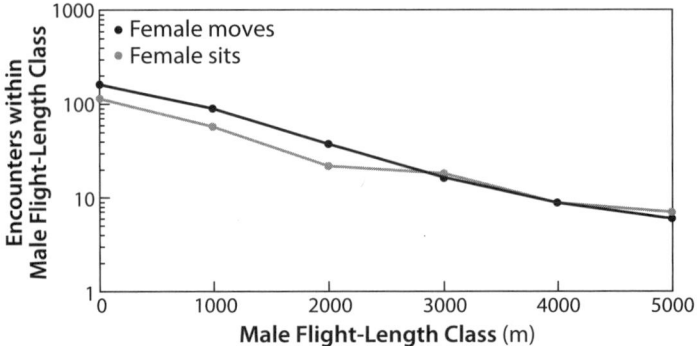

FIGURE 3.12. Frequency distribution of distances that a model Male flies, by the rules of the Forward Vagrancy Model, before encountering a Female. Output is binned in 1 kilometer classes for simulation of 100,000 1-meter steps in a forward-directed random walk with each step at an angle chosen from N(0°, 55°). The curves are respectively green for a female that moves with the same rules as the male until the encounter, and red for a female that sits at a randomly chosen place until encountered by the male. These are essentially survival curves with "death" being an encounter (cf. mating as "le petite mort"). Hence the mean waiting time to encounter is the exact analog of mean lifespan. The waiting times are very close to exponential with fixed exponent, and the maximum likelihood of the exponents are about the same whether the Female moves (Pr [encounter in next km] = 0.51) or does not move (Pr = 0.52).

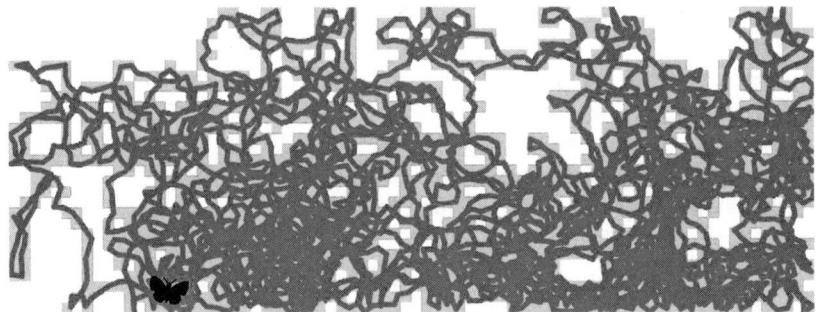

FIGURE 3.13. The Forward Vagrancy Model simulates wanderings of a male Ringlet in a single day in the main study field. See Section 3.4.2 for details and discussion.

study field. The 56 × 150 m field is broken into 28 × 75 patches, each 2 × 2 m. Those patches correspond to the maximum distance within which a stationary female might be able to respond to a flying male (about 1 m), and to the approximate mean free step in a straight line by a male (2 m). Unvisited patches are green; visited patches are light orange. The potential path observed by the male is 1 m wide, corresponding to the maximum distance at which a flying male responded to a female-sized target (0.5 m). The total flight path is 8 km, corresponding to the maximum distance that a male could fly in the typical 8 hours of thermally permissible time on a good day (as in the path of 535 m in 30 minutes recorded in Figure 3.6). The movement model is a forward-biased random walk with headings varied by $N(0°, 55°)$ for each step (as justified in Figures 3.9 and 3.10B and Section 3.3.3). The total coverage of 1,551 patches is 74% of the 2,100 available in the field.

Over the field as a whole, in an average day, the ratio of males to newly emerged females (hence virgins and candidates for mating), is $3 \pm 0.4$, or about 0.33 potential mates per male (Section 3.2.3). With the male covering 75% of the field, this reduces the probability of finding a mate in a continuously searching day to about 0.25. The expected lifespan of a male is only 1.6 days (Section 3.2.2). So the majority of males will not mate, and those few that do will likely have spent all their thermally permissible adult time searching—or rather, flying about to be found by an unmated female.

## 3.5 DISCUSSION: ANDRO- VERSUS GYNOCENTRIC INTERPRETATION

When I began my fieldwork with butterflies, and especially with the Ringlet, I assumed that males were searching for females. All the literature that I had read predisposed me to this androcentric view, and it seemed to fit the fact that males moved about far more obviously than females. My favorite interpretation of this asymmetry in movement is Corner's (1964, p. 86) classic statement: "If two persons wish to find each other it is better for one to wait while the other searches. And if on meeting they must journey, it is better if she who waits should be provisioned while he that searches may travel light and fast. The principle of assignation was worked out long ago by gametes. A well-stocked egg, without flagella and too big to be moved by flagella, is fertilized by a small or undersized but active male flagellate as the spermatozoon." The demonstration that movement by the female does not appreciably increase the likelihood of encounter (Section 3.4.1) reinforces this interpretation.

In addition, Chapter 2 (Section 2.3.4) suggests that a stationary female is more likely to perceive a flying male against the uniform background of open sky than is a moving male likely to perceive a stationary female against a complex background of vegetation that is nearly the same color as she is. And sure enough, two of the three observed Ringlet matings were a result of a perched female flying out to a passing male at about the average distance of detection (Section 3.2.4). The stationary female finding the moving male is consistent with visual physiology and behavioral observation, and suggests a gynocentric correction to my previously androcentric perspective. Who is the searcher and who the quarry is a moot question. It is the interaction per se that is important (Corner's "assignation" above), rather than characterizing the participants as active versus passive.

## 3.6 SUMMARY OF RESULTS *AND SPECULATIONS*

### *3.6.1 Habitat and Lifespan*

Ringlets fly directly over tens of meters in nonpreferred woodland or shrubby habitat, but twist about in their preferred dry open fields. This behavior, plus avoiding wooded boundaries of a field, is sufficient to restrict most of their activity to dry open fields (Rule & Behavior: Section 3.2.1).

Ringlets are short-lived as adults, the mean remaining local lifespan settling to 1.6 days for males and 1.9 days for females (Life History: Section 3.2.2 and Chapter 9).

### *3.6.2 Female Monandry Constrains Male Behavior*

Females tend to mate once, and soon after eclosion, but free-flying virgins are found occasionally. Females enforce monandry; *males may be capable of polygyny*. So males must find virgins, *but females need only find a male that has not very recently mated*. Consequently, there is more urgency for males to find females than for females to find males (Behavior: Section 3.2.2).

Monandry means that males need to pre-eclose females, not only to be ready to mate, but also to expose themselves to as many freshly eclosed females as possible. Accordingly, males can further increase their chances of mating by patrolling. Indeed males pre-emerge females by about two days, their expected local lifespan (Behavior & Life History: Section 3.2.3).

### *3.6.3 Random Vagrancy is Effective and Efficient.*

*A forward-biased random walk can be an efficient tactic for covering the habitat inhabited by virgin females.* In particular, a visually plausible representation of the track of real male Ringlets is a forward-biased random walk with turns to the next move drawn from a normal distribution with mean $0°$ and standard deviation $50°$ (in abbreviated notation, $N[0°, 50°]$), and with reflection from a boundary at the angle of incidence, the "Tactical Forward Vagrancy Model" (Rule & Behavior: Sections 3.2.5, 3.3.1, 3.3.2).

Plots of net distance moved versus path length show high variation for both real flights of males and the output of the reality-tuned Forward Vagrancy Model . Although the overlap between reality and the model is encouraging, the variance of both hinders a critical test (Section 3.3.2).

A forward-biased random walk can be as efficient as a Lévy flight, *and places smaller demands on visual, cognitive, and behavioral capabilities* (Rule & Behavior: Section 3.3.3).

If males patrol, then females *may not need* to wander in order to find, or to be found by, a mate. Females would not appreciably increase their rate of mate encounter by wandering themselves, except when males are rare, as they may well be late in the eclosion season (Behavior: Section 3.4.1).[9]

The Forward Vagrancy Model strongly suggests that a male *may have to* spend his entire weather-permissible adult lifetime on the wing to find a mate, or possibly for a mate to find him (Behavior & Life History: Section 3.4.2).

*3.6.4 Who is Searching for Whom?*

Two of the three Ringlet courtships-to-mating were initiated by a female flying up to a patrolling male. The result *may be interpreted in the light of modeling as:* males wander in order to place themselves in a position to be found by virgin females. Who is "searching" for whom in this scenario is a moot point, and it may be impossible, empirically, conceptually, and even epistemologically, to ascribe: . . . activity versus passivity, . . . male versus female roles, . . . and even cause versus effect. Rather, mate-finding is a matter of interaction between male and female, and should be described as such (Rule & Behavior: Section 3.5).

CHAPTER FOUR

# Fortuitous Site-Fidelity?
## *The Eyed Brown* Satyrodes eurydice

> The butterflies have a very feeble, delicate flight, dancing lazily hither and thither among the herbage, flying generally but two or three feet above the ground; yet when alarmed their movements are more powerful than one would suppose possible in an insect with wings of so delicate a texture.
> —Samuel Hubbard Scudder, 1889, *The Butterflies of the Eastern United States and Canada with Special Reference to New England*, p. 198.

I found that in addition to episodes of powerful flight in interactions between males, Eyed Browns are robust when handled, and they live surprisingly long lives for their delicate appearance and seemingly delicate behavior.

The Eyed Brown stays very local in open sedgy marshes and wet meadows, where the males fly weakly just above or within the sedges, often perching and basking (Clark 1932, Angevine and Brussard 1979, Scott 1986, Opler and Krizek 1984). Adults may sip nectar, but more often feed on sap and decaying vegetation (Cech and Tudor 2005). Females lay eggs on several species of sedges (*Carex* and *Scirpus*, Scott 1986, Cech and Tudor 2005). Both ovipositions that I have observed were of a cluster of three eggs laid on the underside of the blade of a wide-bladed yellow-green *Scirpus*, at the peak where the blade turns over to hang down (Appendix B.3).

The question mark in the title of this chapter is no accident. My story for the Eyed Brown is incomplete because the subpopulation on my study site was small to begin with, but declined to near zero between 1978 and 1980 just as I was beginning to document details of male behavior and to hypothesize about how that behavior might be organized. I episodically visited another subpopulation at a nearby secondary site, at which Jeff Georgia (1978) had made parallel observations in 1977. However, the geometric configurations of the habitats of the two subpopulations were dramatically different, respectively a wet meadow on the study site (about 25 × 25 m; Figure 4.1), versus a marsh on the episodically waterlogged margin of a nearby pond (50 × 12 m around 0.15 ha of open water; Figure 4.2). The meadow site had few clearly defined narrow "trails" through dense

# FORTUITOUS SITE-FIDELITY?

FIGURE 4.1. Main study site for the Eyed Brown and Viceroy. Male Viceroys perched on tall vegetation and patrolled in the open throughout the meadow, and spent some morning time in the shrubbery at the edge. Male Eyed Browns perched and patrolled in spaces between sedges that dominate the wetter sections of the meadow, with episodic patrols above the vegetation (see also Figure 4.2). Female Eyed Browns were usually found within or near the sedges.

vegetation; the marsh site, many interconnected paths through sparse vegetation. The typical behavior of males seemed to differ between the subsites: local movements along the trails with frequent interactions and apparent site-fidelity in the small population of the meadow, contrasting with wide and overlapping ranging and few interactions in the larger but more widely dispersed population of the marsh. So the behavior could be characterized, with some ambiguity and a few exceptions, as apparent vagrancy like the Ringlet but more localized in the marsh, versus apparent territoriality like the Viceroy in the meadow.

This chapter starts by describing this varied behavior in as much detail as my observations allow. I compare and contrast male behavior of Eyed Browns with the more demonstrably territorial Viceroy of Chapter 6. I then present the results of an experiment designed to impose a particular pattern of ranging on male Eyed Browns. The result was not as predicted, but it still confirmed that the butterflies follow channels in the vegetation, for which a hypothetical cue is orientation toward a V-notch in the very near horizon. Such channel-following inspires a model that could produce varying degrees of site-fidelity, without explicit recognition of a territory as particular real estate. Discussion of the behavior of the Eyed Brown in the

FIGURE 4.2. Male Eyed Brown and its habitat. The male is perched overlooking a channel between clumps of sedge shown in the close-up of the habitat. Males and females both navigate through their habitat by flying along such channels, occasionally rising above the vegetation to move from channel to channel. The pond is the secondary site mentioned in the introductory paragraph of this chapter.

context of the model is satisfying, though not definitive. But the real value of the model comes later (Chapter 8), as a contrast to the behavior of the Great Spangled Fritillary (Chapter 5), Viceroy (Chapter 6), and Pearly Eye (Chapter 7), to show that individuals of those three species *do* recognize particular real estate.

## 4.1 NATURAL HISTORY AND BEHAVIOR

### *4.1.1 Habitat and General Movement*

Adult Eyed Browns are found in wet meadows dominated by the presumed larval food plants, particularly sedges (Cyperaceae).[1] Males patrol channels in the vegetation, or perch on the side of a channel, whence they court passing females

(Figure 4.2). Females are found within the sedges or along the channels, and they tend to be relatively inactive unless disturbed.

Both sexes appear fragile and seem to fly weakly, but they are surprisingly rugged and have a local residence time as adults of as much as two weeks for females and nearly a month for males (Table 9.1). Their movements are locally restricted most of the time, with occasional shifts to a distant patch of suitable habitat. Daily movement averaged 15 ± 3 meters for males and 17 ± 6 meters for females (Table 9.1). But of the 104 marked males, two (#77.6 and #77.57) moved from the meadow subpopulation to the pond subpopulation, a distance of more than 300 meters.[2]

### 4.1.2 Female Behavior and Mating

Females may mate multiply; of 15 females dissected, 3 had two spermatophores each, and 12 had one (Table 1.3). Accordingly, any female encountered is at least potentially a mate.

### 4.1.3 Male Ranging and Behavioral Interactions

When densities are high, males seem to be territorial, though the location of a given male's territory may shift from day to day. Males tend to perch on plants that stick out into channels in the vegetation through which the females fly.

Results of interactions are tallied in Table 4.1. There appears to be no tendency to use a set of specific perches, but after confrontation with an "intruder," a male usually returns to a perch near the one from which he launched himself to investigate (16 out of 17 tallied instances). The "intruder" continues flying in a direction that bears no particular relation to his bearing before the encounter (9 on their initial bearing, 15 counter to it). The encounters show elements of territoriality in that the "intruder" is apparently "driven away," and the "territory holder" remains

TABLE 4.1. Movements of Eyed Browns after Interactions by Prior Status

|  | *Return toward Perch or U-turn* | *Fly in Initial. Direction* |
|---|---|---|
| Initially Perched (all males) | 16 | 1 |
| Initially Flying (a few females) | 15 | 9 |

*Note:* See Section 4.1.3 for details and interpretation. Marginal totals differ because some interactions were between two initially flying individuals.

about where he started. However, all that has really happened is that a flying butterfly, which we defined as the "intruder," has continued to fly, and the butterfly that was formerly perched is perched again. This kind of behavioral encounter will be discussed further for the Pearly Eye in Chapter 7, in particular the question of whether the behavior represents defense of a territory (Section 7.2). When population densities are low, males appear to range over the suitable habitat without regard for each others' wanderings, much like the male Ringlets of Chapter 3.

Males interact with each other, but it is difficult to discern whether this interaction is explicitly territorial, as opposed to a mutual investigation of what must appear from a distance to be a potential mate (cf. Scott 1974). Furthermore, the short-term localization of activity for a particular male may be due to the local pattern of channels in the vegetation, rather than to recognition of a particular piece of real estate.

## 4.2 BEHAVIOR OF MALE EYED BROWN COMPARED WITH VICEROY

Chapter 6 will cite detailed evidence that the male Viceroy is truly territorial, in the sense of recognizing and defending particular pieces of real estate. However, here I argue that the male Eyed Brown is certainly less territorial, and perhaps not territorial at all. The Eyed Brown and Viceroy share parts of their habitat, with the Eyed Brown being found less frequently in dry upland fields (Table 1.2 and Figure 8.1). Individual males of both species are often found in the same general location from day to day. Males of both species engage in interactions that often result in one individual returning to a perch near the one that it left. These are all typical of territorial behavior, though not definitive. So here is a comparison of both species with respect to re-sightings, metrics of interactions and the arena in which they occur, and the behaviors during those interactions.

Male Eyed Browns are confined to sedgy depressions, and they perch and fly low along paths in the vegetation, occasionally flying higher to shift to another nearby path. Male Viceroys range over meadow and adjacent shrubbery, perching on high vegetation and flying over the low vegetation. A few individuals of both species are re-sighted repeatedly. However both species are found in a specialized habitat that, at least on my study site, occurs in small patches surrounded by less favorable habitat. So confinement by habitat preference may promote re-sighting of individuals of both species.

Marked male Eyed Browns and Viceroys are re-sighted near their sites of marking with about the same probability per individual (respectively 52% and 48%; Table 4.2). But among such re-sighted individuals, Eyed Browns that are known to be alive are far more likely than Viceroys to be absent, or missed in surveys between re-sightings (respectively 41% versus 27%; Table 4.2). This might

TABLE 4.2. Re-sightings of Marked Eyed Browns and Viceroys

| Number: | Marked A | Re-sighted B | Re-sightings next survey C | Re-sightings not next, but later D | Total Re-sightings E |
|---|---|---|---|---|---|
| Eyed Brown m | 128 | 67 | 109 | + 76 | = 185 |
| Eyed Brown f | 60 | 21 | 23 | + 12 | = 35 |
| Viceroy m | 66 | 32 | 59 | + 22 | = 81 |
| Viceroy f | 38 | 3 | 6 | + 0 | = 6 |
| | | % Re-sighted = B/A | | % Known alive but "hidden" = D/E | |
| Eyed Brown m | | 52% | | 41% | |
| Eyed Brown f | | 35% | | 34% | |
| Viceroy m | | 48% | | 27% | |
| Viceroy f | | 08% | | 00% | |

*Note:* For each species, "m" = male, and "f" = female. The upper entries are tallies from daily records of marking and re-sighting. After marking each re-sighting is tallied if the butterfly was re-sighted on the next census of the study subsite (C), or if it was not sighted on the next visit but was known to be locally alive because of a subsequent sighting (D); so "D" counts the number of censuses in which the butterfly was "hidden," either on or off the subsite. The lower entries are percentages calculated from the upper entries.

be considered evidence that Eyed Browns are less likely to maintain a perch on a fixed piece of real estate. However, the tall perches of male Viceroys are visible from a greater distance than the Eyed Brown's perches within low vegetation, and this may affect the relative likelihoods of my seeing perched individuals of the two species on a given survey.

Changes in location from day to day and the spread of those locations from their centroid are similar for both species (Table 4.3); so the absolute ranging is similar. However the transect of the Viceroy's contiguous habitat is at least twice that of the Eyed Brown, and the Viceroy's typical maximum displacement during a given interaction is at least three times that of the Eyed Brown (Table 4.3). So the Viceroy's degree of "homing" relative to its potential and actual range of wandering is much greater than the Eyed Brown's. This difference suggests that the Eyed Brown is confined more by habitat-fidelity and the Viceroy more by behavior per se.

So far the observations are all in favor of the Eyed Brown's being less territorial than the Viceroy, but the evidence is indirect. Direct observations of interactions between male Viceroys typically involve a male flying from its perch toward a flying male, an interaction that may depart altogether from the favored habitat, and the former percher returning from a great distance to perch very near its former perch. If the initially flying Viceroy returns to perch at all, it is to a perch at an appreciable distance from the initial percher (details are in Chapter 6, Section 6.3.4). The

TABLE 4.3. Scaling of Habitat and Movement of Eyed Browns and Viceroys

|  | Contiguous habitat scale (m) | Daily movement (m/day) | RMS distance to centroid (m) | Displacement in interaction (m) |
| --- | --- | --- | --- | --- |
| Eyed Brown m | ~ 30 | 15 ± 3 | 49 | ~ 20 |
| Viceroy m | > 60 | 16 ± 6 | 46 | > 80 |

*Note:* Contiguous habitat scale for the Eyed Brown is the longest distance within the study site dominated by sedges. Contiguous habitat scale for the Viceroy goes beyond the border of the meadow because its shrubby perimeter is permeable for routine flights. Daily movement "velocity," and observed lifetime root-mean-square distance to centroid, are calculated from daily re-sighting records (see Section 9.2 and Table 9.1). Displacement during interaction is the typical maximum distance that an initially perching male flies during an interaction before returning to near his original perch. Viceroys made several forays out of sight beyond the shrubby border of the meadow, forays exceeding the 80 meters that I could follow and observe.

interactions between male Eyed Browns are not so starkly definitive, but they show a much weaker asymmetry between percher and flier; namely, after an interaction, the former percher tends to return in the general direction whence he came, while the former flier may either continue in his initial direction or turn in an arbitrary direction (Section 4.1.3). So the direct evidence from the geometry of interactions between Eyed Browns supports some degree of site-fidelity, and is consistent with territoriality that is much weaker than that of the Viceroy.

## 4.3 THE "BUTTERFLY CYCLOTRON"

In an attempt to test these ideas, I scythed a pattern of channels into the meadow in the 1983 season. The channels converged to a central circle, entering it at an acute angle. My intent was to create a sort of Butterfly Cyclotron, idealized and explained in Figure 4.3. Unfortunately, the meadow subpopulation of Eyed Browns declined to inadequate numbers in the year of this experiment, and never recovered. Nevertheless, I imported 15 male Eyed Browns from the nearby marsh/pond subpopulation, and released them one by one in the meadow, with results reported in Table 4.4. Ten individuals flew just above the vegetation for a few meters, failed to encounter my laboriously cut paths, and then, uncharacteristically for the species, flew up to about 10 meters above the meadow, and set a course down the valley, back toward the marsh. It is plausible that they did not recognize the meadow as suitable habitat, and flew toward the cue of a lower horizon, and thus toward wetter land—rather than that they were specifically "homing" toward the marsh.

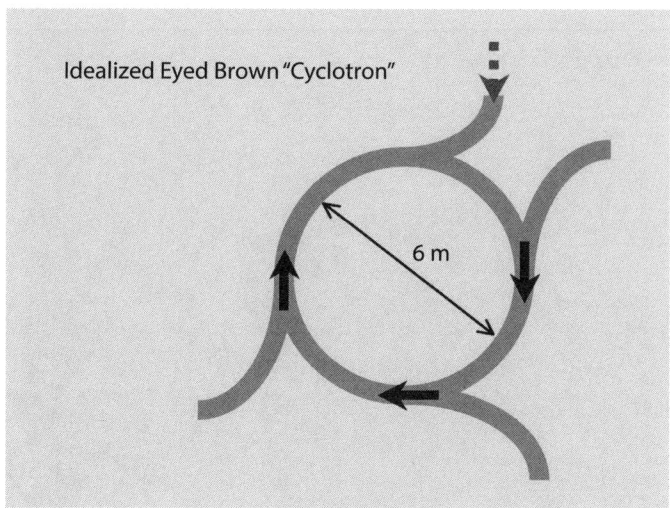

FIGURE 4.3. Idealized Eyed Brown "Cyclotron." Male Eyed Browns perch and patrol along channels in the vegetation, prompting the idea of cutting a traffic circle with entering paths at an acute angle. Ideally, butterflies introduced by the three paths at 3, 6, and 9 o'clock would patrol clockwise. An additional butterfly could be added from the path at 12 o'clock, to turn the "cyclotron" into a "collider." Practicalities prevented the realization of the ideal structure, which is probably a good thing. Figure 4.4 shows the actual structure achieved, and Section 4.3 presents the results.

TABLE 4.4. Behavior of 15 Male Eyed Browns in the "Butterfly Cyclotron"

| *Fly up and away after:* | *Number of males* |
| --- | --- |
| Not encountering a path. | 10 |
| Reaching dead end of a ray path. | 2 |
| Visiting one ray path and one circuit of arena. | 2 |
| Visiting three ray paths and two circuits of arena. | 1 |

The remaining five individuals encountered and variously followed the paths, which are mapped in Figure 4.4. Two reached the dead end of a ray path and then departed. Two visited one ray path, followed it to the arena, made one circuit of the perimeter of the arena, and departed. The most persistent individual visited and followed three ray paths, made two peripheral circuits of the arena and departed. All five departures were like those of the other ten subjects, rising to about 10 meters above the meadow, and flying down the valley toward the marsh.

The experiment was partially successful, demonstrating that channels in the vegetation are a powerful organizing feature of the ranging of male Eyed Browns.

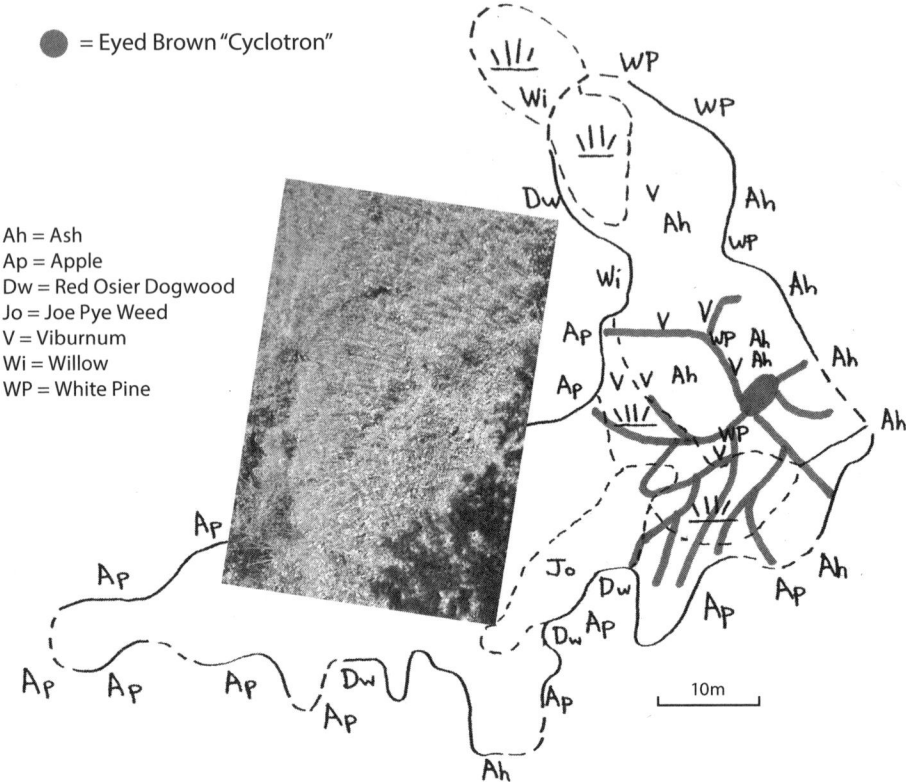

FIGURE 4.4. Realized Eyed Brown "Cyclotron." The photo was taken from about 5 m up in the ash tree nearest the arena. 102 m of paths, about 50 cm wide led into an elliptical arena about 5 × 3 m. 15 male Eyed Browns were released in the area spanned by the convex hull of the paths. Section 4.3 presents the results, particularly the path-following behavior of 5 males.

I was disappointed that none of the imported males settled into my network of paths. It may be that the decline in the meadow subpopulation was caused by reduced habitat quality prior to my experimental manipulation, and that the lower quality was detected by the introduced Eyed Browns. And my experimental landscaping probably further reduced the apparent habitat quality.

## 4.4 A FORTUITOUS SITE-FIDELITY MODEL

Eyed Browns following depressed paths in the vegetation inspires a model, in which simple orientation toward cues in the environment produces behavior that superficially resembles recognition of a particular piece of real estate as a territory, and persistent return to a "home" perch. Suppose that a butterfly in flight orients toward a specific cue to a feature of the local environment. Examples might be the following of channels in the vegetation by the Eyed Brown, orientation toward sunflecks by the Ringlet or the Little Wood Satyr, or a local version of the use of V-shaped notches by the Viceroy and the Fritillary. At each point in space, a vector can be drawn that shows the preferred orientation in flight. If each vector is extended until it contacts the next one, the world is divided into little attractive domains. Within each domain, strings of vectors lead toward a single "home" point, or at most a small loop of head-to-tail vectors. Between adjacent domains is a watershed along their common boundary. A butterfly that has been displaced anywhere within its initial attractive domain will be able to fly "home" by slavishly following fixed orientation features. Figure 4.5 illustrates these ideas, and in particular it shows that even a set of vectors with random orientations will generate a mosaic of attractive domains that look superficially like territories. Any spatial structure to the distribution of such cues in the environment can only increase the apparent regularity of the mosaic.[3]

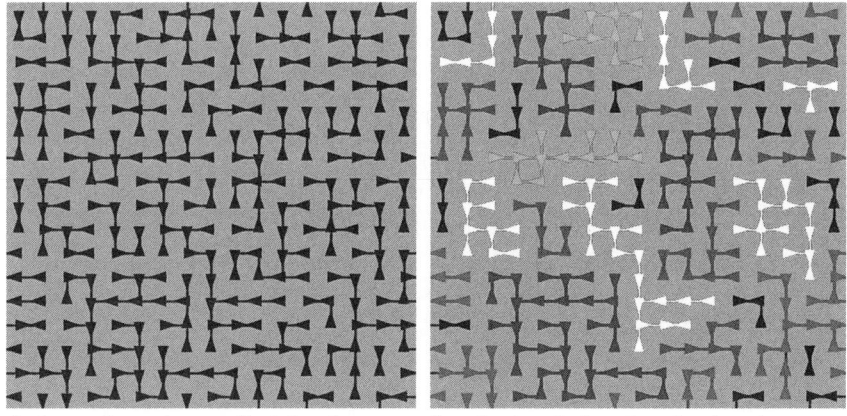

**A.** Arrows in random cardinal directions ...    **B.** ... form exclusive attractive domains

FIGURE 4.5. NetLogo model of Fortuitous Site-Fidelity, inspired by behavior of male Eyed Browns. In each square of a grid, an arrow in a random cardinal direction represents an environmental cue that tells a butterfly to "fly this way." The result is a pattern of small attractive domains each with either a "bow-tie" where two arrows meet or a cycle of four (rarely a couple more) adjacent squares, where the butterfly might stop and perch. Figure 4.4B has neighboring domains given different colors to make the pattern of Figure 4.4A more clear. As discussed in Section 4.4, even in its unrealistic form, the model can mimic the variants of male Eyed Brown behavior. And departures from the model suggest insights into the behavior of other species, specifically the Fritillary, Viceroy, and Pearly Eye.

The wanderings of male Eyed Browns seem superficially to fit this model, but definitive tests have not been done. However, the model provides some strong predictions that are *not* met by other species, and these are expanded in Chapter 8. In particular, an individual butterfly that behaves according to the model will, if it strays into the attractive domain of a second individual, proceed to that second individual's "home." So whenever a butterfly strays into another's attractive domain and then returns to its own initial domain, that is evidence that something more than fixed conventional cues from the external environment are being used.

Individuals show subtle local and temporal variation in characteristic behavior. These variations invite adaptive interpretation, and the interpretations may have some validity, but there are alternative mechanistic interpretations, some involving thermoregulation. In particular, a model of "random" configuration of channels in the vegetation produces attractive domains that can confine most individuals to particular local ranges that vary in size and complexity with the height at which each individual flies. This height of flight may be influenced by thermoregulatory requirements (cf. Tsuji et al. 1986 for Sulphurs *Colias*), or by population density and its influence on the frequency of investigative encounters with conspecifics.

## 4.5 REAL EYED BROWN BEHAVIOR VERSUS MODEL

I admit in advance that the following analysis is equivocal due to inadequate data, but the behavior of male Eyed Browns is ambiguous as well. So the analysis is presented not as a definitive test of the reality of the model, but rather as an example of what could be done with adequate data.

Figure 4.6 records displacements of male Eyed Browns during interactions, most of which lasted 20 seconds or less. The "contest displacement" is the distance between the beginning and the end of a flight path during which two males are interacting within a few centimeters of each other. The "perch displacement" is the distance between the original and the post-contest perches of the perching male. During the periods of observation, most undisturbed males were perched or moving between perches separated by 1–2 meters. I start with the charitable assumption that this approximates a measure of the scale of the attractive domain in a model like Figure 4.5. So displacements greater than about 4 meters represent movements outside of the attractive domain. Accordingly, the two contest displacements of more than 4 meters in Figure 4.6, which result in perch displacements of less than 4 meters, are potential examples of leaving the attractive domain and returning to it, i.e., of "homing" behavior that is not part of the model. Conversely, the two instances of perch displacement greater than 4 meters could

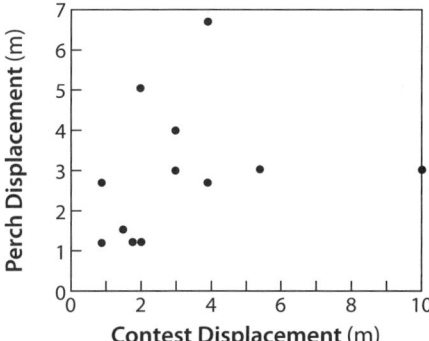

FIGURE 4.6. Displacements of interacting male Eyed Browns. The "contest displacement" is the distance between the beginning and end of a flight path during which two individuals are closely interacting with one another. The "perch displacement" is the distance between perches before and immediately after an interaction. There is a tendency for small perch displacements to be associated with short contest displacements, but the data are too sparse for statistical confidence in that tendency.

be instances of being drawn across the boundary of the initial attractive domain and moving to a perch within its next neighboring attractive domain, obeying the model of site-fidelity by accident. If the tracks followed by male Eyed Browns really do generate environmentally determined attractive domains, their scale is probably larger than the assumed 4 meters on times scales of minutes to hours, but smaller than the daily scale of 16 meters (Table 4.3). Unfortunately, I did not record tracks of Eyed Browns at the appropriate intermediate scale. Figure 4.6 shows no evidence of returns to the original perch, and so counters a prediction of the strict version of the Fortuitous Site-Fidelity Model of Figure 4.5. Consequently the strongest statement that can be made about male Eyed Browns is that they show some behavior that is consistent with the Fortuitous Site-Fidelity Model and some behavior that suggests weak additional homing to a particular piece of real estate.

## 4.6 DISCUSSION: "FORTUITOUS SITE-FIDELITY" AS A NULL MODEL

The Fortuitous Site-Fidelity Model may partly explain the ranging behavior of male Eyed Browns. Such a model is not necessary for the male Ringlet, whose movements, while confined to a preferred habitat, show no evidence of site-fidelity within that habitat.

For the other species, the Fritillary, Viceroy, and Pearly Eye, the Fortuitous Site-Fidelity Model is useful as a null model, in that all three species behave contrary to the model in ways that suggest that their site-fidelity is more than fortuitous. In particular, males of all three species move over ranges that overlap with the ranging of other males, and yet each male tends to return to a perch or a path that differs between individuals. So they are *not* responding mechanically to orientation cues in the environment that are interpreted uniformly by every individual.

## 4.7 SUMMARY OF RESULTS *AND SPECULATIONS*

### *4.7.1 Habitat Guides Male Behavior.*

The Eyed Brown is found in wet meadows, usually dominated by yellow-green sedges (Section 4.1.1). Females may mate several times, so any female encountered is a potential mate (Section 4.1.2). Male Eyed Browns perch and patrol in channels in the vegetation, which are frequented by females. Males follow experimental channels cut into the vegetation, *which may confine them to locally attractive domains* (Cue & Rule & Behavior: Sections 4.1.3 and 4.3).

### *4.7.2 Male Eyed Browns Are Site-faithful but May Not Be Territorial*

Most movements of most male Eyed Browns are local (Section 4.1.3). Males show clear evidence of site-fidelity, *but ambiguous evidence on other requisites for true territoriality:* recognition of particular real estate, and driving away other males. Male Eyed Browns show less evidence of territoriality in all respects than the male Viceroys of Chapter 6 (Behavior: Section 4.2).

### *4.7.3 A "Fortuitous Site-Fidelity Model" Helps to Interpret Behavior.*

A NetLogo Model of Fortuitous Site-Fidelity is proposed (Section 4.4), in which environmental cues, e.g., paths in the vegetation, provide vectors to guide movement. Simply following such vectors could confine individual butterflies to attractive domains. The Fortuitous Site-Fidelity Model as presented in Figure 4.5 has several unrealistic features, in particular movement of fixed distance on a rectilinear lattice, and unidirectional movement in a random direction. However, several important qualitative consequences of the model are independent of these artificialities.

Following such environmental vectors could produce apparent homing to separate domains without explicit territorial behavior (i.e., without defense or recognition of particular real estate). *The site-fidelity of male Eyed Browns could be due partly to such a neutral mechanism* (Cue & Rule & Behavior: Section 4.5).

Following such cues can produce large repetitive cycles only if there is large-scale regular structure in the habitat. This result will help to interpret the apparent traplining of the Fritillary (Behavior: Chapter 5).

An individual following such environmental cues that trespasses on the domain of a second individual should settle at the perch of the second individual's domain, rather than returning "home." If not, something additional or very different is going on, strongly implying individual recognition of particular real estate (Sections 4.5 and 4.6). The Fortuitous Site-Fidelity Model will be used as a null model to show that the Fritillary, Viceroy, and Pearly Eye do indeed recognize particular real estate (Rule & Behavior: Chapters 5–8).

CHAPTER FIVE

# Setting and Running a Trapline
## *The Great Spangled Fritillary* Speyeria cybele

> It is extremely timorous and swift of Wing; and should you, as it flies by you, strike at it with your Nets and miss it, it is vain to pursue it; for, being frightened, it is wild, and will not settle till it be quite out of your Sight.
> —Moses Harris, 1766, *The Aurelian*, describing the European Dark Green Fritillary *Argynnis aglaja*

Harris's notes on the Dark Green Fritillary apply equally to the Great Spangled Fritillary. Indeed, most of my data come from 1977, the year that I was assisted by Jeff Georgia, whose goal during the field season was to catch, mark, and watch as many Fritillaries as possible, employing his considerable skills as a Princeton Varsity Lacrosse letter-winner.[1] I also was lucky to make critical observations in 1982 when an exceptionally dark male Great Spangled Fritillary with an unusually slow wingbeat frequency was individually identifiable over a fortnight without being captured.

The Fritillary flies through moist habitats: meadows, marshes, and open deciduous woodland (all previous authors). Both males and females pause to sip nectar from any flowers that provide a firm landing pad, especially where violets *Viola*, their larval food plant, are present beneath the herbs (Clark 1932 and personal observation). "Males patrol in wide meandering circuits, a few feet above the ground, in locations where they are likely to find females. But dispersing individuals fly quite directly."—in the words of Cech and Tudor (2005, p. 162). Males will often remain in the neighborhood of a single nectaring plant for several hours (Gochfeld and Burger 1997). Males pre-emerge females by a month or more, and most are worn and tattered by the time the females appear (Clark 1932); females wait to lay eggs in early autumn, after many weeks on the wing in summer (Weed 1917). The female crawls about in the vegetation to lay an egg or two on or near violets, then flies on directly to another patch of violets (Clark 1932, Cech & Tudor 2005). Females smell "strong and nauseous" (Clark 1932), and mating apparently involves pheromones (Scott 1986). My own observations are consistent with these patterns. In particular, both sexes early in their seasons fly short circuits from nectaring plants, dipping down and fluttering through vegetation; but later they fly more widely and directly, favoring those patches of nectaring plants that have violets in their understory. My field seasons were not long enough to observe oviposition by females, but I did observe a single Aphrodite *Speyeria aphrodite* behaving as described above, specifically crawling about like a terrestrial insect and tapping various plants with her abdomen (Appendix B.3).

## 5.1 NATURAL HISTORY AND BEHAVIOR

### *5.1.1 General Movement*

Male ranging is traditionally described as strong point-to-point flight, making them difficult to capture, and contrasting with female ranging, which involves fluttering in and out of vegetation apparently searching for oviposition sites (e.g., Clark 1932). That difference in behavior is typical during the time when both sexes are common. However, much of the sexual difference may be due to the timing of observations relative to the adult life cycle. I have observed fresh males fluttering locally early in the season, and well-worn females zooming long distances late.

### *5.1.2 Mating and Female Behavior*

Female Fritillaries tend to mate only once (Table 1.3), and all three pairs that I have found in copula have been near large patches of violets, and have involved very freshly emerged females, one of which had not entirely unfolded her wings. Sims (1984) reports that female Fritillaries mate soon after emergence and then delay oviposition for as much as several weeks. My observations of females are mostly of freshly emerged individuals, which fly short distances between episodes of dipping into the vegetation, and late in August of individuals that fly longer distances between fewer and longer episodes hidden in dense vegetation where violets were obvious in the spring. These observations suggest that the older females repeatedly visit a trapline of patches of the larval food plant, dispersing their eggs singly in time as well as in space. Such behavior has been directly observed for the Viceroy by Platt (1984) and by Scudder (1889, pp. 261–266), who discusses the broadcast of single eggs as a means of averting parasitism and predation (discussion reappears in Section 9.3.5).

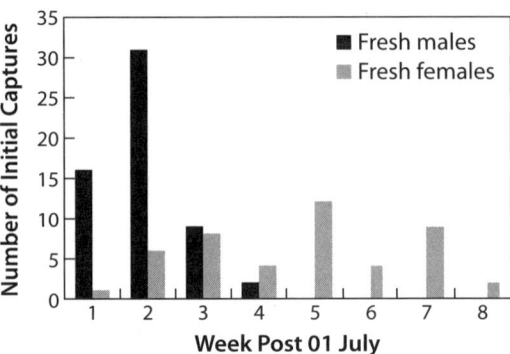

FIGURE 5.1. Dates of initial capture of fresh or slightly worn Great Spangled Fritillaries (all years combined). The emergence of most males predates that of most females by several weeks.

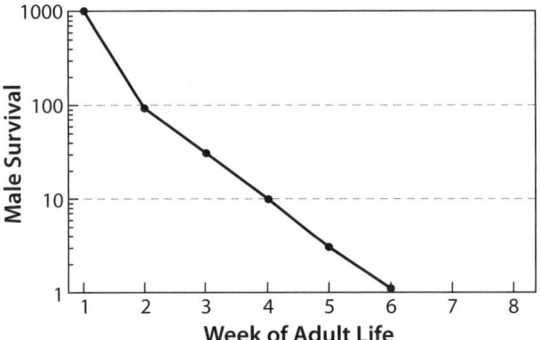

FIGURE 5.2. Average survival for male Fritillaries (literally "local residency" = number known to be alive ; see Section 1.6.4). Only one in a thousand is expected to be locally alive by week 6. Mortality, or equivalently local loss, is an extreme cost for the benefit of emerging early to be exposed to many fresh females.

### 5.1.3 Life History

Male Fritillaries emerge nearly a month before females (Figure 5.1) so that by the time fresh females are abundant, most of the males are worn and many are departed, tattered, or even dead (Figure 5.2; the reason for such "protandry" is given in Section 3.2.3).

### 5.1.4 Male Ranging and Behavioral Interactions

Evidence for male Fritillaries running traplines is of four kinds, all indirect, but very strongly suggestive, and discussed below. In day-to-day sightings, particular numbered individuals were observed repeatedly at the same spot, with some extensive absences, and occasional intercalated visits to distant spots (Figure 5.3). Within a day, a few individuals were observed to repeat, in detail and on a regular schedule, a segment of their presumed traplines (Figure 5.3). By simply mapping

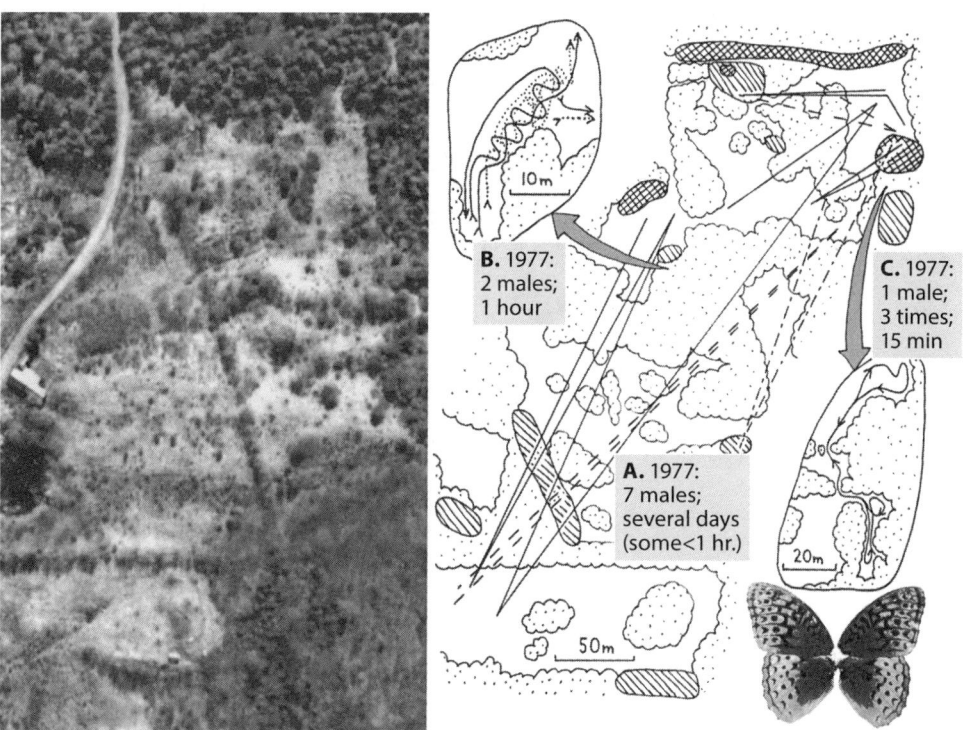

FIGURE 5.3. Sightings and repeated paths of male Fritillaries. A. Straight lines connect captures and re-sightings over several days for 7 male Great Spangled Fritillaries. They repeatedly visit several distant sites. Insets B and C. Movements at a finer scale (see Section 5.1.4 for detailed descriptions).

vectors for all traveling large Fritillaries, I uncovered a network of preferred flyways (Figures 5.4 and 5.8). In July of 1982 I had the good fortune to mark a freshly eclosed male and to map his local explorations over the next 10 days. In an attempt to discover his presumed trapline, our daughter Jennifer and I made several attempts to follow his wanderings beyond a particular hedgerow, each of us on opposite sides and communicating by CB radios; he thwarted us every time, presumably by taking diagonal paths that we never discovered.

Figure 5.3A maps sequential sightings of several males in 1977, connected by straight lines of 50 meters or more. This gives clear evidence of repeated visits to a site with long movements in between, though the temporal pattern of repeats cannot be estimated because of the episodic nature of my own visits to the sites. The insets of Figures 5.3B and 5.3C record detailed tracks for three individual males at two sites, over periods of minutes to an hour. Entries to these sites, as

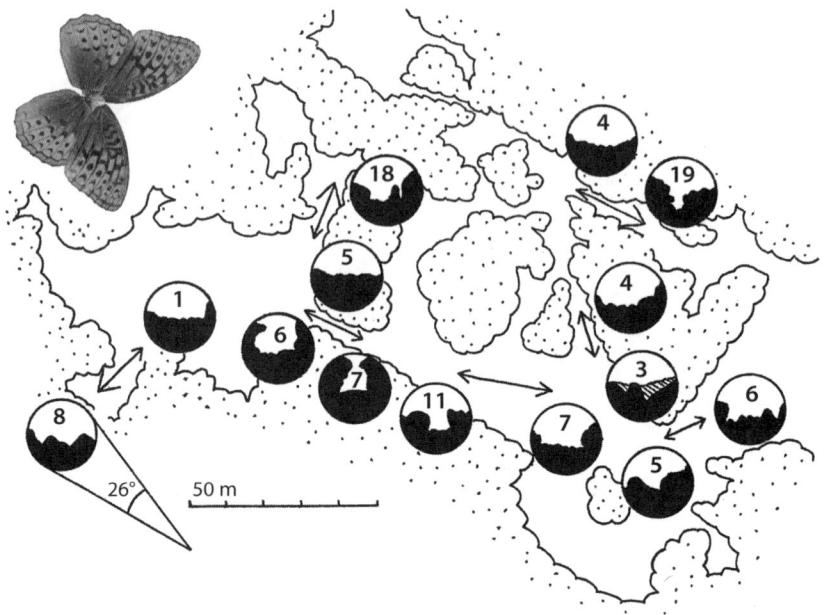

FIGURE 5.4. Flyways (arrows) and horizons (26° views) for 104 sightings of free-flying Fritillaries. The observations are mostly of male Great Spangled Fritillaries, but may include a few females and a few Aphrodite Fritillaries. About 50 additional observations between the alleys were oriented in arbitrary directions. One-way flyways tend to have a V-notch in the distant horizon in the predominant direction of movement, and two-way alleys tend to have V-notches in both directions, though the number of all flyways is too small to show statistical significance to this pattern.

well as exits from and paths within, were via V-shaped gaps in the shrubbery, a larger-scale version of the following of V-paths in sedges by the Eyed Brown (Chapter 4, Section 4.4).

In some years I simply mapped every observation of a traveling Fritillary (mostly males because it was early in their season), recording a small arrow to represent its mean direction of flight. The vast majority of the arrows were concentrated into particular flyways, some of them with flights in both directions, some with flights predominantly in one direction. The flyways were through local V-shaped "valleys," literal valleys in the landscape itself, or vegetative channels: between crowns of forest canopy trees, over shrubs among trees, or over grasses and sedges among shrubs. Two-way flyways had a V-notch in the horizon in both directions; one-way flyways had a V-notch in the horizon in the predominant flight direction, but a flat or bulging horizon in the opposite direction (cf. the smaller-scale Vs followed by the Eyed Brown of Chapter 4, Section 4.4.)

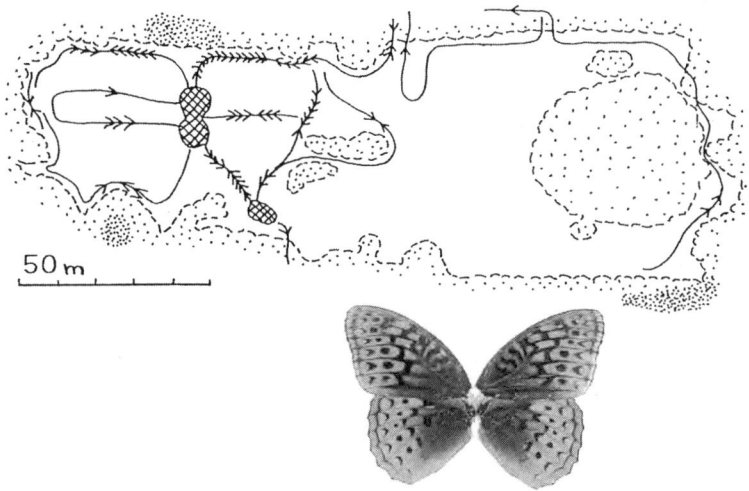

FIGURE 5.5. Paths of Great Spangled Fritillary #82.00 "Main Man" as he developed his circuit. The paths and arrows map 67 independent observations in 3.5 hours over 7 days in July 1982.

Many individuals share common flyways, but different individuals may diverge at the end of a flyway. Thus the orientation of male Fritillaries is consistent with a mechanism suggested by Cartwright and Collett (1987), in which Honeybees *Apis mellifera* make "snapshot decisions" between guiding vectors established by fixed responses to the geometry of the landscape (see also Wehner 1992).

On 10 August 1982 I found an unusually dark male Great Spangled Fritillary nectaring assiduously on milkweed. He initially showed no signs of wear on the scales of wing or body; so I infer that he had emerged within a day or two. Because he was so distinctive I decided to try to observe him without capture and marking, as "Main Man" GSF #82:00. I mapped his movements in detail over the following seven days, and saw him episodically until August 21st. His recorded movements in the main study field are mapped in Figure 5.5. During the week of mapping, he was seen discontinuously at three locations as far as 200 meters SE of the study field.

Figure 5.6 represents the dramatic change in Main Man's behavior between the first and fifth days of observation. On the first day he traveled many short circuits with stops at a central patch of milkweeds where he sometimes nectared (Figure 5.6A). On these circuits he flew close to, and dipped down into, vegetation near the edge of the field, behaving very like females do much later in the season when they are looking for sites to lay eggs. Four days later his circuits were longer, and his flights were more direct, faster, and up to a meter above the vegetation (Figure 5.6B).

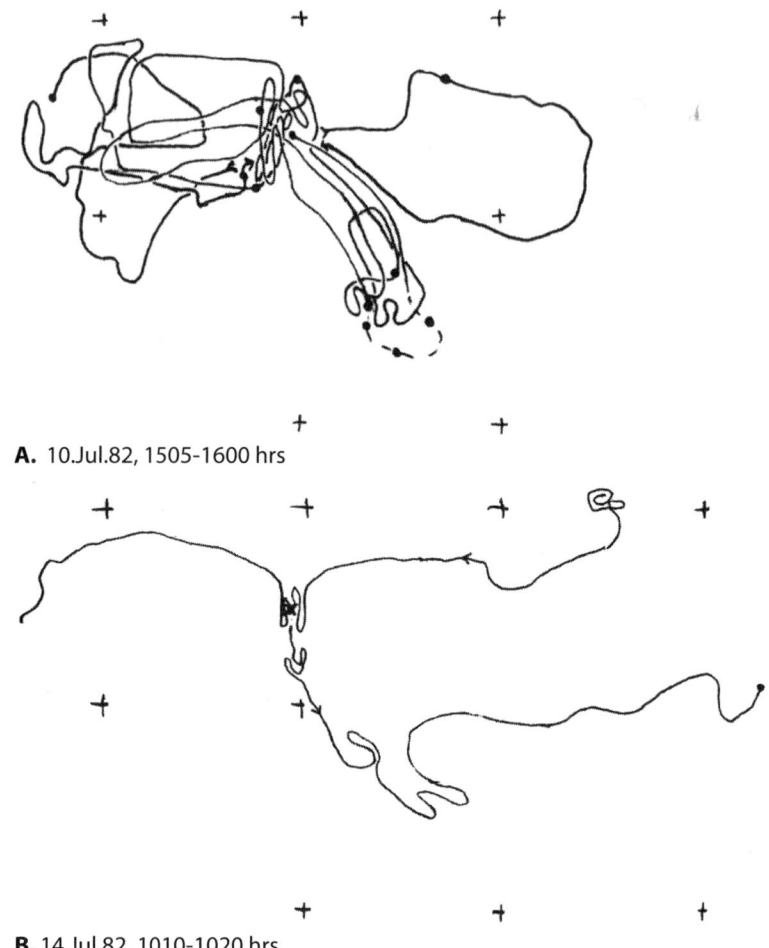

FIGURE 5.6. Tracks of Great Spangled Fritillary #82.00 "Main Man." A. One hour of movement on the first day observed. The central nexus was a patch of milkweed on which the male nectared. On the longer tracks, he often dipped into the vegetation like an ovipositing female. B. Ten minutes of movement four days later. Tracks in B were more direct than in A, and as much as a meter above the vegetation.

Further notes on male ranging and interactions are presented as generalities, rather than as maps or compilations of data from numerous individuals. They are based on observations of marked and unmarked individuals, and the latter are assumed to be males on the basis of consistency of season and behavior with known males. A few of them may be Aphrodites, rather than Great Spangled Fritillaries.

When marked males, or males with idiosyncratic color or pattern of wear, repeated visits to a particular nectaring site or flyway, those visits were often at regular intervals, 10–15 minutes for local circuits, and half an hour to an hour if they traveled through a distant gap in vegetation. Males on a flyway often flew parallel to a forest edge by roughly following the shadow of the top of the nearby forest canopy (A potential cue and rule for doing this would be to fly so as to keep a constant difference in illuminance between right and left ventrolateral fields of view; see Section 2.3.4).

Males were drawn to inspect Fritillary-colored-and-sized objects from distances of a meter or two (30 observations; Table 2.1). Once a male has "found" a female, he can re-approach her directly from more than 3 meters. In July of 1975 three recently emerged females[2] were confined to a screened cage near a patch of violets that was along a male flyway, but not near nectaring plants. The females hung upside-down from the top of the cage with wings spread. At least three males passed within two meters of the cage without diverting course toward it. A tentative conclusion is that if an attractive pheromone is used, it is emitted under the female's control, and likely broadcast only when the female is freshly emerged. Recall from Chapter 2 Section 2.4.1 that human detection of a strong odor from female Fritillaries depends on squeezing them until they evert an abdominal process.

As mentioned in the discussion of Figure 5.4, different males may share flyways along a circuit, but they often diverge to different paths at the end of a flyway. The sharing seems to be a result of independent responses to a common cue, and the divergence may be a response to idiosyncratic cues. However, there is some evidence to suggest contagion in behavior. Several times one male Fritillary was observed to follow the track of another for tens of meters with a following distance of 2–3 meters, the upper limit of reaction distance. One male followed another at the prodigious distance of 10 meters, but that was likely due to their both tracking environmental cues along the edge of the field, rather than to the follower continuously perceiving the leader.

Males set up circuits of places where nectar and larval food plants are available. They then patrol these circuits on a regular schedule and will investigate any objects that are the approximate color of the female. The circuits of different males are different, but they may share a common path for part of their respective circuits. A prominent feature of the common paths is a distinct V-notch in the distant horizon in the direction of travel. Marked male Fritillaries have been observed taking long flights and yet revisiting the same site on a regular schedule. They seem to be running a "trapline" of the kind documented by Ehrlich and Gilbert (1973) for nectaring Longwings *Heliconius ethilla*, and by Alcock (et al. 1977) for mate-seeking anthophorid bees. The Fritillaries are indeed collecting nectar at their stopping points, but they seem to spend more time at nectaring sites like *Asclepias* and *Eupatorium* that also have the larval food plant violets *Viola* nearby. This behavior with respect to nectaring and larval food plants is also shared with some species of Longwings *Heliconius* (Gilbert 1976).

## 5.2 INTERPRETING MALE AND FEMALE BEHAVIOR

It is as though males locate several patches of larval food plants, and set up a trapline that they regularly traverse, looking for freshly eclosed virgin females. Analogously, the Hackberry Butterfly *Asterocampa leila* and the Checkerspot *Euphydryas chalcedona* defend perches near larval food (Rutowski and Gilchrist 1988 and Rutowski et al. 1988). For their part, female Fritillaries, by the almost null behavior of sitting still at the site of eclosion until they allow only a single mating, exert very strong selective pressure for males that are good at finding larval food, who will pass on whatever genetic disposition they have to daughters who will need to find oviposition sites. This interpretation is entirely speculative, but such a scenario has been documented by Watt (et al. 1986) for Sulphurs *Colias*.

A measure of the strength of this selection is that males emerge nearly a month before the females to set up their traplines. By the time that most of the females eclose, many of the males are either dead or departed. Violets are very patchy in their distribution, and though they may be out in the open in the spring, they are difficult to locate when the adult butterflies are searching because they are then deep in the "understory" of lush field-edge vegetation. A possible alternative "strategy" for males is to emerge late and to follow another male to adopt his diligently acquired trapline. Although some following behavior has been observed (Section 5.1.4), the evidence is too weak to be more than tantalizing.

The few females that I have observed repeat the characteristic behavior of short open flights and fluttering into dense vegetation early on, and flying longer and more direct paths between flutters into more widespread patches of vegetation later in their season. By that time, leaves of violets are dried and decaying, though there is often a secondary growth phase during a warm fall. Clark (1932) has observed that ovipositing females move long distances between patches of violets in each of which they lay a few eggs, dispersing eggs both in space and in time, presumably to thwart egg predators and parasites (Section 9.3.5).

## 5.3 FRITILLARIES AS TRAVELING SALESFOLK

It is tempting to analogize the task faced by a freshly emerged male Great Spangled Fritillary to the famous traveling salesman problem;[3] the male has to find an efficient route by which he travels among freshly emerged females, attempting to sell them a spermatophore. For any appreciable number of customer locations, the combinatorics of possible connecting routes make calculation of the most efficient route formidable even for fast computers (see, e.g., Lewis and Papadimitriou 1978 and Cook 2012).

Adding an additional salesman complicates the calculation, especially if the first one to arrive at an appropriate site makes the sale (Fekete et al. 2004), a condition that the female Fritillary may enforce by accepting only a single mating soon after she emerges. However, adding travel costs to the model puts a premium for a given male on those females to which he is already closest, which may foster small and potentially exclusive sales territories, within which there are few enough customer locations to make finding an efficient circuit more tractable. Averbakh (et al. 2008) shows that separate domains on a network that share a common starting point favor a tractable solution that neither of two competitors can best. In addition, adopting separate domains reduces the number of customers in each domain. For a small number of sales-sites, Anderson (1983) argues that simply moving among near-neighbor sites comes close to an optimal route, particularly if customers are renewing—as they are if additional female Fritillaries continue to emerge in a given patch of violets.

It is possible to estimate the order of magnitude of a male Fritillary's traveling sales problem, and this estimation makes plausible the argument for an efficient circuit. The number of males known to be alive in the study area can be estimated from the maximum number over the 1977 season, when Jeff's and my study of this species was most intense. Figure 5.7 shows a maximum of 19 males known

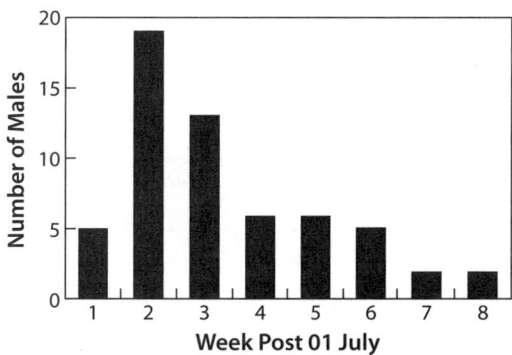

FIGURE 5.7. Great Spangled Fritillary males known to be alive by date. The data are from 1977, when the greatest number of Fritillaries were captured, marked, and observed, thanks to help from Jeffrey Georgia. This pattern of seasonal presence suggests the relative number of males competing for females through the season, and hints that males are, at best, integrating over the emergence times of all females (see Figure 5.1).

alive in the second week of July 1977. The total number of females marked during July 1977 was 35, 26 of them marked during the third and fourth weeks. So to a first approximation, over the 8 hectare (440 × 180 m) study area, during an 8-week season, as many as 20 males are competing for as few as 25 freshly emerging females from about 20 patches of violets about 75 m apart (Figure 5.8). If the females were highly aggregated in space, it might pay an individual male to find and defend the aggregation, but it would also pay other males to do the same at the same patches. If females were randomly distributed with uniform probability per patch (25/20 = mean of 1.25 females per patch), the binomial number expected in each patch would be most often be 1 (7 patches), followed by 0 (6 patches), 2 (5 patches), 3 (2 patches). For either aggregated or random distributions of females, not knowing in advance which patch is which, it would pay a male to visit every patch repeatedly. But the only male who can mate is the first to find a receptive female at a given patch. Accordingly travel between patches imposes a cost in time, energy, and opportunity lost to competitors. These costs place a premium on a short and exclusive circuit, perhaps visiting fewer patches, more often and spread over a smaller area—rather than all patches in a large area with a longer, less frequent circuit.

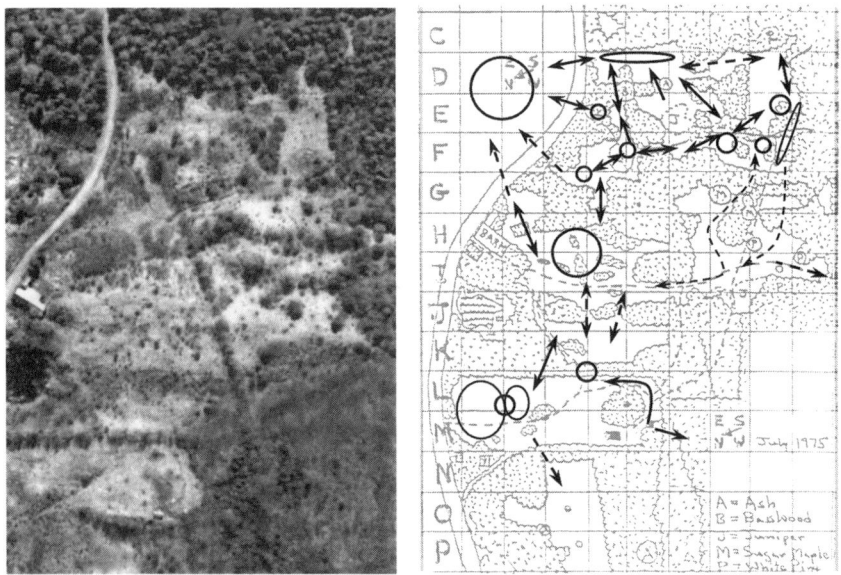

FIGURE 5.8. Larval food plant nodes and connecting paths for Fritillaries. Circles and ellipses represent places where patches of the larval food plant, violets (*Viola* spp.) are common, though during the period of Fritillary flights, the violets are visually hidden in the understory of taller herbs and grasses. Solid arrows are frequently observed paths; dotted lines, more rarely observed paths.

MacGregor and Ormerod (1996) analyzed the apparent algorithm by which humans attack a traveling salesman problem with enough customer locations to present intractable calculations. Their behavior is consistent with visiting neighboring sites on the convex hull of the locations of all sites, and then gradually adding nearby interior sites. This process is close to maximally efficient because all peripheral sites must be visited, leaving only the potentially problematic addition of the interior sites, which, if they are few enough, pose tractable combinatorics (see, e.g., Cook 2012).

Figure 5.8 shows 6 peripheral patches of violets, and 7 interior ones, 2 of which are so near a peripheral patch that their inclusion in a circuit is trivial. This leaves only 5 non-trivially located patches, one of which is well connected by frequently used paths between two groups of other patches. So a simple near-neighbor joining rule such as Anderson (1983) posits would allow a male Fritillary to finesse the traveling salesman problem, even if not explicitly to solve it.

Although I do not have sufficient data to do more than speculate about female Fritillaries, they might similarly be analogized to saleswomen selling eggs to violets. Adopting a circuit of patches of violets disperses the eggs in space, and repeating the circuit at intervals disperses them in time. Both of these dispersals may thwart predators and parasitoids of eggs and larvae, or at least increase the difficulty of these enemies' finding the eggs and larvae.[4] I do not have direct evidence that females repeat circuits. However, the similarity of their flight phenology to that of males is suggestive. Their early flights tend to flutter within low vegetation and seem to be exploratory; later flight have many long segments that seem to be directed toward goals that they have already found.[5]

## 5.4 DISCUSSION: HABITAT MAY FOSTER CLOSED CIRCUITS

The Fortuitous Site-Fidelity Model in Chapter 4, in addition to producing attractive domains, sometimes yields closed circuits. With the randomly oriented cues of the literal model, such circuits would be rare and very small. Regular patterns in the environment at a large scale could produce environmentally determined closed circuits. Indeed, this is part of what is going on in the case of Fritillaries, where the environmental cues are the topography of the landscape, both geological rumples in the land and variations in the height of the vegetative canopy.

Whether there is more to the story is unclear. The developing circuits of "Main Man" in Figure 5.6 could be interpreted as condensing to a perimeter circuit with interior visits, just as humans address an empirical traveling salesman problem (MacGregor and Ormerod 1996). I'm not saying "are," just "could be." However, the details of Figures 5.4–5.6 and 5.8 depend crucially on the geometric configuration of successional vegetation following agriculture that is patchily established and patchily abandoned. This cannot be taken seriously as the natural setting within which the behavior evolved.

I had a brief opportunity in early August 2000 to observe a population of Great Spangled Fritillary in a locale with an entirely different configuration of vegetation. This was the subspecies *Speyeria cybele leto* at the Yellowstone Bighorn Research Association in Red Lodge, Montana. Instead of small patchy fields set in a matrix of deciduous woodland, the Montana population moved among large patches of coniferous forest set in a matrix of subalpine meadow. Violets in both locales were concentrated at woodland edges. The subspecies *leto* has the convenient property that the males look like male Fritillaries elsewhere, but the females are almost black and white. The local behavior of Montana *leto* seemed very like my Fritillaries in New York; the males followed woodland edges and V-notched paths through the forest. On 06 Aug. 2000, I marked 15 males, most of them very worn and slightly tattered, and counted 68 unmarked males in the vicinity (an area of about 150 × 100 m plus about 10 m on either side of 250 m of a dirt road = 2 hectares). On the following day I marked 7 more males, but only encountered one of the 15 marked the day before. Over the next 5 days I saw many male *leto*, but none of them were marked. So in the Montana population I had no evidence of traplining, though both the absolute number and the proportion of males marked were smaller than in the New York population. However, in New York V-notch vectors among trees were reliable paths connecting patches of violets, but in Montana V-notch vectors did not help to locate violets as much as using the trees themselves as landmarks. That means that Montana Fritillaries might travel irregularly between clumps of trees, and if they did establish traplines, they would only be between violet glades within sufficiently large patches of forest.

## 5.5 SUMMARY OF RESULTS *AND SPECULATIONS*

### *5.5.1 Traditional Natural History*

The Fritillary frequents meadows with flowering plants as a source of nectar, often with violets, the larval food plant nearby. Most general accounts describe the male as flying strongly and directly, often in repeated circuits, while the female flutters in and out of vegetation (Behavior: Section 5.1.1). Females tend to mate once, and the three that I found mating were freshly eclosed and near violets (Behavior: Section 5.1.2). Males pre-emerge females by about a month; females fly for about a month before laying eggs. (Behavior & Life History: Section 5.1.3).

### 5.5.2 Males Patrol Repeated Circuits.

At my field site, male Fritillaries revisit particular spots, often on a regular schedule and by repeated paths. Favored paths involve V-notches in the vegetation, openings in tall vegetation, and/or boundaries defined by the shadow of a high canopy on lower vegetation. Favored spots involve nectar "filling stations," and violets nearby (Behavior: Section 5.1.4).

*Individual males may share a favored path, but at some point, usually diverge to follow different paths.* This divergence contradicts the Fortuitous Site-Fidelity Model of Chapter 4, which predicts convergence by fixed response to vegetative cues within each attractive domain (Cue & Behavior: Section 5.1.4).

### 5.5.3 Interpreting Circuits as Traplines for Prospective Mates

Traditional lore has males flying strongly and directly in the open while females flutter in and out of low vegetation. I find that individuals of both sexes flutter in vegetation early after eclosion and fly directly later. Since males pre-eclose females by about a month, it is likely that traditional lore is based on observations made in the middle of the flight season, when most females are freshly eclosed and most males are veteran adults (Behavior & Life History: Sections 5.1.1 and 5.2).

*A plausible interpretation is that both sexes are prospecting for and locating larval food plants early on, the males as potential sites for emergence of virgin females, the females as potential sites for oviposition. I suggest that males are later traplining among the found sites for prospective mates, and females are later traplining to spread eggs in time over food plants that are patchy in space. And I speculate that both are selected to detect and efficiently and repeatedly to visit violets* (Cue & Behavior: Section 5.2).

### 5.5.4 Have Fritillaries Solved the Traveling Salesfolk Problem?

At my study site, patches of violets are widely dispersed, and the number of virgin females emerging in a given patch on a given day is of order one or zero. *Hence to have an appreciable likelihood of mating, a male must travel among several patches*, and efficiently doing so is the famous traveling salesman problem. *The possibility of being beaten to a given female by another male imposes opportunity costs to travel between patches, and favors adopting a reduced and ideally exclusive sales area.* With a small number of patches to be visited repeatedly, the traveling salesman problem may become tractable. I cannot claim that Fritillaries have solved the problem, but the behavior of one male included elements of one of the effective algorithms for coming close (Rule & Behavior: Sections 5.1.4 and 5.3).

### 5.5.5 Are Closed Circuits Idiosyncrasies of Habitat?

The Fortuitous Site-Fidelity Model of Chapter 4 shows that random orientation cues would generate very few and small closed circuits. *That suggests that cues for larger circuits, if they exist, are generated by regularities in the environment at the geometric scale of those circuits. So closed circuits may be easy to achieve in some landscape configurations, but difficult or impossible in others* (Cue & Behavior: Section 5.4).

CHAPTER SIX

# Defining and Defending a Territory

## *The Viceroy* Limenitis archippus

> All these butterflies live a considerable time, and indeed the eggs do not mature in the bodies of the females until they have been a fortnight on the wing; and then they do not lay all their eggs at once, or even within a few days, but prolong the operation over many days or even several weeks.
> —Samuel Hubbard Scudder, 1889, *The Butterflies of the Eastern United States and Canada with Special Reference to New England*, p. 265, describing the whole genus *Limenitis*.

Because of academic commitments, I was unable to remain late enough in the season to repeat Scudder's observations, other than confirming the long lifetime of adult males, and seeing a few females laying eggs on willows *Salix*, aspens *Populus*, and apple *Malus*. But most of my other observations confirm what is in previous literature, and add interpretive details, especially for the males.

### 6.1 BEHAVIORAL POSTLUDE AS PRELUDE

The behavior of Viceroys in my study site was so consistent that I shall start with an interpretive summary and then present the observations on which that interpretation is based. Female Viceroys prefer to lay their eggs on willow. One way to find willows is to follow streams. One way to follow streams is to consistently fly toward the visual cue of low spots in the horizon. If female Viceroys do so, then they may enter a meadow from any direction, but would tend to leave via the lowest spot in the horizon. The part of the field adjacent to the low spot would then be the place where females occur at highest density in space and time. Consequently, the quadrant of the field nearest the low horizon would be the optimal place for a male to perch and to watch for females flying against the relatively uncluttered background of sky and distant vegetation.

All my observations of sexual interactions fit the following description of behavior.[1] A male flies out from its perch and approaches the female, which quickly drops to land within the vegetation. The male hovers nearby for a short time, and then returns to one of its perches. I never saw explicit behavior on the part of the female that could be interpreted as either rejection or acceptance, nor did I see a male land nearby. So I must depend on circumstantial evidence to suggest that free-flying females are potential mates.

Male behavior in response to other males is dramatically different. When a male flies from its perch and approaches another male, the two interact in a long flight that can leave the meadow entirely, and the original percher is always the first to return, usually to his original perch or to one nearby. Thus the behavior of male-male interactions is qualitatively different from pre-courtship investigation, a crucial requirement for an agonistic interpretation of such behavior (Waage 1974 for the Black-winged Damselfly *Calopteryx maculata*, Bitzer and Shaw 1979(80) for the Red Admiral *Vanessa atalanta*).

However, to show that this behavior is explicitly territorial it is necessary to demonstrate: (1) Males choose habitats and perches with the potential for observable females. (2) Individual males recognize particular pieces of real estate. (3) Perched males drive other males away.

## 6.2 NATURAL HISTORY AND BEHAVIOR

### *6.2.1 Habitat and General Movement*

Scott (1986) reports adults feeding on nectar, honeydew, and plant ferments. I noticed a diurnal pattern in feeding behavior. Neither sex was obvious in the morning, but the few seen were nectaring or attending slime fluxes on willow *Salix* bark. In the afternoon there were some instances of nectaring, but most females were exploring or flying between clumps of the larval food plants, and most males attended perches within a restricted area.

In the afternoon, when Viceroys are most active and most obvious to me, they are in and around the same wet meadows as the Eyed Brown (Figures 4.1, 6.1, and 6.2). Females move through the meadow, particularly along the edges. And each male moves episodically from one perch to another within a particular piece of real estate.

### *6.2.2 Female Behavior and Mating*

The Viceroy inhabits wet meadows with shrubs, and especially with young willows *Salix* and poplars or aspens *Populus*, which are the preferred larval food plants, and hence the oviposition sites for females (Clark 1932, Cech and Tudor 2005).

FIGURE 6.1. Viceroy meadow with locations of typical perches used in 1982. This is not a map of perches actually used, but is rather a photograph on which plants that stick out of the general tangle are marked. Nevertheless, the photo gives an accurate representation of the relative density of real perches over the field of view. The photo was taken from 6 m up in an ash tree at the right edge of Figure 6.2, and looks toward territories A, B, and C of that figure. The primary V-notch in the horizon (mentioned several times in text and other figures) is to the right of center. It was created in 1977 by zealous use of a chainsaw to extend the margin of Baker/Dunbar Road. Previously, perches in the foreground had been favored by males; after 1977, perches in the background were favored.

According to Scudder (1889), female Viceroys will fly for two weeks before laying eggs, and then they distribute the egg-laying over several more weeks—which he suggests is a way of thwarting predators and parasitoids of eggs and larvae. It is the flying females that males encounter from their perches. Unfortunately, all 12 free-flying females that I caught and dissected had a single spermatophore. So I infer that most female Viceroys found by a territorial male will already have mated and may not need a second mating. However, free-flying females of other species in the Viceroy's genus *Limenitis* have been found without spermatophores and with more than one, though the majority still have one (Table 1.3). So it is likely that some small percentage of free-flying female Viceroys may be receptive to mating.

## 6.2.3 Male Ranging and Behavior

The low probability that free-flying females are receptive is consistent with the lack of male persistence in encounters with females, but also favors behaviors that lead males to habitats where females are abundant, and behaviors that ensure a given male of exclusive access to those females.

Male Viceroys spend their mornings either hiding from me or attending slime fluxes on willows, which, being the most favored larval host species, are likely places to find recently eclosed virgin females. During the afternoon, they defend territories in meadows along flyways frequented by females that are probably either laying eggs or exploring potential places to do so. Clark (1932) notes that males are sedentary when rare, but active when abundant. My males were always rare, and almost always relatively sedentary.

Table 6.3 shows that the perching areas most favored by males are those in Figure 6.2 that offer the clearest view of the lowest spot in the horizon. Males perch no higher than about 1.5 meters, with periodic patrols and inspections of other flying insects, and then return to the same or a similar perch. Males may

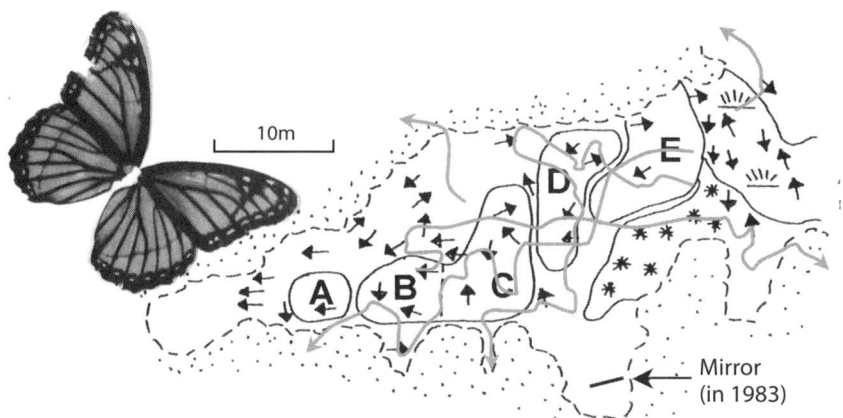

FIGURE 6.2. Territories held by males listed in Table 6.3 and vectors showing movement of females. This is the wet meadow of Figure 6.1, surrounded by shrubby woodland (stippled). The data for territories and short movements by females were gathered from 30 July to 15 Aug. 1981. Table 6.3 suggests that the prime territory was B+C, from which the primary V-notch (on the left, and in the background of Figure 6.1), toward which most females were moving, was most distinctly visible. An indistinct secondary V-notch was locally visible along the drainage through the wet area on the right. The longer tracks are of unharassed females, recorded in 1983. Where each of these females left the habitat, a local V-notch in the horizon was apparent. In 1983 a giant mirror was emplaced as shown to generate an artificial low spot in the horizon (see Section 6.3.1 and Figure 6.3).

TABLE 6.1. Male Viceroy Behavior in 33 Approaches to a Potential Perch

| Change in behavior in 33 independent close approaches: | Settle on perch | Continue flying |
|---|---|---|
| If horizon remains clear from perch | 21 | 1 |
| If horizon is blocked by intervening vegetation | 2 | 9 |

*Note:* Males prefer perches that allow an unobstructed view of the lowest spot in the horizon.

favor a given area, or even a particular perch for as much as 3 weeks, defending against other males, with the resident usually prevailing, though some males change favored areas frequently. My observations of male Viceroy behavior are very like observations of its congeners *Limenitis weidemeyerii* (Rosenberg 1989) and *L. arthemis* (Lederhouse 1993). Lederhouse even cites details like perches being associated with flyways of females, rather than with female abundance per se, or with larval host plants, or with nectaring sites. He also documents male defense of a sequence of perches, a pattern that he calls "dynamic territories."

Males indeed station themselves in favored places to discover females, and they tend to remain faithful to a particular set of perches that present an unobstructed view of the horizon, especially of the low spot in the horizon. Each male appears to defend a specific territory against incursions by other males. Although initial responses to intruders are not noticeably different from flights to investigate females, there is a special behavioral pattern that characterizes male-male interactions that strongly implies assessment, conflict, and driving competitors from a particular territory. There is a hierarchy of defended territories that corresponds roughly to their inferred values for efficiently encountering females. Some males alternate between defending a highly preferred and hotly contested locale and defending a less contested area of lower quality.

Males dart out from these perches toward flying objects, and their subsequent behavior depends on the identity of the object. During this activity, the males may nectar infrequently, especially on Joe-Pye Weed *Eupatorium* early in the season and Goldenrod *Solidago* later.

## 6.3 MALES DEFEND TERRITORY AGAINST OTHER MALES

### 6.3.1 Cues for Male Choice of Habitat and Perches Foster Seeing Females

Favored male perches are typically concentrated in one section of a meadow, near a low spot in the horizon, which the perched males face more often than not.

The insight about the importance of the horizon came from an unwitting manipulative experiment by the road crew of Whitehall Township in 1977. In zealous defense of a municipal right-of-way, they felled a sufficient number of trees to shift the configuration of the horizon from the study meadow. Puzzling over a sudden change in the favored perches for male Viceroys, and remembering the importance of "V"-notches for Fritillaries, I found that the Viceroys had moved from near an old notch, gradually filling in by growth of shrubbery, to near the newly created notch.

In 1983 I confirmed this insight by making a giant mirror (12 feet wide by 3 feet high) of aluminized Mylar on Masonite, and putting it up on the edge of a Viceroy's territory orthogonal to the natural notch in the horizon (Figures 6.2 and 6.3). When the mirror was angled to reflect sky toward the favored perches, the male faced the mirror or the natural notch about 30% of the time. When the mirror reflected vegetation, the male faced almost exclusively toward the natural notch (Table 6.2).

Within the suitable habitat, males prefer an area with perches from which they can watch the sky near low spots in the horizon. The ideal perch has such a view, and is high enough so that the sky is not blocked by nearby vegetation, but low enough that low-flying females will still appear against uncluttered sky and therefore be detectable as a visual cue at a great distance. An additional requirement of

FIGURE 6.3. Male Viceroy watches low horizon simulated by giant mirror (4 × 12 feet). The mirror could be oriented as shown to reflect sky toward potential perches, or it could be tilted forward to reflect vegetation. The different responses of male Viceroys are compiled in Table 6.2.

TABLE 6.2. Male Viceroy Orientation toward Artificial versus Real V-notch in Horizon

| Male orients ±15° toward (109 independent observations) | Mirror | Real V-notch | Neither |
| --- | --- | --- | --- |
| If mirror reflects sky | 21 | 48 | 4 |
| If mirror reflects vegetation | 0 | 33 | 3 |

a suitable perch is that there is some vegetation above it on the same plant. This awning above sometimes wholly or partially shaded the perch, yet sometimes left it in full sunlight. I could not discern an interpretable pattern of the shade n = beneath the awning with respect to time of day, ambient temperature, or behavior of the perched male, and so I have no evidence of its potential importance in thermoregulation (cf. Douglas 1979, Kingsolver 1985a, 1985b, Heinrich 1986a, 1985b, Tsuji et al. 1986). It is conceivable that it provides a fixed object in the nearby field of view against which to calibrate movement of small objects elsewhere. Another speculative function of the behavior is to avoid perching on terminal inflorescences wherein potential predators lurk, specifically crab spiders (Thomasidae) or ambush bugs (Reduviidae); this will be discussed further in Section 9.3.4.

### 6.3.2 A Male's Ability to Detect Visual Cues is Equal to His Tasks

Figure 6.4 records the frequency of male orientation and flight toward potential conspecifics and potential perches at varied distances. Measurement and interpretation of responses to conspecifics is straightforward. The perch-to-object distance was recorded for males leaving a perch and flying directly toward the potential conspecific. Male Viceroys are the champions for this maximum detection distance, 7 meters, versus 3 meters for the closest rival (Figure 2.1 and Table 2.1). For all the responses beyond 3.5 meters, the male saw the potential conspecific as a tiny object of high contrast against a simple, unpatterned background. Measurement and interpretation of movements toward a perch were more problematic. The measure that I used was that distance from which a flying male approached and settled on a perch after making an abrupt turn of 90 degrees or more from a previous, more twisty path. My reasoning was that a perch could better be distinguished from its background by differential optic flow when flying normal to the vector toward the perch than when flying more nearly toward it (see Figure 2.14 for the relevant geometry). My other reason was that I wanted to avoid spurious inference of distant detection from instances in which a male was flying more or less linearly and settled on the first perch that it almost bumped into; hence my insistence on a turn to redirect flight from a previously twisty path.

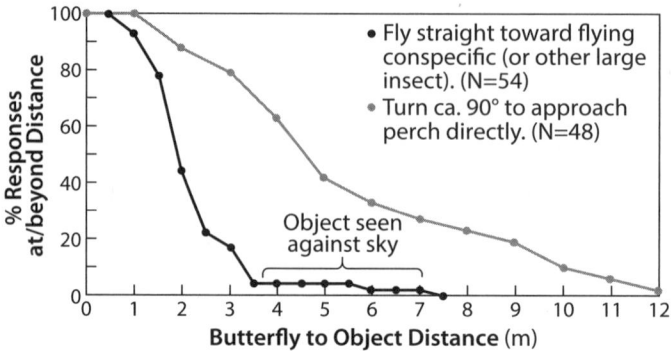

FIGURE 6.4. Frequency of responses of male Viceroys to objects at or beyond given distances. Flights toward conspecifics or other large insects were launches from a perch directly toward an object flying across the field of view; objects at extreme distances (≥4 m) were all against open sky as viewed from the perch. Turns to approach a perch were measured from detailed maps of tracks like those behind Figure 6.6; interpretations and cautions are discussed in Section 6.3.2.

Figure 6.4 shows that a male can respond to a conspecific up to 3 meters away if it is seen against a complex background of vegetation, but as much as 7 meters against a featureless background of open sky. Orientation and approach toward potential perches can be from a distance of 10 meters, with a median of 5 meters. These ranges of response are diagrammed in Figure 6.5, along with a map of perches that males used in 1983. Many pairs of perches are within the 2–5 m at which a male can potentially detect a rival, and individual perches within the most favored territory (Figure 6.2 and Table 6.3) are no further distant from the horizon determined by the edge of the meadow, favoring the detection of females. However, the territories are sufficiently large that movement among perches is necessary to view all of the horizon to detect females, or all of the perches to detect other males. Perches themselves are inter-visible among multiple neighbors and over more than half of a typical territory, as can be seen by superimposing the ranges of visibility diagrammed in Figure 6.5 onto the territories mapped in Figure 6.2. So the demonstrated ranges of response to visual cues are adequate to permit moving among perches within a typical territory, efficiently spotting flying females, and detecting other males encroaching on the territory.

### 6.3.3 Males Occupy Individual Real Estate.

There are two curious features of the male Viceroys' territories in this study. Particular blocks of real estate are defended as units by different individuals at

DEFINING AND DEFENDING A TERRITORY 133

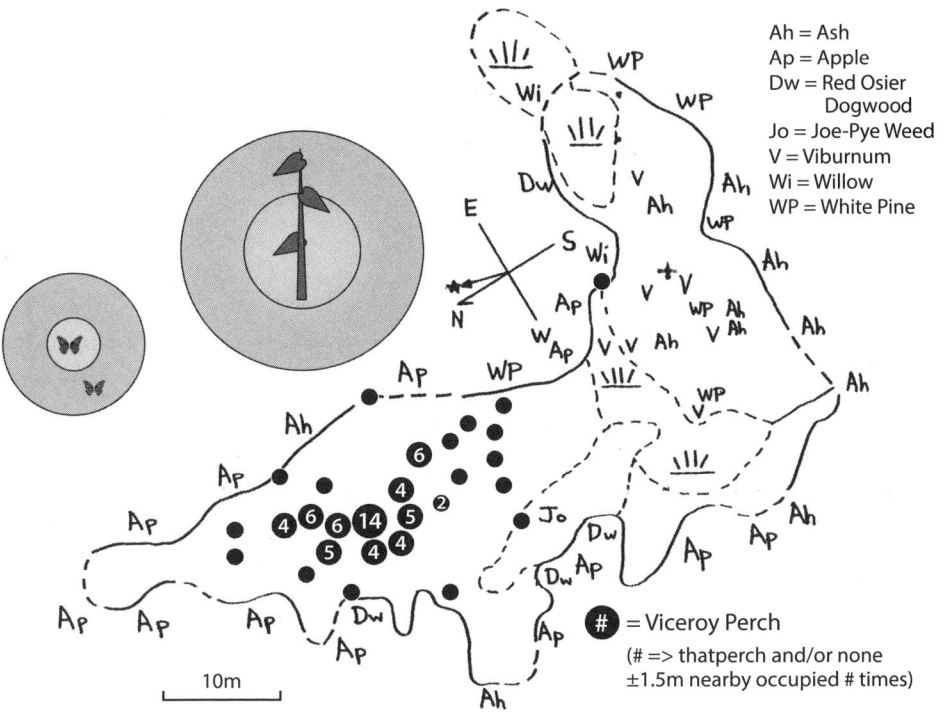

FIGURE 6.5. Scaled median (inner circle) and extreme (outer circle) distances of response of male Viceroys to conspecifics and potential perches (data from Figure 6.4). Potential intervisibility of perches spans about the size of a defended territory of patrolled perches. And most favored perches are concentrated where a large proportion of the horizon at the edge of the habitat is visible and within the extreme distance of response to a conspecific.

different times (Table 6.3). Individuals sometimes alternate between a highly favored but highly contested block and a less favored and less contested block, "leapfrogging" the intervening territory. The boundaries between these blocks are characterized by a low density of a particularly favored type of perch, namely vegetation taller than the surrounding meadow but with an awning above.[2]

*6.3.4 Males Defend Particular Real Estate against One Another.*

Territorial male Viceroys fly from their perches to investigate moving stimuli at prodigious distances compared with the other species studied—as much as

TABLE 6.3. Male Viceroys Holding Specific Territories of Figure 6.2 in 1981

| Date & Time | Holder of Territory Section: | | | | |
|---|---|---|---|---|---|
| | A | B | C | D | E |
| 30 Jul.81 | | | #81.3 Adam | #81.4 | |
| 1 Aug. | | #81.8 John | | #81.7 | |
| 2 Aug. A.M. | | #81.9 | | #81.10 Charlie | |
| 2 Aug. P.M. | #81.9 | #81.8 John | | #81.3 Adam | |
| 3 Aug. | | *#81.3 Adam | | #81.8 John | *#81.3 Adam |
| 4 Aug. | | *#81.7 | #81.10 Charlie | *#81.7 | |
| 5 Aug. | *#81.8 John | #81.10 Charlie | | *#81.8 John | |
| 6 Aug. | | | #81.10 Charlie | | |
| 9 Aug. | | | #81.10 Charlie | #82.3 Adam | |
| 10 Aug. | | #81.10 Charlie | | #81.18 | |
| 12 Aug. A.M. | *#81.15 | #81.10 Charlie | | *#81.15 | |
| 12 Aug. P.M. | | #81.10 Charlie | | #81.17 Arthur | |
| 13 Aug. | | | | #81.10 Charlie | #81.17 Arthur |
| 14 Aug. | | #81.10 Charlie | | | #81.17 Arthur |
| 15 Aug. | | | #81.10 Charlie | #81.17 Arthur | |

\* Note instances in which individuals switch to defend distinct areas, flying across another's territory without settling. By metrics of number of days held or preference by the dominant individual, Charlie, the most favored territory was a combination of B and C, followed by D.

4–7 meters for an object the size of a conspecific if it is presented against a hazy sky (a subtended visual angle of only 1° to 0.6°). Behavior upon close approach is very different in response to a conspecific female versus a conspecific male. A female quickly drops to the vegetation, and the male hovers nearby for a short time. A male flies after another male, the two typically exchanging status as leader and trailer by the leader flying up into a loop and half-roll to descend and approach from the side of the trailer. This tactic is the classic Immelmann turn of aerial warfare (Sims 1972), which has also been documented in flies (Wilkerson and Butler 1984). The new leader then loops and rolls to "get on the tail" of the follower. In a typical encounter, this behavior of trading places will proceed faster and faster until the two males complete a loop and roll in which the follower maintains its status behind the leader. Then the flight becomes a fast and nearly linear chase out of the territory, and often out of the habitat. During this chase, the territory holder is sometimes the chaser and sometimes the chased, but invariably the first to return to the territory when the chase is broken off.

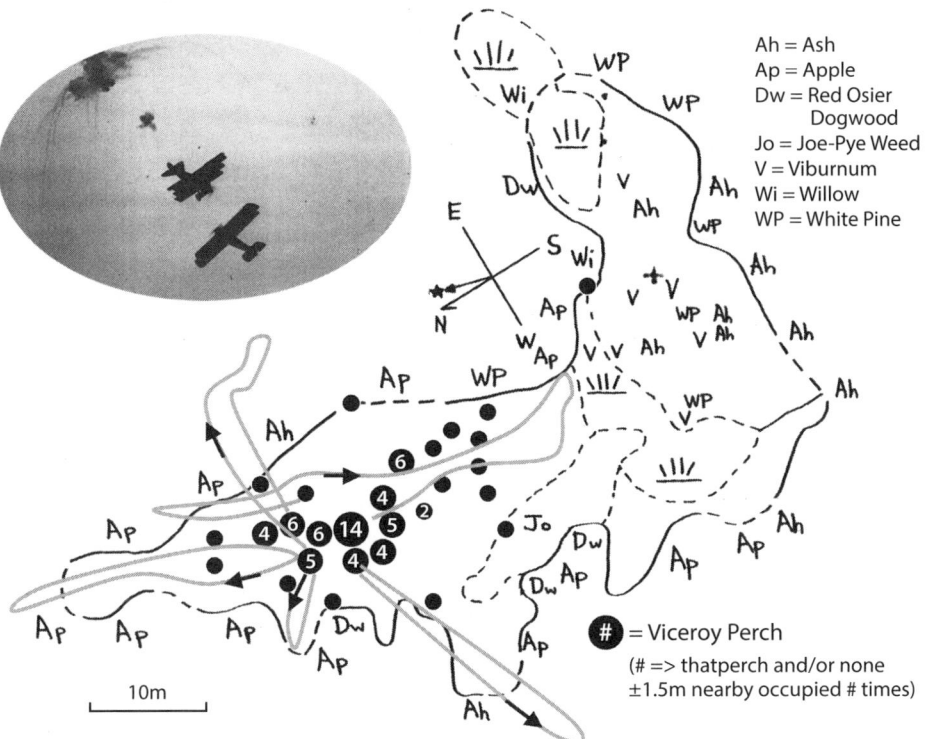

FIGURE 6.6. Perches used by male Viceroys in 1983 and short-term territorial excursions by one male in the meadow of Figures 6.1–6.3. 77 independent instances of perching were mapped for at least 5 males, including #83:2 and #83.5, from 4 to 10 Aug. 1983. Note that many pairs of perches are within the 2–5 m at which a male can potentially detect a rival (Table 2.1). The excursions of a single unmarked male were mapped during a period of only 2 minutes (14:35–14:37 on 7 Aug. 1983)! The continuous intensity of activity of this male was unusual, but the natures and lengths of individual excursions are typical of many males over many years. Track (A) was a spontaneous patrol of part of the territory and a change of perch. Track (B) was an approach to another flying male at 1.5 m, an intense interaction for 2–3 m, and a chase out of the meadow. Track (C) was a launch to another male at 1.0 m, an intense interaction for 1.5 m, and a chase out of the meadow. Track (D) was a launch to another male at 2.0 m and an immediate chase. Track (E) was a launch in response to a dragonfly, and a return to the perch. The intense interactions often share elements of WWI aerial "dogfights," illustrated in the inset, which shows a Sopwith Camel biplane, having accomplished an Immelmann turn, approaching a Fokker Dr I triplane (taken at Rhinebeck Aerodrome, NY, summer 1986).

## 6.4 DISCUSSION: DEFENSE OF PERCH VERSUS TERRITORY

Daily and seasonal changes in male behavior are described and discussed, as are missing data that are crucial to speculative interpretation. The elements of my interpretation of the behavior of Viceroys are found in other species (Baker 1972, Baughman et al. 1990, Bitzer and Shaw 1983, 1995, Daily et al. 1991, Ehrlich and Wheye 1986, Forsberg and Wiklund 1989, Ross 1963,[3] Rutowski 1978, and others already referred to).

Cues involved in the behavior of males may be simple. Once in the right habitat, males appear to select a territory mainly on the basis of visibility of a low patch of open sky. Within the territory, they select perches largely by the same criterion. In turn the availability of appropriate perches helps to set the boundaries of the defended area. Male Viceroys initially respond to other butterflies at distances at which the targets can, at most, be stimulating a few adjacent ommatidia. The distance of response is greatest when the target is seen against an uncluttered background of open sky (Figure 6.4).

## 6.5 SUMMARY OF RESULTS *AND SPECULATIONS*

### *6.5.1 Habitat*

The Viceroy is found in shrubby meadows, especially near the larval foods willow (*Salix*) and aspen or poplar (*Populus*). Males occupy territories where females are found (Behavior: Section 6.2.1).

### *6.5.2 How Females May Find Sites to Lay Eggs, and Males May Find Females*

*Females may follow stream courses by orienting toward low spots in the horizon* (Figure 6.2 and Section 6.2.2). Males choose perches that allow detection of females at a great distance (Section 6.3.2), and preferentially *look for females* by orienting toward the low spots in the horizon (Cue & Rule & Behavior: Sections 6.2.3 and 6.3.1).

### *6.5.3 Males Appear to Defend Territories Against Each Other.*

Males appear to defend their territories against other males, ranging far from them but returning, and engaging in interactions that are qualitatively different from those with females (Behavior: Section 6.3.4). *Individual males may alternate between a contested high-quality territory and a middling one, in terms of potential number of visible females* (Behavior: Section 6.3.3).

*Specific behavior of male-male interactions implies conflict and site-related defense* (Behavior: Sections 6.3.4 and Section 4.2 of Chapter 4).

CHAPTER SEVEN

# Sociology at a Singles Bar

## *The Pearly Eye* Enodia anthedon

> It is interesting from its social and gamesome habits. A particular individual will frequent the foot of a particular tree for many successive days, contrary to the roaming habits of butterflies in general. Hence he will sally out on any other passing butterfly, either of his own or of another species; and, after performing sundry circumvolutions, retire to his chosen post of observation again. Occasionally I have seen another butterfly of the same species, after having had his amicable tustle, take likewise a stand on a neighboring spot: and after a few minutes' rest, both would simultaneously rush to the conflict, like knights at a tournament, and wheel and roll about in the air as before. Then each would return to his own place with the utmost precision, and presently renew the "passage of arms" with the same result, for very many times in succession.
>
> —Philip Henry Gosse, 1859, *Letters from Alabama (U.S.), Chiefly Relating to Natural History*, p. 122, describing the male Southern Pearly Eye.[1]

Gosse's description of behavior fits my male Pearly Eyes in every detail. Part of my task in this chapter is to see how much of his metaphor of knightly conflict is supported unequivocally.

### 7.1 NATURAL HISTORY AND BEHAVIOR

#### *7.1.1. Habitat and General Movement*

Pearly Eyes favor woodland edges and open glades with wide-bladed grasses dominating the ground cover. Both sexes congregate at slime fluxes on wounded trees, where exuding sap has become colonized by yeast to form a beery nutrient broth (Figure 7.1A). Females go there for the nutrients; males for the nutrients

FIGURE 7.1A. Nine or ten Pearly Eyes, mostly males, at a natural wound on an apple tree (*Pyrus malus*). The leftmost, lighter individual is probably a female. Such dense aggregations are common in the late afternoon and early evening. Figure 7.1B. Thirteen Pearly Eyes at a sponge baited with a ferment of sugar and red raspberries (*Rubus strigosus*). Pearly Eye #50 is a female, being courted from the side, "flank-butted," by a male.

and the females. Thus their social system is literally a singles bar, or more technically a resource-based lek ("lek" = an aggregation of males in which the primary activity is mate-finding, but to which females are drawn by an additional resource, usually food; Bradbury 1985, Höglund and Alatalo 1995).

Female Pearly Eyes live long lives (Chapters 1 and 9) and lay clusters of 3–7 large eggs on the undersides of wide-bladed grasses.[2] The females presumably need extra nutrients beyond what they had stored during the larval period. Females appear to aggregate independently at slime fluxes.

Male Pearly Eyes gather at such sites, sometimes tolerating each others' presence in a tight aggregation, sometimes appearing to vie for exclusive access to a given slime flux and the attracted females.

Both sexes apparently find slime fluxes initially by flying along woodland edges[3] and tacking upwind when they encounter a plume of odor from the ferment. Once a male has found a suitable flux, he apparently learns the vicinity and tends to stay nearby, flying out to investigate all butterfly-like objects. Eligible females tend to explore for fluxes in the early afternoon, when light, laminar breezes from woods to field are usual (Section 2.4.2).

To explore this behavior, I brewed my own ferment, soaked cellulose sponges in it (Figure 7.1B; see recipe in Appendix C.8), and set them at 5 stations across a 12 meter gradient from field to woodland. Figure 7.2 shows the result of this experiment. Ideally I would like to measure the Pearly Eye's preference for degrees of shade along this gradient. However, a given Pearly Eye cannot make independent comparisons among the baits, because the separation between stations, and indeed the length of the whole experiment is less than the average attractive radius of each station, which depends on the velocity and direction of light breezes. Furthermore, the presence of other individuals at a given bait increases its attractiveness (cf. Section 7.1.2 and Chapter 8). The best I can do is to record "presence" rather than "preference," and to ensure some semblance of independence, I tally individuals that were initially captured and marked at a given station along the gradient. Figure 7.2 shows that both sexes were initially found throughout the gradient, but more in woods than in field. Figure 7.2 also shows the accumulated number of "butterfly-days" that marked Pearly Eyes were found at a given station on the gradient, over the 12 days of the experiment, during which

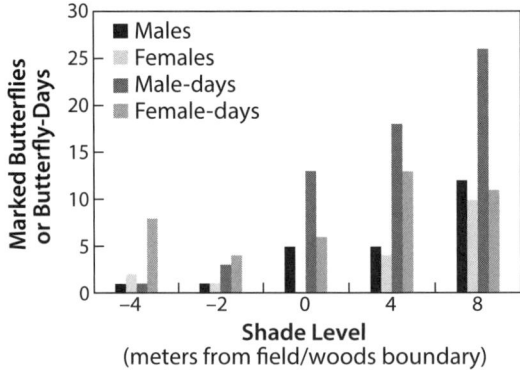

FIGURE 7.2. Pearly Eye presence at baits across a field/woodland boundary, 6–17 Aug. 1980. Baits were set at the given distances from the boundary. The solid bars record the number of Pearly Eyes that were initially captured and marked at each bait (totals: 24 males; 17 females). Both sexes were initially found more in woods than in field. The dotted bars record the cumulative number of individual-days that the marked Pearly Eyes were found on each bait. The butterfly-days data are highly heterogeneous and do not permit statistical tests, but they still suggest that males sort toward the woods and females sort toward the field. See Section 7.1.1 for further details and discussion.

new individuals continued to appear and be marked. The data on butterfly-days are even more heterogeneous and interdependent than those from initial captures, and do not permit statistical tests. Nevertheless, they give some indication of the re-sorting of individuals along the gradient after initial marking, and suggest that males sort toward the woods and females toward the field. This fits with observations of the behavior of both males and females in this experiment when released after marking. When released from the station 4 meters into the woodland, all 17 females flew toward the field, but only one male did; the other 23 males flew either directly into the woods or parallel to the boundary for several meters before turning toward the woods.

### 7.1.2 Male Ranging and Behavioral Interactions

Pearly Eyes seem to be more patchily distributed than their already patchy optimal habitat. By distributing baited sponges in varied spatial designs, I could influence the distribution and behavior of Pearly Eyes. When baits were distantly spaced, males were distributed either at random or in a hyper-dispersed fashion. Males showed varied but individually characteristic loyalty to particular home baits (Figure 7.3). When baits were clumped, males showed a contagious distribution among baits within a clump. Casual observation suggests that this distribution may be generated simply by settling soon after interaction with another individual (for theoretical discussion, see Chandrasekhar 1943, Odendaal et al. 1988, Reed and Dobson 1993, and Turchin 1998, pp. 109–125). In a randomized choice experiment between two otherwise identical baits, males settled much more often on the one that had lightly fluttering Pearly-Eye-colored pinwheels around it (Figure 7.4).

Males gather at wounds and "compete" to intercept females. There they engage in "contests" with other males, superficially displaying all the signs of "the bourgeois strategy," in which the owner wins and the interloper retreats (Maynard Smith 1978, Maynard Smith and Parker 1976). Males engage in "sexual chases" of females, which may turn into "conga lines." It is conceivable that females can exert some "mate choice" in these chases. Matings have been observed only near tree wounds, and at particular times of day (midafternoon).

### 7.1.3 Artificial Baits Foster Observations and Experiments.

By setting out many artificial baits, I initially increased the local abundance of both male and female Pearly Eyes, and increased the rate at which I could observe their interactions. Some of the observations led to specific hypotheses that I

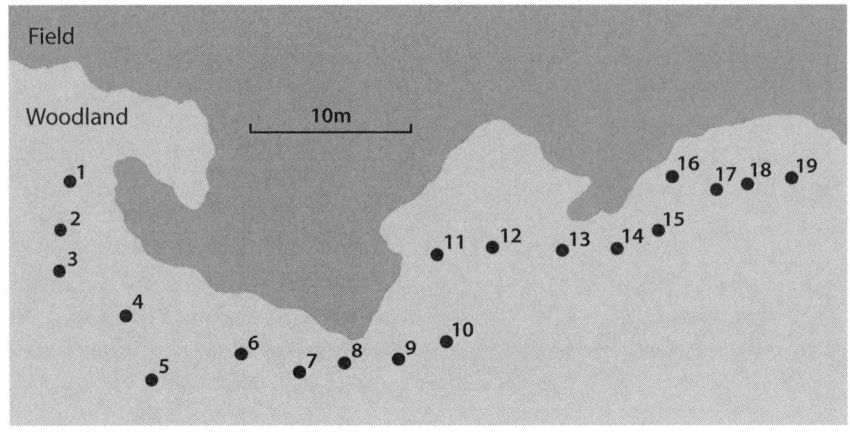

FIGURE 7.3. Long-term perching (x) or brief visits (o) by individual male Pearly Eyes at each of 19 numbered baits (July 1981). Each row is an individual, in attendance for the number of days noted. Although a few males attended several baits, most were faithful to a particular location, even though their flights often took them within the attractive radii of many other baits.

explored in more detail with an appropriately designed set of baits. For those activities that take place on or very near the bait, I had the opportunity to record them on videotape, though lighting in the forest, especially in the later afternoon, is far from ideal. So I can manipulate this species and observe its natural behavior closely in a natural setting. Accordingly, I know more about the social behavior of this species than any of the others.

My observations of Pearly Eyes are of four kinds: (1) ad hoc distant observations of unconstrained individuals, (2) ad hoc observations of individuals after

FIGURE 7.4. Pearly Eye laboratory for manipulative experiments. Baits on saplings are sponges soaked in sugar and raspberry (*Rubus strigosus*) ferment. Pinwheels made from 2-inch squares of kraft paper are suspended by threads from movable branches. A breeze as light as 0.9 m/s can turn one so as to simulate a fluttering Pearly Eye. The woods provide an effectively dark background for flash and strobe photography. Photo of the author was taken before he started to dye his hair gray. In the prime experiment, six pinwheels were suspended from a branch that was randomly placed on one of two baits 3 meters apart. Twenty one Pearly Eyes (mostly males) were tallied as they came to the baits. Twenty settled at the bait with pinwheels, and only one at the bait without, though they were evenly divided between the baits per se. The pinwheels clearly help to attract and/or to induce settling. Several males briefly inspected a spinning pinwheel on the way to the bait.

marking some with permanent numbers (Chapter 1; Horn 1976b), (3) ad hoc and structured observations at artificial baits arranged in a spatial configuration that favors interactions, (4) manipulative experiments with artificial lures. Each of these poses problems of interpretation and statistical analysis, problems that are exacerbated by particular features of Pearly Eye behavior. Pearly Eyes are found singly or in aggregations. The aggregations may be due partly to independent choices of a particular microhabitat in a patchy environment, and due partly to attraction between individual butterflies. Particular butterflies may be active or inactive at any given time. The nature of activities, including their daily schedule, and their changes with age, can vary from one individual to another. Accordingly, the statistics of unconstrained behavioral observations can be as much influenced by my ability to notice butterflies, interacting with my schedule and coverage of their habitat, as by their behavior under observation. Conversely, the outcomes of manipulative experiments and structured observations are very sensitive to idiosyncrasies of the individual butterflies observed.

### 7.1.4 Exploratory and Courtship Behavior of Males

It is hard not to think of some male Pearly Eyes as exhibiting "personalities," describable with terms like: nervous, inquisitive, feisty, placid, timid, mercurial, systematic, picky, indiscriminate, and dogged. In most of my initial encounters with a group of males near a natural tree wound or an artificial bait, one or more males flew out and circled me several times in what seemed to be inspection of a novel object. In later encounters with marked butterflies, it was usually the same individuals that approached and circled me. Some individuals attended so much to my presence and movements that I could not get useful observations about their behavior relative to conspecifics or other features of their environment. Conversely, many individual males became habituated to my presence and activities, while still responding to each other and to potential mates. All males, even the highly active ones, sometimes go into an inactive mode in which they respond only to physical disturbance.[4]

Conversely, the "courtship" behaviors of male-female dyads are consistent enough to support a general description. A female pursued closely by a male will usually dip down and then fly upward, often along a tree trunk to land on either a vertical or a horizontal surface. The male follows to land and court from the female's side with some nudging from behind. He may vibrate his wings, but does not flutter and bash, the way he does with other settled males on a bait. He may flutter and dislodge the female from a perch that provides no dance floor, or from a bait on which there are other settled males. A pursued female may alternatively dive suddenly to a landing or to a continued flight. Most "chases" are broken by the male losing her as she starts her dive, though a male who dives quickly can keep up with her. If male and female both land on a bait, the "chase" can be broken by the male turning to feeding or launching toward another flying individual. When both sexes are settled, the female may apparently "reject" by tilting so far toward the male that he cannot get close to her body unless he crawls beneath the shade of her wings. When a male has lost a female, he does what appears to be a short local search and then dives low and flies a helical path for several turns up or down the nearest tree. He then does the same for two to four trees within 2–3 m, then heads back to the bait. I have seen a final revisit to the actual spot at which contact was lost with the female often enough to make me suspect that it is part of an organized search, but not often enough to be sure that it is not simply accidental behavior to which I give a wishful purpose.

## 7.2 MALES SEEM TO REPEL MALES AND COURT FEMALES

Male Pearly Eye behavior at a slime flux shows all the elements that Davies (1978) has documented for the Speckled Wood butterfly *Pararge aegeria* at sunflecks in a woodland. Male Pearly Eyes investigate and chase passing females (Figures 7.5 and 7.6). A single female may be pursued by a "conga line" of three to five or more males (Figure 7.7), which invites interpretation as a mechanism by which females incite sexual competition among the males to foster mating with only the best among potential mates. Sitting males drive away passing males; i.e., there are male-male "contests" with a "home court advantage"; that is, the males appear to assess each other's prospects in a contest, and to settle it according to the convention that the resident wins and the intruder loses.[5] If a sitting male is experimentally removed, another takes his place and wins subsequent contests with passing males. When the original male is released, the contest between him and the new site-owner is longer than usual, and the outcome may be indeterminate; i.e., there is "escalation" in even contests when both individuals have the status of former site-owner.

These assertions are consistent with outcomes of male-male interactions recorded in Table 7.1. These interactions are all "contests" or "spinning wheel flights" between males of known identities and known initial statuses: "Resident" = at or near the bait for a full day or more; "Vagrant" = newly arrived in the vicinity of the bait; "Percher" = initially on or within 1.5 meters of the bait; "Flier" = initially flying. A butterfly is called a "Winner" of the contest if, after interaction, it settles within 1.5 meters of the bait (the typical distance of response to another butterfly) while its rival either flies away or settles further than 1.5 meters. The interaction is declared an "Equal Draw" if both males settle within 1.5 meters of the bait, and an "Ambiguous Draw" if both settle further than 1.5 meters.

Table 7.1 supports the following conclusions:

- When there is a winner, Resident tends to win over Vagrant (home court advantage or "bourgeois strategy" of Maynard Smith 1978), but this could be simply because Percher wins over Flier.
- When both contestants are Resident, Percher still tends to win over Flier, but statistics are not strong enough to reject an indeterminate outcome.
- Some instances of long contests could be sequences of repeated reencounters, i.e., multiple short contests, rather than "escalation" (footnotes to Table 7.1).

FIGURE 7.5. Stroboscopic photo of a male Pearly Eye (four leftmost images) sallying to investigate one of two passing Pearly Eyes (respectively two images above and one below). The initial reaction distance was about 15–20 cm, and the following distance is about 25 cm.

FIGURE 7.6. Stroboscopic photo (four flashes) of a male-male encounter (left) versus a male-female encounter (right). On the left, the two males are engaged in what is often described as a "territorial contest;" Wickman and Wiklund (1983) refer to it as the more neutrally descriptive, but equally poetic, "spinning wheel flight." On the right a female comes to a landing, "chased," or more neutrally, "followed" by a male.

FIGURE 7.7. Stroboscopic photo (four flashes) of a "conga line." On the far left is a female. She is being followed by a line of four males. Each butterfly is about 15–20 cm behind its predecessor. The geometry and dynamics of this behavior is qualitatively the same as a "sexual chase" in birds, which has traditionally been interpreted as a female's inciting competition among males to sort out who is best able to keep up with her. The implication is that she will thus end up with the fittest mate. See Section 7.3 for an alternative interpretation.

Davies' (1978) study of the Speckled Wood quickly became a contested classic of optimal game-theoretic hermeneutics (Maynard Smith 1978, criticism by Austad et al. 1979, rejoinder by Davies 1979). In particular, the home court advantage was christened the "bourgeois strategy," in which prior ownership of a territory is an asymmetry that favors the eventual victory of the owner, and therefore settles the contest most efficiently by having the interloper retreat if challenged. Data that superficially supported the home court advantage were given an alternative explanation by Wickman and Wiklund (1983). They suggested that at any given time given males of the Speckled Wood tend to remain either basking perchers or flying patrollers.[6] Since perching defines both ownership and victory, and flying defines both intruding and defeat, the supposed home court advantage may represent no more than constancy of behavior. Baker (1983), reviewing the evidence, concluded that the evidence for behavioral contests between males was strong, but the evidence for settling those contests by arbitrary conventions or behavioral biases was equivocal. Shreeve (1984) noted the difference in thermal physiology between a heated Speckled Wood sitting in the sun and a cool butterfly flying in the shade (cf. Jacobs and Watt 1994, who have the opposite situation, Sulphurs *Colias* flying in the sun and sitting in the shade). Shreeve also noted that the thermal difference may reinforce alternative behaviors, and differences in thermal environment might explain some of the differences among studies of the Speckled Wood (after Wickman and Wiklund 1983).

## 7.3 SIMPLE BEHAVIORAL RULES FOR COMPLEX ENCOUNTERS

### 7.3.1 Experiments

Much of this apparently complex behavior can be produced by a few simple rules of behavior in response to simple cues. The attempt to discover such rules can be viewed as reductionism, mechanism, strong inference (Platt 1964), or any of a number of "-isms." However, my ultimate intent is to use the rules as tentative "explanations" of as much of the behavior as possible, and then to search for interesting behavior that goes beyond the predictions of the simple rules (cf., my use of the Fortuitous Site-Fidelity Model as a null model in Sections 4.6 and 8.4.2).

In a series of experiments modeled after the classical work of Tinbergen (et al. 1942), I tested the responses of male and female butterflies to tethered individuals of their own or other species, and to various artificial lures. The fake lure that most successfully elicited responses that mimic those to a tethered live conspecific was christened the "Butterfly Rotisserie" (Figure 7.8 and Appendix C.6). The lure itself is a rotating 5 cm-diameter circle of brown kraft paper.

Here are the results of presenting the lure to Pearly Eyes. Flying or perched female Pearly Eyes ignored or actively avoided a rotating lure. Perched males often ignored the lure, but sometimes flew out to inspect it closely. Flying males more often inspected than ignored. After bumping into the rotating lure, most males became wary of it, but some continued to approach and/or follow it for several seconds.

My experiments with tethered butterflies and artificial lures have confirmed or suggested a number of simple behavioral "rules" that together could produce much of the Pearly Eye's complex behavior.

### 7.3.2 Rule #1: Males investigate, but females ignore, fluttering objects.

The first rule concerns the response to a fluttering, butterfly-like object. Female Pearly Eyes usually avoid it, and sometimes ignore it. Males show "off" and "on" behavior. When "off," they ignore. When "on," the usual state, they invariably approach and inspect it (cf. Scott 1974, and Silberglied 1977). This behavior alone is sufficient to generate male-female chases, and even male-male-male-female conga lines, as well as male-male "contests" (respectively, male-female

# SOCIOLOGY AT A SINGLES BAR 149

FIGURE 7.8. The Butterfly Rotisserie. A lure, a circle of brown kraft paper or other colored cardstock, is held in a clip at one end of an automotive speedometer cable. It is rotated by a fishing rod spinning reel at the other end of the cable. The whole is mounted on a gunstock with a camera aimed at the lure. So the lure can be rotated and translated with considerable freedom of movement both before and after presenting it to a sitting or flying Pearly Eye. Our daughter, Jennifer, does the honors here. See Figures 7.9C–D for examples of the rotisserie in action, and see Sections 7.3 and 7.4 for results.

"horizontal flight pursuit" and male-male "spinning wheel" of Wickman and Wiklund 1983).

Here is how the rule works. If a male is perched and a female flies past, the male flies out to investigate, the female turns to avoid, and the male pursues in investigation. If a male is perched and another male flies by, the sitting male flies out to investigate, the passing male investigates the newly launched one, and the two bump around in a mutual investigation until each discovers that the other is not a female. Thus a simple behavioral rule can simulate the same pattern that has previously been interpreted as intersexual courtship versus intrasexual competition, with the male making a decision as to which to do based on prior assessment. The rule can be extended to interpret more complex patterns, like the "conga lines" or "sexual chases" of the Pearly Eye. The first individual in the line is a female, avoiding what is behind her. The second is a male, investigating what is in front of him. The third is a male, investigating what is in front of him; and so on. Occasionally a natural experiment confirms this interpretation when a long conga line is disrupted by bumping into vegetation. The front section continues on as a line, but the rear section immediately degenerates into a set of one-on-one bumpings. Note that the male is not required to change his cue or rule of behavior in order to generate any of these varied patterns of interaction.

This rule is also sufficient to promote and reinforce male spatial aggregation by the mechanism most clearly explained by Turchin (1998, p. 113 ff.). If a male traveling in a straight line encounters and interacts with another male he may leave the encounter in a new direction. So if a male encounters a cluster of other males, his rate of turning increases and he is less likely to depart the vicinity of those other males; the other males are less likely to disperse by the same mechanism (cf. "Turn to stay!" in Section 8.2.4).

### 7.3.3 Rule #2: Males return to previous behavior (flying versus sitting).

The second rule is "Butterflies in motion tend to remain in motion; butterflies initially at rest tend to go back to rest" (Newton 1687).[7] As Wickman and Wiklund (1983) observe, this generates an apparent home court advantage in male-male encounters. The owner is defined as the butterfly that is initially at rest. By the rule, the butterfly at rest goes back to rest following an investigative encounter, which in turn defines the winner. The interloper is defined as the individual that initially flies in. If, by the rule, this interloper continues to fly following the encounter, then by definition it retreats. There is no need to invoke a sophisticated, game theoretic "bourgeois strategy" (Davies 1978 and Maynard Smith 1978) to interpret the behavior of an owner winning against an interloper. Each individual merely returns to what it was doing before the encounter.

### 7.3.4 Rule #3: Males initially sitting tend to return to their previous perch.

The third rule argues that two males initially at rest tend to return to their rest sites after an aerial encounter. If the two initial rest sites are close to one another, it is likely that the males will pass within mutual detection distance of one another. If so, they may reencounter several times, each time adding to the total time before they settle again. Table 7.1 records some instances of long contests that are indeed sequences of repeated reencounters. So much for "escalation in even contests."

The return to a perch is usually more direct than the exploratory or interactive path away from the perch. This suggests that residents "know" the detailed topography of their territories and have maneuvering advantages over intruders. Davies also noted this for the Speckled Wood and suggested that it was part of the intrinsic advantage of a resident in a contest with an intruder.

TABLE 7.1. Outcomes of Male-Male Pearly Eye Interactions by Prior Status of Participants

| Symmetry | Result = Draw or | | Clear Winner of Known Status | | | |
| --- | --- | --- | --- | --- | --- | --- |
| | Ambiguous | Equal | Percher | Flier | Resident | Vagrant |
| Resident versus Resident | 4 | 7 | 5[a] | 1 | — | — |
| Resident versus Vagrant | 16 | 6 | 10[b] | 1 | 12[b] | 5 |
| Vagrant versus Vagrant | | 2 | | | | |

A "Resident" had been on or near the bait for at least a day, and a "Vagrant" was newly arrived in the vicinity of the bait. The "Percher" was initially within 1.5 m of the bait, and the "Flier" was flying. The outcome is an "Ambiguous Draw" if both males end up sitting more than 1.5 m from the bait, and an "Equal Draw" if both males settle within 1.5 m. A "Winner" was defined by sitting on or within 1.5 m of the bait at the end while the other male was further away, either settled or flying. See Section 7.2 for conclusions.

a. One instance of Resident vs Resident with initial Percher winning was repeated 20 times in succession as the initial Flier re-approached the bait.

b. One instance of Resident vs Flier with Resident/Percher winning was similarly repeated 7 times.

## 7.4 A FIGHT-OR-FOLLOW MODEL FOR PEARLY EYE INTERACTIONS

I have incorporated these rules into a "Fight-or-Follow" Model, using the programming language NetLogo (Wilensky, 1999, http://ccl.northwestern.edu/netlogo/). The model for the Pearly Eye is based on a correlated random walk, in which a butterfly repeatedly moves a unit length in a direction that differs from its current heading by a random Gaussian deviate of 0° mean and an adjustable angular standard deviation (i.e., the Forward Vagrancy Model of Section 3.3.3). A male encountering another butterfly within a fixed detection distance turns toward it, but then chooses the usual random direction before moving. A female encountering another butterfly within a fixed detection distance turns away from it, but then chooses the usual random direction before moving. Other than the reorientation on encounter, flying males and females initially move with the usual forward-biased random walk. A perching, "guarding" male sits at a particular location on its "perch," until another butterfly comes within the fixed detection distance, at which point the percher adopts the behavior of a flying male. When the percher gets further than a set distance from its perch, it returns to the perch using a correlated random walk with a smaller variance in the change from forward direction at each move.

Even in the primitive form just described, the Fight-or-Follow Model produces "movies" that look frighteningly like my videotapes of the behavior of real Pearly

Eye. Still frames from a model run are shown with trails for 50 steps of the movements for a male-male interaction (Figure 7.9E) and a male-female interaction (Figure 7.9F). The former records a two-episode spinning wheel flight, and the latter records a continuous chase. Figures 7.9A and 7.9B show corresponding natural interactions, and Figures 7.9C and 7.9D show interactions of Pearly Eyes

**A.** Male sallies to investigate male

**B.** Male chases male chasing female

**C.** Male investigates rotating lure

**D.** Male chases male chasing lure

**E.** Model males investigate each other

**F.** Model male investigates; female avoids

FIGURE 7.9. Pearly Eye interactions: A–B. Natural behavior, C–D. Experimentally induced behavior, and E–F. Modeled behavior. All are consistent with males behaving the same way, simply investigating another butterfly, in "contests" (left column) versus "chases" (right column). The presence and avoidance behavior of the female generates the difference between "contest" and "chase."

with the Butterfly Rotisserie, C with the lure rotating in place, D with the lure beginning a backward sweep. Nature, experiment, and model all show the difference between a contest and a chase without any change in the rules governing the behavior of real or robotic males.

## 7.5 DIURNAL CHANGES IN MALE BEHAVIOR MAY BE ADAPTIVE

There is daily variation in the behavior of a typical male Pearly Eye, and in the extent to which a given male remains in the alternative states of patroller or percher.[8] In the morning, most males feed on a bait or bask in the sun away from it (Figure 7.10). Each bait typically has only one male on it, but often there is another male 0.75 m to 1.5 m above it (Figure 7.11A). As the morning passes, an increasing proportion of males sit alertly on or near a bait, flying out to any butterfly-like object, chasing it briefly if it is a female, and quickly returning to the bait if it is not (Figure 7.10). During the brief period in the early afternoon when new females are approaching the baits (Figure 7.12), males engage in lengthy chases of those females, and begin to engage in lengthy "contests" with other males (Figure 7.13). In the later afternoon, chases of females become short again, but male-male interactions remain lengthy (Figure 7.14). Characteristic differences between individual males seem to become exaggerated in the later afternoon and evening, as some carry on as usual, some settle on baits to feed, and others bask in the setting sun (Figure 7.11B).

Individual interactions between males become more intense and longer at just the time when eligible females appear, and they continue during the times when

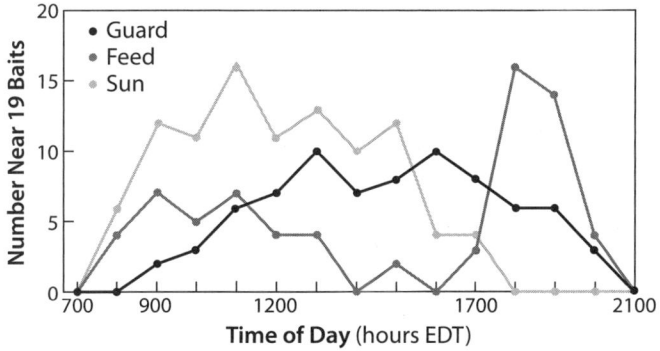

FIGURE 7.10. Male Pearly Eye stationary behavior in a survey of 19 baits during the day of 28 Aug. 1983. "Guarding" is characterized by a male's perching on or near a baited trysting place with its proboscis retracted so that it is clearly not feeding. Such guarding increases before, at, and after the time that new females are most likely to appear at the bait (cf. Figure 7.12).

**A.** Late morning     **B.** Late afternoon, early evening

FIGURE 7.11. Diurnal pattern of attendance by Pearly Eyes at trysting places. This pattern typifies a large local population at the height of the flying season. A. During the morning one or two males perch right at the ferment, with perhaps one or two males distant by 0.5–1.5 m (about the mutual detection distance), usually above the bait. B. In the late afternoon and early evening, many males and a few females are active at and around the ferment, in this case 6 or 7 perching and 9 in flight.

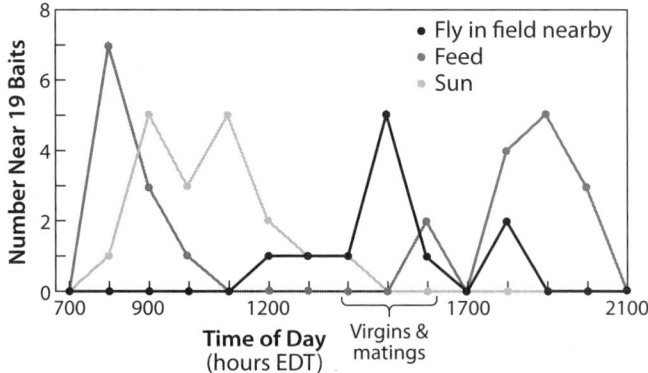

FIGURE 7.12. Female Pearly Eye activities in a survey of 19 baits during the day of 28 Aug. 1983. "Feeding" involves attendance at a bait or a natural ferment with the proboscis extended. Feeding females may be courted from the side by males, but without any success that I have seen. "Sunning" females are apparently thermoregulating in an exposed microsite. "Flying" females appear to be tacking upwind toward baits. Numbers are small, but there is increased activity during the midafternoon period, when a few virgin females and pairs in copula were found.

SOCIOLOGY AT A SINGLES BAR         155

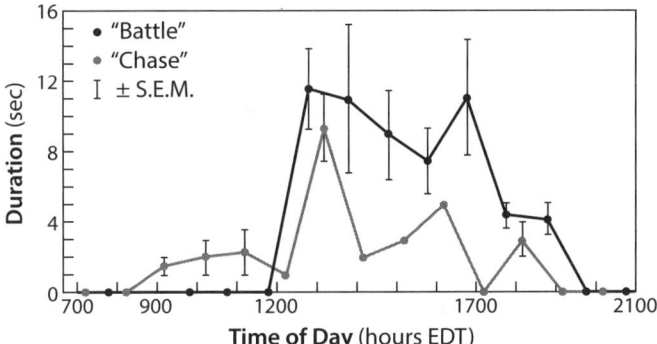

FIGURE 7.13. Duration of male Pearly Eye interactions in a survey of 19 baits during the day of 28 Aug. 1983. "Battles" are symmetric irregularly spinning mutual investigations; when the sex of the other individual could be determined, it was always a male. "Chases" are irregularly linear flights, following another individual; when the sex of the leading individual could be determined, it was always a female. The presence and duration of battles rises dramatically during the period when eligible females are most likely in attendance at the baits. Data on chases are too sparse and variable to support much speculation, though longer chases occur when eligible females first appear.

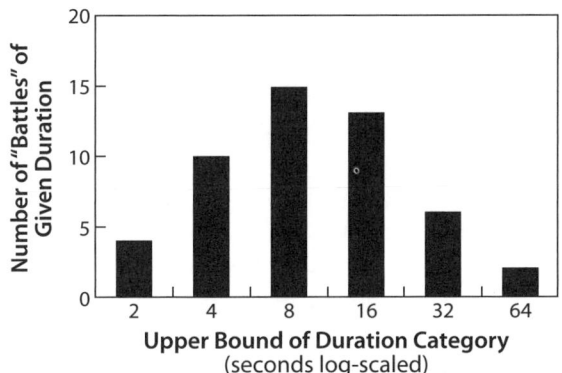

FIGURE 7.14. Duration of 50 male-male Pearly Eye "battles" during the prime period for such activity, 1300–1700 on 23 Aug. 1983.

matings have been observed. This is strong, albeit circumstantial, evidence that these male-male interactions are indeed potentially adaptive contests for positions with greater access to females.

What bothers me about this argument is that if I had found exactly the opposite temporal pattern of male behavior, I could invent an equally plausible adaptive interpretation. If the interactions had been strongest and longest in the morning, I could argue that the males were staking out territory within which to have exclusive access to females, and that once the females appeared, the males would be wiser to chase them than to engage other males.

## 7.6 DOES AN INVESTIGATING MALE DISTINGUISH THE SEXES?

The major differences described so far between male-male and male-female interactions can be interpreted as due to the behavior of the female, rather than that of the male. Some detailed differences can be seen in slow-motion review of videotapes, but even these differences are equivocal. They can be plausibly interpreted as Newtonian reactions to being buffeted by puffy vortices shed with each wingbeat by an interacting partner (cf. Section 2.2.2).

Nevertheless, it is very likely that males have sufficient sensory input to distinguish the sex of an interacting partner. The optical flow of the visual field is very different between a linear flight chasing a female and a spinning-wheel flight interacting with another male (Section 2.3.4). The typical behavior at the end of an investigatory flight differs with the sex of the object of investigation. When a formerly perched male leaves a male-male encounter, he almost always heads directly toward the nearest bait or his former perch, though he may be diverted by another encounter before he gets "home." A male's aerial encounter with a female may end with the female diving to perch on the ground or fluttering to perch on a baited tree. If the male follows successfully, he will land nearby and engage in courting behavior, typically moving to the side of the female and "flank-butting" with wings partly or extensively spread (see Figure 7.1B). If the male is unable to follow, or if he otherwise loses the female in flight,[9] he may fly back to bait or to perch as in a broken male-male encounter, but he often engages in the following additional behavior. The male approaches the nearest tree and performs a descending helical flight about the trunk; he may fly up and repeat this maneuver and/or fly to a nearby tree and repeat it, and so on, up to as many as five times, before perching again. Every time that I have noticed this special behavior, it has followed the break-up of a chase, either of a female or of an encounter-avoiding male. So a male may make a special effort that looks like a systematic search after it has lost contact with a female, an effort that it never makes after losing contact with a "contesting" male.

## 7.7 DISCUSSION: STRATEGIC OR MECHANISTIC BEHAVIOR—OR BOTH?

Figure 7.15 records the ultimate goal of all this behavior, a 20 July 1983 mating between Male #83.31 and a recently eclosed female. Male #31 was initially

FIGURE 7.15. An early afternoon, 20 July 1983 mating, between Male #83.31 and a recently eclosed, unmarked female. Couples can remain coupled for several hours, and the female can make short flights (2–4 m) towing the male.

marked on 17 July, at which time his wings had lost enough scales for me to estimate his age since eclosion as 2–3 days. Later on the day of marking, he assiduously courted a recently eclosed female, following her through 5 episodes of short flight (2–4 m) and landing. When he left her, I assumed that she had rejected him because she was already mated, but I collected and dissected her anyway. She had no spermatophores; she was a virgin! Three days later he achieved the goal of all the previous behavior, the mating recorded in Figure 7.15. He was next seen on 25 July, and then again on 26 July, 02 Aug., 03 Aug., 06 Aug., and 07 Aug., by which time his wings were so worn that I could not be sure of reading his number without recapturing him, which I chose not to do in deference to his elder status.

Male #83.31 was observed often as a percher, and he must have patrolled between perches. Did he defend his territories? Did he escalate when likely to win contests? Did he engage in many chases? Did he leave his first virgin love or did she rebuff him? Or did he just investigate butterfly-like objects until one either recourted him or simply sat still long enough for mating? Had either of the females observed with him incited a chase? Had she induced a conga line and sorted out a champion? Or had she avoided or ignored until near the moment of mating? Were the butterflies engaged in evolved strategic games, or were they mechanically following simple rules of behavior in response to simple cues?

Much of the observed behavior is equally consistent with either extreme of strategic behavior or mechanical behavior. There is even an intermediate philosophical position in which a strategic outcome is achieved by natural selection acting on simple mechanical responses to simple cues. And there is also the possibility (which I hopefully doubt) that the whole of mating behavior is orchestrated by olfactory cues that I cannot sense.

I don't know the answers. For me, some aspects of a seemingly complex behavior are now perhaps less mysterious. Others have become even more mysterious: the details of male behavior when he has lost contact with a chased female, whether either sex is choosing versus simply finding a mate, how males map and navigate individually characteristic regions of the landscape, and what is the right interpretation of the diurnal variation in behavior—of individuals and among individuals. The Pearly Eye deserves and will repay much more study, and if you can find a stable population, it is great fun to work with.

## 7.8 SUMMARY OF RESULTS *AND SPECULATIONS*

### *7.8.1 General Natural History*

Pearly Eyes congregate at fermenting slime fluxes on wounded trees. *They apparently navigate there initially by following odor plumes* (Section 7.1.1). *Later they navigate to particular sites by using other cues* (Cue & Behavior: Sections 7.1.1 and 7.1.2).

*Females apparently go there for nutrients needed to lay large eggs in clusters of 3–7 each.* Males go there for both nutrients and access to females (Behavior & Life History: Section 7.1.1).

### *7.8.2 Simple Cues and a "Fight-or-Follow" Model Can Produce Complex Male Behavior.*

Males are attracted to aggregations of other males, but often act to repel each other individually (Cue & Rule & Behavior: Sections 7.1.2 and 7.2).

The details of male behavior have been interpreted strategically. *They court females but contest with other males. They engage in competitive chases of possible mates, perhaps at the instigation of the female.* In asymmetrical contests with an interloping male, there is a home court advantage. *Even contests between two local males are escalated to greater length* (Behavior: Section 7.2).

Simple tactical rules can generate those details of behavior without the need for strategic interpretation. Males investigate, but females ignore, fluttering objects. Interactions result in locally convoluted flight paths, reinforcing aggregation of males. After aerial interaction, individual males return to previous behavior (flying = patrolling versus sitting = perching). Males initially sitting tend to return to their previous perch (Cue & Rule & Behavior: Section 7.3).

A NetLogo Fight-or-Follow Model based on those simple tactical rules can generate the supposedly "strategic" behavior (Cue & Rule & Behavior: Section 7.4).

### 7.8.3 Is the Behavior Really that Complex?

Diurnal changes in male behavior suggest an adaptive interpretation, but ambiguously (Section 7.5). Individual male Pearly Eyes differ dramatically and characteristically in the same settings.

Male-male versus male-female interactions *should produce perceptibly different signals*, and end in characteristically different behaviors, a broken encounter with a female often leading to searching behavior not seen after an encounter with another male (Behavior: Section 7.6)

### 7.8.5 Is Male Behavior Strategic or Robotic?

*I suspect both!* (Behavior: Section 7.7)

CHAPTER EIGHT

# Do Butterflies Make Decisions?

> I am much more interested in the way that nature shapes
> theory than in the constraints that theory imposes on nature.
> –Henry S. Horn, 1971, *The Adaptive
> Geometry of Trees*, p viii

It would be easy to state that butterflies *must* make decisions: to favor the right habitat but avoid the wrong habitat, to approach conspecifics but to avoid enemies, to court the opposite sex but to repel their own sex, and even to decide when they have sufficient information to make the other decisions. Alternatively, it may be that what look like decisions are instead fixed and robotic changes in behavior in response to simple and dependable cues that appear when changed behavior is adaptive. So the organization of this chapter is a local version of the organization of previous chapters. I start with the interaction of each species' ranging with the spatial configuration of its habitat, which requires description of what constitutes the favored habitat for each species.[1] I then review cues and rules of behavior that might be used by automata to find, preferentially settle in, and move about the right habitat. And I simulate those behaviors with individual-based models, shamelessly favoring models that impose little or no cognitive burden on a butterfly. I shall show that interspecific differences in habitat use by real butterflies can be interpreted in terms of automatic rules of behavior given cues that differ between species. I then ask what aspects of behavior require more than fixed responses to fixed cues. Specifically, when do individuals exposed to the same cues behave differently? Such situations suggest individual decisions, but do not necessarily require them.

## 8.1 FAVORED HABITAT, MATRIX, AND RANGING

### 8.1.1 What Distinguishes Favored Habitat from Matrix?

A favorable habitat must provide food for larvae (= oviposition sites for adult females and nearby emergence sites for both males and females), food for adults

TABLE 8.1. Resources and Conditions Provided by Habitats

| Characteristic: | Dry Field | Wet Meadow | Edge | Woodland | Forest Glade |
|---|---|---|---|---|---|
| Larval Food | Ringlet | Eyed Brown Viceroy | Fritillary Viceroy Pearly Eye | Pearly Eye | Fritillary Pearly Eye |
| Adult Food | Ringlet Fritillary Viceroy | Eyed Brown Fritillary Viceroy | Eyed Brown Fritillary Viceroy Pearly Eye | | |
| Likely Mates | Ringlet | Eyed Brown Viceroy | Fritillary Viceroy Pearly Eye | Pearly Eye | Fritillary Pearly Eye |
| Predators Frequently Observed at Study Site | Dragonflies Spiders Wasps, Ants | Dragonflies Spiders Wasps, Ants | Dragonflies Spiders Hornets Ants, Birds | Hornets Birds | Birds |
| Temperature Humidity | Usually Hot and Dry | Moderate and Moist | Varied and Patchy | Varied and Patchy | Cool and Varied |

(usually nectar, ferments, and/or decaying organic matter), likely mates, safety or opportunities to avoid predation, and an appropriate range of temperature and humidity for basic physiology (after Dennis 2010). At the study site, these resources are spread among habitats in ways that differ for each of the main species, as listed in Table 8.1.

Table 8.1 shows that all the resources that the Ringlet needs are found in the dry upland fields that it favors. It is likewise for the Eyed Brown and Viceroy that favor wet meadows and their edges, sedgy for the Eyed Brown and potentially shrubby for the Viceroy. The situation is more complicated for the Fritillary and the Pearly Eye.

The ideal Fritillary habitat may be patches of violets, the larval food plant, with connecting paths through unfavorable habitat. Or the habitat may be where patches of violets are closely dispersed in a slightly less favored matrix that provides navigable connecting paths. Nectaring plants as filling stations may be either along the paths or associated with the violets. Different locales could provide different distributions of primary resources and pathways among them.

The ideal Pearly Eye habitat is woodland/field or woodland glade. A singular resource within it, fermenting sap, is attractive from a distance.

## 8.1.2 Spatial Extent of Ranging Behavior and of Habitats

Figure 8.1 compares the habitats and spatial ranging of the five intensively studied species with the distribution of their habitats in the landscape of the study site. In Figure 8.1A, daily sightings of marked butterflies are tallied in pie-diagrams whose radius is the root-mean-square dispersion of all sightings of each individual. The pie is essentially a measure of the area over which an individual ranges from the centroid (think "center of gravity") of observations. So the size of each diagram measures the ranging of each species on the same scale as the map. The map, Figure 8.1B, shows that the study area has irregular patches of dry field and shrubby meadow, enclosed and dissected by woodland (trees forming an irregular canopy), which is further enclosed by forest (trees forming a closed canopy).

The Ringlet, Eyed Brown, and Viceroy are usually confined to patches of field or meadow (Figure 8.1A). Overlaying their diagrams on the map shows that individual males range over an area of about the same size as patches of their favored habitats. Accordingly, the behavior at habitat boundaries is important for all three species.

Fritillaries are apparently more catholic in their strong presence among all habitats except closed-canopy forest (Figure 8.1A). The size of their diagram shows that Fritillaries also range more widely than the other species, over an area larger than the patches of open habitat, and about half the size of the main study area. Accordingly, the apparent repetition of circuits by individuals becomes even more intriguing (Section 5.1.4 and Figures 5.3, 5.5, and 5.6).

The Pearly Eye favors woodland edges and glades, and the fields nearby. The small size of its ranging diagram (Figure 8.1A) is misleading, because wanderings of males are badly underestimated by the root-mean-square distance to the centroid of observations of marked individuals. A number of factors interact to make observations of Pearly Eyes spatially aggregated. The butterflies are attracted to fermenting wounds on trees, whose intrinsic patchiness is compounded by the patchiness of the special woodland habitat configurations and wind conditions that allow the ferments to be found readily (Figure 8.1B, and Sections 2.4.2 and 7.1.1). The attraction of males to other males further reinforces aggregation (Section 7.1.2). And because I worked with Pearly Eyes where they were numerous, my observations were even more spatially aggregated than the butterflies, and in addition my observations were seasonally concentrated into the periods when Pearly Eyes were most numerous and most densely aggregated. So the local dispersion of my observations badly underestimates the true dispersion of individual Pearly Eyes.[2]

## A.
**Habitats of Sightings**

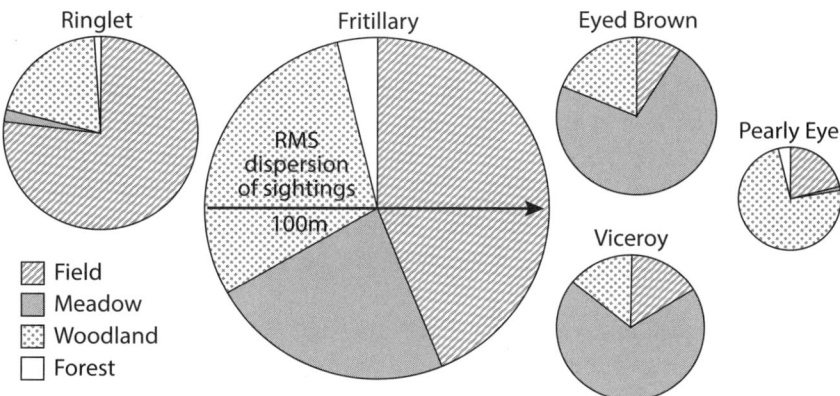

## B.
**Map of Habitats**

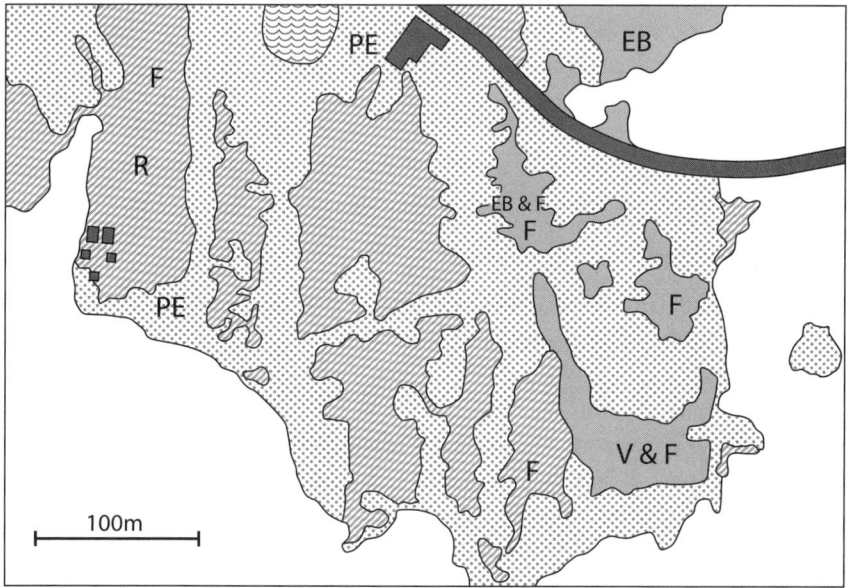

FIGURE 8.1. Habitats and sightings of butterflies. A. Daily sightings of marked male butterflies are tallied in pie diagrams whose radius is the root-mean-square dispersion of all sightings of each individual (a measure of ranging). B. Map of habitat types to the same scale (spatial distribution of habitats), with main study areas initialed for each species. See Section 8.1.2 for further details and discussion of the matches between movements of butterflies and spatial configuration of favored habitat (including the fact that Figure 8.1A underestimates the wanderings of male Pearly Eyes).

## 8.2 CUES AND RULES TO FIND AND STAY IN HABITAT

I approach cues and rules from both theoretical and empirical directions. In proposing theoretical models, I favor models that are based on plausible simple cues and on behaviors that entail little or no cognitive burden. I start by generalizing the rules of movement in response to particular cues that I have already proposed for individual species. The rules are deliberately stated in anthropomorphic terms to emphasize the apparent complexity of behavior that could be observed despite the simplicity of the cues and rules themselves. The compound eye has the ability to encode very subtle information about relative degrees of dappling of light and shade as dynamic patterns of flicker and flow (Section 2.3). Butterflies behave as though they use such information, especially recognizing open habitats, perceiving (and/or ignoring) habitat boundaries, following edges, and orienting to singular points along those edges.

Such cues and rules can generate automatic behaviors that will direct individuals to and through the right habitat. The empirical examples given here are expanded in the following subsections. Moving toward well-lit habitats keeps the Ringlet within its favored upland fields (Section 3.2.1). The favored meadows of the Eyed Brown and Viceroy are found in depressions, so flying toward low spots in the horizon will help (Section 6.5.2). Additionally, meadows have odors that I can detect, and the Eyed Brown and Viceroy are likely better at it. Flying toward V-notches in the far horizon also helps to guide the Fritillary between patches of larval and adult resources (Section 5.1.4). The Pearly Eye, by flying along edges and then upwind to odors of ferment, can more likely find potential mates; attraction to other fluttering objects will help additionally, by aggregating both males and females (Sections 7.1.1 and 7.1.2).

The Tactical Vagrancy Model of Chapter 3 suggests that habitat preference could be achieved by a simple rule of "Turn to stay; proceed to explore."[3] However, I shall show below that such a rule, when explicitly enacted in a heterogeneous environment, is surprisingly ineffective, and even capricious. Conversely, even a modest tendency to recognize and respond explicitly to habitat boundaries effectively enforces habitat preference.

### *8.2.1 Rule: Run for Daylight!*

Both male and female Ringlets follow this rule. By strictly obeying it they can turn away from wooded boundaries of their preferred open field habitat, but also penetrate openings in hedgerows to move to neighboring open fields (Section 3.2.1 and Figure 3.2B). This rule is the basis of my protocol for taking stroboscopic photos of butterflies in flight (Section 2.2.6). Butterflies released in an enclosed building flew toward an open window, a lighted area initially spanning a solid angle of about 45° of their field of view, within which the stroboscope was tripped and the photos were taken. All species initially flew toward the opening, though some normally open-flying species became spectacularly disoriented (Fritillary, Cabbage White, and Monarch, Figure 2.8A–C).[4] In the field, I routinely observe several species moving toward open glades within woodlands, including the Pearly Eye, Great Spangled Fritillary, Mourning Cloak, and Red-spotted Purple. In dappled shade, the Little Wood Satyr explicitly dives toward sunflecks (Section 2.3.2).

Within open habitats all my species tend to fly at or within a few meters of their nearby canopy top, be it grass and herbs, shrubs, or trees. Srygley and Oliveira (2001) argue that this is necessary to remain within the boundary layer of light breezes in relatively still air.[5] Potential mechanisms to orient such flight are mentioned in Section 2.3.5.

*8.2.2 Cues: High Spots and V-Notches Provide Guidance: Paths in Vegetation and Beacons in Landscape*

Depressed paths in the vegetation are detectable as sharp V-notches or shallow U-notches in the horizon, at scales from the near horizon for paths between herbs and graminoids, and spaces between canopy trees, to the distant horizon for stream corridors and depressions in the landscape. Movements that seem to follow paths defined by such notches include the Eyed Brown (Section 4.3), Viceroy (Section 6.3.1), and Fritillary (Section 5.1.4; Figure 5.4). Experiments with a mirror (Figure 6.3; Table 6.2) confirm that a male Viceroy will orient toward a low spot in the horizon, and territorial male Viceroys both leave and return to the habitat of their territories through pathways that are indicated by V-notches (Figure 6.6).

I have watched other members of the Fritillary's genus *Speyeria*, and the Viceroy's *Limenitis*, throughout the United States as they followed streams and woodland edges. V-notches were an almost invariable feature of their paths when viewed from the positions of the butterflies, including the points at which they rose at a woodland edge.

"Hilltopping" is a classically named phenomenon, in which many species of butterflies tend to collect at higher densities on the tops of hills (Shields 1967, Scott 1968, and many subsequent studies). Turner (1990) has shown that Tiger Swallowtails tend to congregate at the top of the canopy on hilltops.[6] And Baughman et al. (1990) have shown that Edith's Checkerspot *Euphydryas editha* reaches higher integrated numbers in high spots of very low relief (15 m rise in a 300 m run = 5% grade).[7]

Conversely, the Viceroy (Chapter 6), and to some extent the Eyed Brown seem to be "Valleybottomers" (cf. Scott 1973). Male Eyed Browns #77.06 and #77.57 were marked and resident for three days in the central wet meadow of my study area, and were rediscovered two days later on a bog 200 m ENE and 8 m lower in elevation.

### 8.2.3 Rule: Follow Your Nose and Join the Party!

Having already anthropomorphized the social behavior of the Pearly Eye as a "singles bar," I have no difficulty analogizing its demonstrated ability to follow edges and then fly upwind to fermenting tree sap (Sections 2.4.2 and 7.1.1) as "following your nose," and the tendency to settle near other flutterers (Section 7.1.2) as a mechanism that "joins the party." Furthermore, if a male butterfly flying a more or less direct route interacts with others so as to turn from his course, that is sufficient to foster local congregation (Odendaal et al. 1988, Turchin 1998, pp. 113–115). Indeed, that is a special case of the following rule.

### 8.2.4 Rule for Habitat Preference: Turn to Stay; Proceed to Explore!

The turning bias model was originally framed in the context of foraging for aggregated prey (Smith 1974) or aggregating for efficiency of mate-finding (though Odendaal et al. 1988 argue that such aggregations may, indeed sometimes do, decrease individual male mating success due to losses to aggregated competitors). More recent models provide statistical and analytic flexibility for copious and detailed tracking data, and theoretical flexibility to span qualitatively different behaviors (e.g., Fleming et al. 2014, 2015, Calabrese et al. 2016). However, it is not clear how a creature could encode and realize them.[8] Perhaps more important is the fact that these models are intended to uncover and to parameterize heterogeneity hidden in a long track recorded from georeferenced radiotelemetry. I am in just the opposite position: I know my heterogeneity of habitat, I can map it, and I can look directly for responsive and/or adaptive heterogeneity in ranging behavior of my butterflies. My approach is less elegant, but I know more about what my critters are doing. Most of the preceding theory initially assumes a uniform arena and derives implicit spatial heterogeneity from varied movement by the individual agents. In this context, an increased rate of turning both defines the favored habitat and helps to confine wandering to it. This circularity inflates the prospect of a turning bias model for general habitat preference.

Nevertheless, simply turning more frequently and/or at greater angles in the preferred habitat than in the less-preferred matrix should be a sufficient behavior to bias presence in the preferred habitat (classically Cody 1971 and Smith 1974, and in more detail and generality Levin 1986, Johnson et al. 1992, and Turchin 1998). Chapter 3 presented a Tactical Vagrancy Model, which showed that a Forward-biased Random Walk effectively and efficiently explores a uniform area, and that a Brownian Walk, with its more frequent turns, restricts the area explored (Section 3.3.3 and Figure 3.11). Figure 8.2 generalizes this result, and shows that turning tends to maintain a local and self-crossing path, while straighter lines explore newer territory.

This is part of the mechanism for habitat preference in the Ringlet, as argued qualitatively in Section 3.1 and inferred from the model in Section 3.3. And it is clearly adaptive for the Ringlet to fly directly in the less favored matrix in order to traverse it rapidly to the next patch of favored habitat (Section 3.2.1). However, there is a delicate balance to be achieved. Turns within the more favored habitat help the butterfly to remain in it, but turns also hinder systematic exploration of the favored habitat (Figures 3.11 and 8.2).[9]

It is tempting to infer that preference for a favored habitat over a less favored matrix can be achieved by turning more often in the favored habitat than in the matrix, or equivalently by forward-biasing a random walk more tightly in the matrix than in the favored habitat. So, it is instructive to frame an explicit Turning Bias Model for Habitat Selection, whose output is shown in Figure 8.3. This type of NetLogo model has already been introduced in Section 3.3. On a $31 \times 31 = 961$ patch toroidal lattice, the butterfly takes a random walk of 960 steps with a specified forward bias in the central $21 \times 21 = 441$ patches. In the 520 peripheral patches, outside the central area, the butterfly moves in a nearly straight line, a random walk with forward bias $N(0°, 5°)$, and movements wrap from top to bottom and side to side. Figure 8.3 records the number of visits to the central patches, along with a percentage bias toward the central area from its "fair share" of visits, which with no bias would be 441 visits. Figures 8.3A–C were chosen from 10 runs at each rate of turning to show a typically strong favoring of the preferred habitat. Comparing Figures 8.3A, B, and C suggests that an intermediate rate of turning provides the strongest discrimination. Figures 8.3C and D are different runs of the Turning Bias Model with identical parameters, specifically a central forward bias with standard deviation equal to that of a uniform distribution between 0° and 359°.[10]

Two results of Figure 8.3 differ from my prior expectation, but are understandable in retrospect. One result is that an intermediate forward bias, equivalently an intermediate tendency to turn in the preferred habitat, produces a stronger

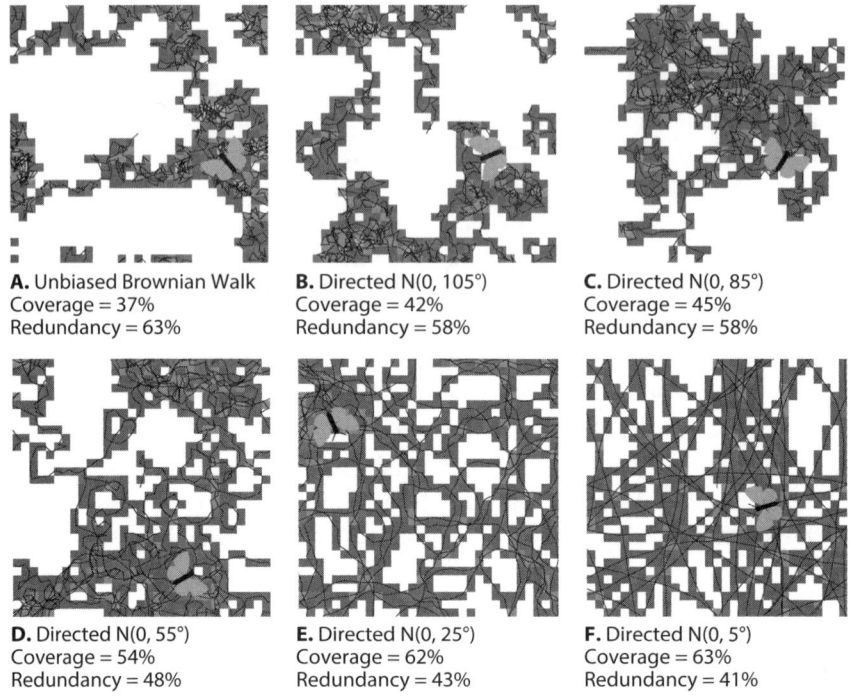

**A.** Unbiased Brownian Walk
Coverage = 37%
Redundancy = 63%

**B.** Directed N(0, 105°)
Coverage = 42%
Redundancy = 58%

**C.** Directed N(0, 85°)
Coverage = 45%
Redundancy = 58%

**D.** Directed N(0, 55°)
Coverage = 54%
Redundancy = 48%

**E.** Directed N(0, 25°)
Coverage = 62%
Redundancy = 43%

**F.** Directed N(0, 5°)
Coverage = 63%
Redundancy = 41%

FIGURE 8.2. Effect of turning on local versus exploratory movement. As in the Random Vagrancy Models of Chapter 3 (Figure 3.11), the butterfly takes 960 steps with a variously forward-biased random walk, on a 31 × 31 = 961 patch toroidal lattice. 8.2A. In the Brownian Walk, a turn is made in each patch in a uniformly random direction. 8.2B. The vaguely Directed Walk is achieved by choosing, in each patch, a forward-biased direction from a normal distribution of headings with mean 0 (straight ahead) and standard deviation 105° (chosen to equal the standard deviation of the Brownian Walk), denoted N(0°, 105°). 8.2B–F. The walks are increasingly forward-biased, as the turns are chosen from a narrowing distribution of angles. Coverage is the percentage of patches visited, and increases with forward-bias from 8.2A to D, but then seems to level off for straighter paths (D–F). Redundancy, as measured by the percentage of visited patches that are visited more than once, decreases for straighter paths, but the multiplicity of redundancy is markedly greater for more twisty paths (A–C). Turning tends to keep explorations local and repeated; straighter lines explore newer territory.

preference than a high rate of turning (cf. Figures 8.3A–C). The other is that a high rate of turning produces extreme variability in preference from moderate (Figure 8.3C) to negative (Figure 8.3D). The heuristic reason for these results is that without explicit behavior to recognize and respond to the boundary, a move near the boundary is equally likely to take the model butterfly out of the favored area as to keep it within, or vice versa, no matter what the difference in forward

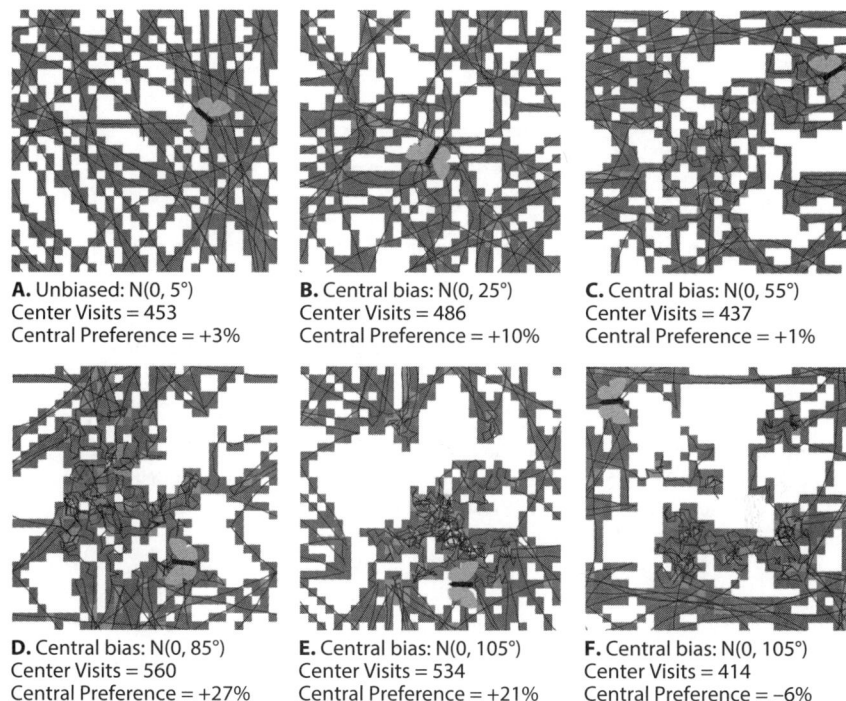

**A.** Unbiased: N(0, 5°)
Center Visits = 453
Central Preference = +3%

**B.** Central bias: N(0, 25°)
Center Visits = 486
Central Preference = +10%

**C.** Central bias: N(0, 55°)
Center Visits = 437
Central Preference = +1%

**D.** Central bias: N(0, 85°)
Center Visits = 560
Central Preference = +27%

**E.** Central bias: N(0, 105°)
Center Visits = 534
Central Preference = +21%

**F.** Central bias: N(0, 105°)
Center Visits = 414
Central Preference = −6%

FIGURE 8.3. Turning Bias Model for habitat preference. This is a modification of the model in Figure 8.2 (q.v. for details). In the central $21 \times 21 = 441$ patches the forward bias is as given; outside the central area the bias is nearly straight, N(0°, 5°). Visits to the Central 441 patches are tallied. Central Preference is the excess of that tally over a "fair share" of 441 visits to Central Patches. Comparing A–E, an intermediate forward bias most favors the preferred habitat, though by a surprisingly small amount. E and F are different runs with identical parameters, and vary from preference to avoidance of the "preferred" habitat. Multiple runs with the same parameters produce highly varied results (see Figure 8.4). See Section 8.2.4 for details and interpretation of why turning per se is so weak as a mechanism for habitat selection.

bias between habitat and matrix. Furthermore, for a butterfly entering the favored habitat, an increased rate of turning keeps it near the boundary, hence subject to a chance to leave, whereas for a butterfly leaving the favored habitat, its increased forward bias takes it further away from the boundary. So in the vicinity of the boundary there is a stochastic bias against the favored habitat.[11]

Figure 8.4 plots the number of central visits, and its variation over 10 runs of the model for each of the full range of forward biases portrayed in the examples of Figure 8.3. Figure 8.4 paints an inauspicious picture for the pure Turning Bias Model for Habitat Preference. At best the model produces only a weak average preference, and even then, it can result in weak avoidance in any given instance.

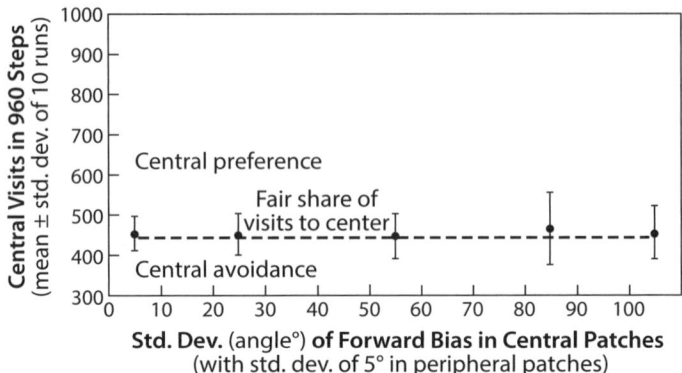

FIGURE 8.4. Inauspicious statistics of Turning Bias Model for Habitat Preference. The model of Figure 8.3 was run 10 times with the same parameters as each of Figures 8.3A–F. Paths were nearly straight, N(0°, 5°), in the Peripheral Patches, and varying from that to very twisty, N(0°, 105°), in the Central Patches. The results are graphed as mean ± standard deviation of the number of Central Visits. The graph undercuts the inference from Figure 8.3, that intermediate forward bias (mildly twisty paths) in the Central Patches produce somewhat more Central Preference than either very twisty or nearly straight paths. Indeed, the range of variability overlaps no preference at all for all degrees of forward bias modeled. Reasons for such variability and the absence of an average effect are discussed in Section 8.2.4.

## 8.2.5 Rule: Reflect From Boundary to Stay in Habitat!

Strong boundary behavior has been directly recorded for the Ringlet (Figure 3.8), observed during the "Cyclotron" experiments with the Eyed Brown (Section 4.3), and strongly inferred as the Viceroy returns to its territory from brief but distant excursions (Figure 6.6). For these three species the boundary behavior helps to confine them to patches of favorable habitat. The Fritillary ranges widely over a more catholic habitat (Figures 5.3 and 8.1). I infer, but have not directly observed, that the wanderings of Fritillaries are ultimately bounded by the configuration of flyways among patches of preferred habitat (Figures 5.4 and 5.8). Aggregations of Pearly Eyes are explicitly along the boundaries between field or open glade and woodland or forest (Sections 7.1.1 and 7.1.2 and Figure 7.3). Furthermore, captured males released at an edge tend strongly to fly initially toward the woods, to stay within view of large open patches, to interact with other butterflies that they encounter, and eventually to settle on baits that they find (Section 7.1.1). Females tend to fly toward the field, settle a short distance away, and stay perched for a while (Figure 7.2).

Figure 8.5 presents the results of a model that incorporates boundary behavior as an adjustable tendency within the favored habitat to reflect, at the angle of incidence, from the boundary with the less favored matrix. At each of 960 steps,

# DO BUTTERFLIES MAKE DECISIONS?

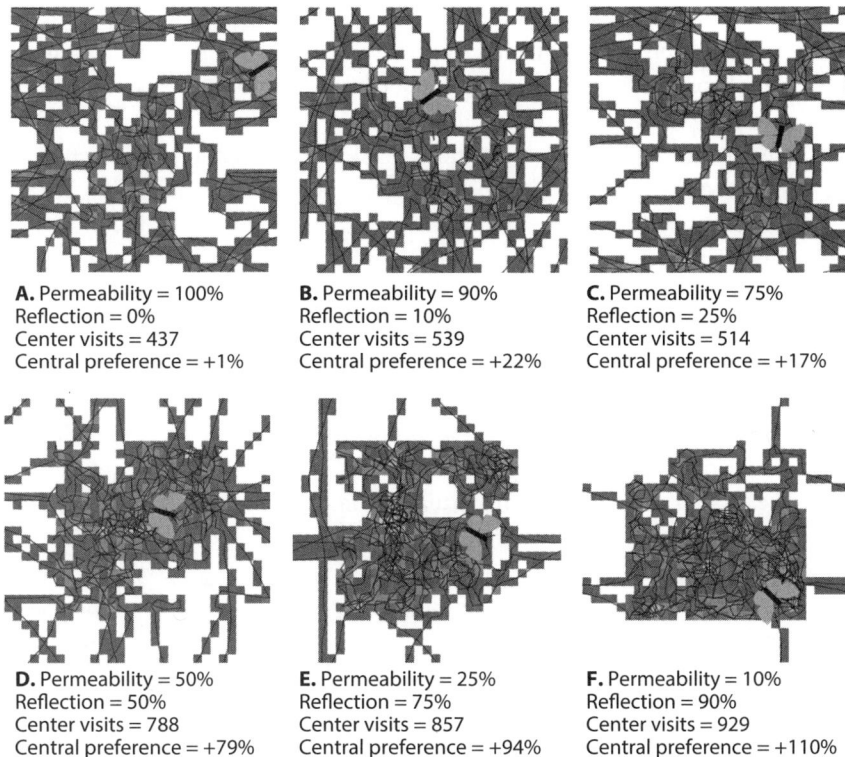

FIGURE 8.5. Adding boundary behavior to Turning Bias Model of Habitat Preference. Figures 8.5A–F show progressively increasing reflection from the boundary between the 441 favored central patches and the remaining peripheral patches. Permeability of 100% denotes no explicit boundary behavior. Were the boundary impermeable (0%), a butterfly within the central area would reflect from peripheral patches at its angle of incidence with the boundary. Intermediate permeability is modeled by a biased random choice between ignoring the boundary versus reflecting from it. A forward bias of N(0°, 55°) was chosen as standard for the favored central habitat, because it matches the observed behavior of both Ringlet (Section 3.3.2) and Pearly Eye (Section 7.4). Figure 8.5C shows that 75% permeability, or only 25% reflection at the boundary, yields a substantial concentration in the preferred habitat, and Figure F shows that 90% reflection virtually confines the butterfly to the preferred habitat. Contrasting Figures 8.5 and 8.2, small changes in boundary behavior have much greater effect than large changes in turning angle. (Fig. 8.5A is a repetition of Fig. 8.3C since the parameters of both are the same, though different runs produce somewhat different figures.)

the model butterfly moves in a random direction, with forward bias of N(0°, 5°), i.e., nearly straight ahead in the matrix, but with forward bias N(0°, 55°) within the preferred habitat. N(0°, 55°) was chosen for the forward bias in the preferred habitat for several reasons. It produces some degree of habitat preference (Figure 8.3), albeit undependably (Figure 8.4). It gives roughly the same variation in forward progression upon reflection from the boundary as in the open (Figure 3.8 and Table 3.2). It produces paths that visually match the paths of male Ringlets, and that, perhaps fortuitously, result in model animations that are very like videos of male Pearly Eyes, alone or interacting.[12]

Simple patterns of optic flow (Sections 2.3.4 and 2.3.5) can provide cues and criteria for the types of boundary behavior incorporated in the model of Figure 8.5.

The model encodes a varied tendency to engage in boundary behavior as a probability of reflecting versus ignoring the boundary from within the favored habitat (the central half of the patches in Figure 8.5). Figures 8.5A–F show the effects of progressively increasing the probability of reflecting from the boundary (or equivalently decreasing the permeability of the boundary). As in Figure 8.3, central Preference is the excess of the tally of Central Visits over a "fair share" of visits to Central Patches (441). As Figure 8.5 shows, 25% reflection at the boundary produces a substantial preference for the favored central habitat, and reflection of 75% or more virtually confines the model butterfly to the favored habitat.

Figure 8.6 plots the number of central visits, and its variation over 10 runs of the model for each of the full range of boundary permeabilities that are portrayed in the examples of Figure 8.5. The message of Figure 8.5 is confirmed; more than a 25% tendency to reflect produces increasing and dependable preference for the favored habitat.

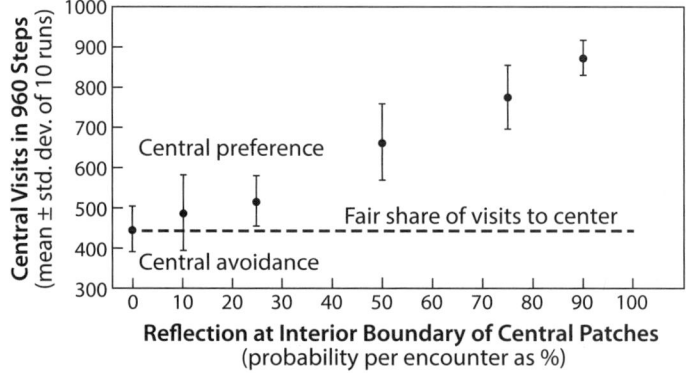

FIGURE 8.6. Auspicious statistics for habitat preference by internal reflection from boundary. The model of Figure 8.5 was run 10 times for the same parameters as each of Fig.8.5A–F. Paths were nearly straight, N(0°, 5°), and mildly twisty, N(0°, 55°), in the Central Patches. The results are graphed as mean ± standard deviation of the number of Central Visits. The graph confirms the messages of Figure 8.5: modest Reflection provides sure Central Preference, and the degree of Preference rises with degree of Reflection. Comparing Figures 8.6 and 8.4, Boundary Behavior is effective in generating and ensuring habitat preference, but Turning Behavior alone is not.

## 8.3 MOVING WITHIN THE FAVORED HABITAT

The models of movement for all species are some kind of graph in the technical sense, if one allows the degenerate egocentric path of the Ringlet (Figures 3.6 and 3.13) to be called a graph. A butterfly's perches define the nodes or vertices of the graph, and the flights between perches define edges or lines. The movements of males, previously caricatured in Figure 1.2, can be further idealized in Figure 8.7 as technical qualitative graphs, amenable to analyses and insights from graph theory (Benjamin et al. 2015). The theoretical representation as a graph does not necessarily imply that butterflies are good at technical graph theory. But it allows the discussions of different species, particularly the Eyed Brown, Fritillary, and Viceroy, to have a common ground in the configuration of nodes and edges. Furthermore, if the butterflies do behave as suggested, characteristic differences between the species, and variations between individuals of the same

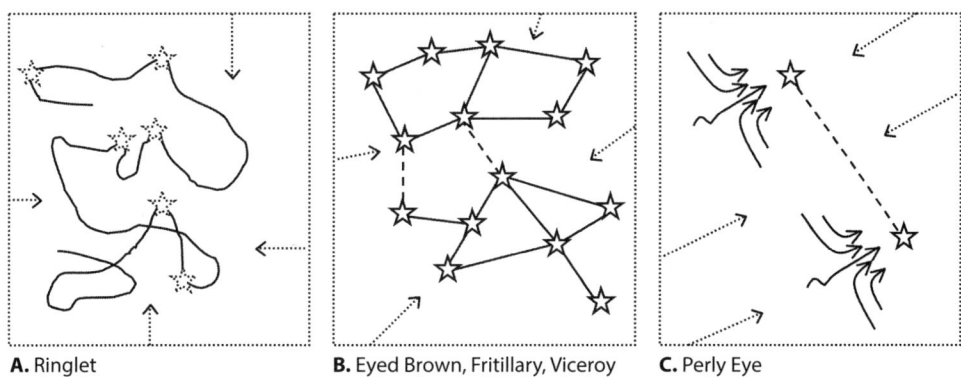

**A.** Ringlet     **B.** Eyed Brown, Fritillary, Viceroy     **C.** Perly Eye

FIGURE 8.7. Male movements as qualitative graphs. The favored habitat is bounded by the dotted line; the dotted arrows represent reflection from boundaries with unfavored habitat. Potential perches are represented by stars, and paths between them by lines. Dashed lines represent paths that are taken infrequently, not at all, or perhaps routinely but by few individuals. A. Within its strongly bounded habitat, the Ringlet wanders with no detectable pattern and may or may not stop at ad hoc perches (Section 3.2.5). B. Three species can be characterized as entering the favored habitat or reflecting from its boundaries to traverse networks of paths between favored perches, with differences of detail among the species. The Eyed Brown seems to be guided by both the paths (V-notches in nearby vegetation) and the perches (Section 4.3). The Fritillary must be guided by potentially tortuous paths (V-notches in near and distant vegetation) because their perches (nectaring sites near larval food plant) are not intervisible (Section 5.1.4). The Viceroy is guided in relatively straight lines between intervisible neighboring perches (Section 6.3.1). C. Within its generally favored habitat, the Pearly Eye orients toward edges (Section 7.1.1), flies along them, and then approaches its favored perches, fermenting tree sap, by tacking upwind (Section 2.4.1). The behavior described here is consistent with robotic responses to simple fixed cues in the vegetation; see Section 8.3 for discussion of further complexities.

species, as well as temporal variations in the behavior of individuals, need not require complex decision-making. And further speculative insights can be drawn from characteristic properties of the various types of graphs, and from whether given species exemplify them or violate them.

### 8.3.1 Rules for the Ringlet

Virgin female Ringlets may be rare, and slow to "renew" as a resource, but they are distributed in a fine-grained way over a fine-grained habitat, and so the best way to search for them may be to find the appropriate habitat and cruise it in a way that covers as much territory as possible. The Ringlet's "graph," Figure 8.7A, is an egocentric forward-biased random walk, and the only "decision" needed is to turn back at the boundary of the appropriate habitat and to increase the forward bias when in an inappropriate habitat (Sections 8.2.4–5). Both of these rules can be achieved by context-dependent fixed behavior, without explicit decisions.

### 8.3.2 Mystery for the Eyed Brown

Female Eyed Browns are found on paths through the vegetation that are effectively the edges of a graph. Ideally, I would like to map the graph in detail, and it would include many edges that are two-way. So neighboring perches and junctions form a highly connected network in Figure 8.7B, which poses the problem of just how many circuits and attractive domains there might be (cf. Figure 4.5).[13] The behavior of males is sometimes locally restricted and sometimes not. The Fortuitous Site-Fidelity Model is essentially a set of one-way directed graphs on a Cartesian grid (Figure 4.5). The graphs are disjunct trees, each spanning the attractive domain of a single point or a small cycle of edges, and they produce a large part of the range of variation that individual males show. But they also predict that episodes that draw a male beyond the boundary of its initial attractive domain will bring it into a neighboring domain with a new settling point or cycle. Whether the Eyed Brown really behaves according to the model is a subject for further study.

### 8.3.3 Rules for the Fritillary

The nodes of the graph for the Fritillary, Figure 8.7B (a caricature of Figure 5.8), are patches of violet, the larval food plant. The edges are alleyways bordered by patches of tall vegetation. The nodes are not intervisible, but the edges are characterized by V-notches in the horizon at the end or ends of the alleyway toward which movement is directed (Figure 5.4). So the graph is navigated by moving along the edges. Different individuals share edges from which they may depart in different directions, strongly implying that decisions are being made. The number of nodes within the study area is small (Figure 5.8), and the number of nodes within the convex hull of the whole lot is even smaller, small enough that touring the perimeter and nearest-neighbor-joining within the perimeter can generate a highly efficient trapline (Section 5.3).

### 8.3.4 Behavior of the Viceroy

In Figure 8.7B for the male Viceroy, the nodes are perches, and the edges are flight paths between perches. Because perches are intervisible over distances greater than those between next neighbor perches, the perches are highly connected to each other, and navigation within the territory is perch to perch (cf. Figure 6.5). Boundaries between territories of different individuals are in part defined by the outcomes of mutual contests, but are biased toward sections of the habitat with lower potential perch density, i.e., places where fewer perches are intervisible. Other things being equal (which they will not be in the next sentence), Viceroys should be drawn toward parts of the territory with higher density of potential perches. However, different perches will have different potential presence and detectability of images of potential mates, and these properties will shift among different perches during the day. Individual male Viceroys shift orientation on perches (Table 6.2), shift preferred perches (Figure 6.6), and even shift territories (Table 6.3). I have the overall impression that such shifts are the result of adaptive decisions, but I have no explicit evidence that they are more than automatic responses to changing visual cues as the males fly between nodes (Figure 6.5).

### 8.3.5 Cues and Rules for the Pearly Eye

The graph of Figure 8.7C for the Pearly Eye is idiosyncratic because the nodes are initially attractive from a considerable distance. They can be found by traveling along a boundary between field or glade and woodland, which can be achieved automatically by maintaining equal optic flow velocities of different views for right and left eyes (Section 2.3.4). A fermenting food source can be found by tacking upwind (Section 2.4.2). Contagious settling can be fostered by automatically moving toward fluttering objects and landing soon afterward (Section 8.2.3).

## 8.4 DO BUTTERFLIES DECIDE WHERE TO GO?

### 8.4.1 Butterflies as Cue-&-Rule Automata

All the species show evidence of responding to environmental cues like those envisioned in the simple models. The Ringlet turns to avoid a boundary between field and woodland. The Eyed Brown flies along channels in the vegetation, possibly viewed as V-notches in the near horizon. The Fritillary uses V-notches in the far horizon to orient large-scale movements, and often uses the shadow of trees to orient parallel to the edge of a woodland. The Viceroy uses V-notches in the horizon to locate and orient a territory; it may use channels in the vegetation on return to its territory following a chase; and it uses the configuration of sky and shade in its choice of perches. The Pearly Eye flies along woodland edges, potentially exposing itself to odor plumes, and tacks upwind when it finds one. And captured male Pearly Eyes tend to fly to the woods when released at the edge of a field, whereas females show no preference.

### 8.4.2 Fortuitous Site-Fidelity versus Site-Recognition

So far the behavior of all five species has been described in ways that allow interpretation as though they were automata following fixed rules in response to simple visual cues. To infer a decision, it is necessary to show that different individuals behave differently when exposed to the same set of cues. The Fortuitous Site-Fidelity Model of Chapter 4 provides a standard against which to inquire whether butterflies make spatial decisions, or more narrowly, whether they recognize particular pieces of real estate. The model argues that fixed cues guiding movement will divide the environment into attractive domains that would look like exclusive home ranges to an observer ignorant of the cues. The configuration of the attractive domains might be complex in three dimensions because the direction indicated by visual cues might change with changes in the height from which they are observed, including heights from which the cues cannot be perceived.

The Fortuitous Site-Fidelity Model of Chapter 4 (Figure 4.5) is based on fixed responses to simple spatially varying cues. And it makes a particularly useful prediction about the consistency of behaviors among individuals, namely that two individuals released within the same attractive domain will stay within it. Accordingly, a test of whether two individuals are treating the same space differently is whether their ranges overlap in part and yet differ drastically in total. Here is how each species answers that test.

The Ringlet is confined to fields by potentially automatic behavior at the boundary, but within those fields it shows a large component of random walk, specifically the Tactical Forward Vagrancy Model (Chapter 3; Figures 3.9–3.10). The wanderings of each male can potentially span a whole field, given enough time, overlapping completely with the tracks of all other males. Thus the Ringlet fails the test and shows no evidence for recognition of particular sites, and no evidence of making spatial decisions.

The case for the Eyed Brown is unclear, due to a paucity of data from a population of high local abundance. This is unfortunate because I invented the Fortuitous Site-Fidelity Model of Chapter 4 to explore my suspicion that the spatial behavior of male Eyed Browns actually fits the model. If it does, then there would be no need to invoke decisions in the ranging of male Eyed Browns (Section 4.5).

But the following species, the Fritillary, Viceroy, and Pearly Eye, all violate the primary prediction of the Eyed Brown's Fortuitous Site Fidelity Model which encodes a twig-fugal directed tree graph. At least in the short term, these three species routinely transgress the attractive domains of other individuals, and yet return to their own individual "home" ranges. So individuals must recognize particular parts of their landscape and make context-dependent decisions about how to return.

The Fortuitous Site-Fidelity Model gives two insights into the ranging of the Fritillary (Chapter 5). First, random small-scale one-way cues rarely produce closed loops, and even more rarely do they produce large closed loops. So the large scale of the Fritillaries' traplines must be due either to large-scale cues or to prodigious memory. Notches in the horizon provide large-scale cues (Figure 5.4), lessening the potential role of memory. Larger-scale patterns among these clues may be partly responsible for closing the loops, but the pattern of spatial contiguity among visited sites may also play a role (Anderson 1983 and Figure 5.8). Second, Fritillaries pass the test of departure from the null Fortuitous Site-Fidelity Model of Chapter 4. Several males share nodes and alleyways and yet characteristically diverge to complete different, and apparently characteristic, circuits. This fact also argues that large-scale patterns among cues like notches in the horizon cannot be solely responsible for the closure of loops. The picture that emerges is of guiding environmental vectors for much of the trapline, but decisions must be made at crucial locations to close the loop (Cartwright and Collett 1982 have demonstrated this kind of navigation explicitly for honey bees).

The male Viceroy leaves its favored habitat in chases, but returns, and so apparently "knows" the lay of the land. Often the "chases" are up cul-de-sacs, with a return from the end that often brings them within measured response distance of a traditionally used perch (cf. Figures 6.5 and 6.6). When going further afield, they usually make the approach back to the home field though what turns out to be a V-notch in the near horizon.

Male Viceroys also pass the test for decision-making by departing from the Fortuitous Site-Fidelity Model. When territorial, they patrol a piece of real estate with defined boundaries, returning to it after chases that transgress nearby territories, and even after chases that leave the territorial habitat altogether (Chapter 6; Figure 6.6). There is perhaps a component of the null model in their behavior, however. Extraterritorial flights are typically at a higher altitude than territorial patrols, exposing the flyer to a different set of cues from the local territory holder, and low-altitude trespassers often perch briefly within someone else's domain. The overall pattern is exclusively defended "home" territories, but broadly overlapping wanderings that include the territories of other individuals (Similar behavior is shown by Cartwright and Collett 1987 for honey bees returning to a food source).

Male Pearly Eyes vary in their behavior, some of them passing the test. When I set out many baits at high local density, a few males never attended more than one bait, and a few others attended many baits with no detectable pattern. Most, however, attended one or two mainly and repeatedly, with intervening visits to others (Figure 7.3). I had a few males that repeated visits to several baits that were separated by other baits with no recorded visits. I could not detect evidence for traplining by these males, nor for distinctive visual configurations of the canopies that consistently separated visited from unvisited baits.

In summary, I have yet to find evidence for recognizing a particular site by the Ringlet or the Eyed Brown. The other species, the Fritillary, the Viceroy, and the Pearly Eye, all show evidence of individually specific site-recognition. Because different individuals of these three species traverse a common space, and are exposed to the same spatial cues, before going their separate ways, I tentatively infer that they are making individualistic decisions.

## 8.5 DO MALE BUTTERFLIES DECIDE WHAT TO PURSUE AND WHEN TO STOP?

### 8.5.1 Ringlet

The initial question needs to be turned around, because for the Ringlet it may be the female who does the pursuing (Sections 3.4 and 3.5), and in the two such cases observed, the pursuit ended in copulation within about two seconds. Interactions between males are of variable length but usually so short (1 or 2 meters over 1 or 2 seconds) that it is likely that their termination is a result of the males drifting beyond the distance or orientation that favors detection. In neither case is there a reason to invoke a decision, as opposed to fixed, context-dependent behavior.[14]

### 8.5.2 Eyed Brown and Fritillary

Eyed Brown male interactions can be distinctly longer (ca. 3 meters over 6–8 seconds) than those of the Ringlet, but they are not qualitatively different. Given that the interaction distance of the Eyed Brown is nearly four times that of the Ringlet (Table 2.1), the length of interaction could be that order of magnitude greater for the Eyed Brown, and still be consistent with a purely visual break-up when both fortuitously lose sight of each other.

For male Fritillaries, unambiguous turn-and-initial-approach are limited to less than 3 m (Section 5.1.4). There are several returns and re-approaches from greater than 3 m, and several instances of "long-following" at 3–10 m, which, given the maximum unambiguous detection distance of 3 m (Table 2.1), may be due to common following of physical environmental cues (Section 5.1.4). Experiments with caged females produced no detectable attraction of males flying within 2 m, suggesting that pheromones are either not attractive at distances greater than vision, *or* that females do not release attractive pheromones much after eclosion (Sections 2.4.1 and 5.1.4).

So for the Eyed Brown and the Fritillary, I have no observations that demand an explicit decision by a male to cease "pursuit" of another male.

### 8.5.3 Viceroy

Male Viceroys must decide what to pursue, or at least whether to continue pursuit, given that their behavior is qualitatively different when pursuing other males compared with investigating females or other flying creatures (Section 6.1). In addition, they break off pursuit when they leave familiar territory and/or the favored habitat (Figure 6.6). In my main study area, I did not see whether they navigated using known landmarks, though mapped returns to the favored habitat were along pathways from which I could find V-notches in the horizon leading back. However, in other places where the habitat per se is not confining, they turn around and head more directly back in the general direction whence they came. Since both departure and return follow habitat channels or boundaries, both require only direct responses to environmental cues.

### 8.5.4 Pearly Eye

Simply watching the behavior of interacting male Pearly Eyes, it is hard to believe that they are *not* continually responding to changing and ambiguous information by making decisions. However, the rules outlined in Section 7.3 are fixed responses to simple stimuli, and the burden of Chapter 2 is that some surprisingly complex data can be filtered to provide a simple stimulus. In turn, the rules of Chapter 7 produce behaviors that seem complex and purposive, especially when mapping their patterns in time and space. I must admit that even with all the hard evidence pointing to Pearly Eyes being mechanical automata, I deeply believe that there is more going on, and I intend to keep searching for it.

The difference between the male-male spinning wheel encounter and the male-female horizontal flight pursuit seems a good starting point. The external appearance of the two kinds of encounters is so different that at first I thought that they were qualitatively different behaviors, like the different encounter behaviors of the Viceroy (Chapter 6) or of damselflies of the genus *Calopteryx* (Pajunen 1966 and Waage 1988). But the behavior of the male Pearly Eye is the same in both cases, investigating a butterfly-like object. It is the different and fixed behavior of the object, not necessarily a decision by the focal male, that creates the difference in behavior of the interacting pair. A fleeing female produces a horizontal pursuit; a symmetrically investigating male produces a spinning wheel (Section 7.6). It is harder to explain the breaking off of an encounter as a fixed response to a particular stimulus. The pursuit and the spinning wheel surely give different stimuli to a male, respectively linear translation and rotating of the whole visual field except the object under inspection (Section 2.3.4). The male is often close enough to the object or to its wake for olfactory cues to be involved as well. Yet the typical duration of the spinning wheel is four times that of the horizontal pursuit (Figure 7.13). This suggests that the male-male behavior is important for its own sake, not just as a mistaken investigation of a female-like stimulus.

The daily temporal pattern of male Pearly Eye activity is subject to adaptive interpretation. The following summarizes the results of Section 7.6. Males interact more with each other in the afternoon when receptive females are common than in the morning when they are rare; i.e., they "fight" only when there is something worth fighting about (Figures 7.10 and 7.12). An alternative explanation of this pattern is that males interact more when the local density of males is high, which it is when females are common. The problem with this exercise is that I could have given a plausible adaptive interpretation had I found exactly the opposite pattern. Males would have congregated and fought to gain exclusive access to a female-attracting site before the females got there. Once the females were there, males would have had better things to do than to engage in long interactions with other males.

## 8.6 *SICUT ERAT IN PRINCIPIO:* LANDSCAPE CONFIGURATION MATTERS

I wish that I could end with a discussion of how my observations and experiments with butterflies put my study into the long line of ideas about how simple behavior can generate complex behavior and achieve subtle ends, and even strategic goals.[15] But even for the species for which I have the most extensive information, the Pearly Eye (Chapter 7), I cannot cleanly interpret its behavior as strategic versus mechanical (Sections 7.4–7.7). Moreover, Section 8.2.5 shows that boundary behavior is a crucial part of habitat preference, and my most explicit empirical information on boundary behavior is rudimentary, even for the Ringlet (Figures 3.2, 3.3, and 3.8). Finally all my NetLogo models caricaturing mechanical behavior assume simple configurations of boundaries, dominated by straight lines, already a caricature of the real landscape (Figure 8.1). In addition, real landscapes (Figure 1.2) have gradations between favored habitats and unfavored matrix, a wide range of consistent patch sizes of favored habitat, and highly varied path lengths, both across patches of favored habitat and through the matrix between adjacent patches. And because the favored habitats differ among my study species, so do the configurations of their landscapes.

The behavior of my five species all played out on the landscape of the study area, and so was both constrained and facilitated by the habitats of that landscape and especially by the geometry of boundaries between habitats. Abandoned farmland growing back to woodland and forest is not the landscape in which the butterflies have spent most of their evolutionary history. So it is appropriate to ask whether their behavior should be expected to represent "natural" behavior. The following answers are all speculative.

### 8.6.1 Ringlet

The behavior of male Ringlets in my study may be representative, because twists and turns occur within the scale of fields of uniform structure. The "natural" habitat was probably edaphically dry northern prairie and semimontane meadow with a similar, or even larger spatial scale. The species has expanded its range into postagricultural land relatively recently, becoming two-brooded in the process (Wiernasz 1983, 1989). Ringlets have shown the same range of behaviors where I have observed them in New York, New England, and the northern Great Plains, and in montane meadows of the Rockies. So I am confident that the studied behavior is typical of the species.

### 8.6.2 Eyed Brown

"Natural" patches of the habitat of the Eyed Brown, wet sedgy meadow, have a wide range of sizes that depend on local topography and drainage. The wet meadow of my study site is likely at the smaller end of that range. Furthermore, my observed behavior of individual Eyed Browns may well be constrained to the small scale of paths within the favored vegetation, paths determined in part by the patchy establishment and growth of sedges, and in part by the wanderings of mammals, including me. Whether such paths are more broadly characteristic of their favored habitat I do not know.

The small size of the patches of appropriate habitat may have contributed to the extirpation of my study subpopulation. And of 15 males that I imported for the "Butterfly Cyclotron" experiment (Section 4.3), 10 departed the habitat immediately, and the remaining 5 departed after brief exploration.

### 8.6.3 Great Spangled Fritillary

My observed closure of long paths for individual Great Spangled Fritillaries may be an artifact of the postagricultural landscape. But anecdotal observations of regular returns to the same spot and circuits have been reported by many observers, from Scudder (1889) to Opler and Krizek (1984), Gochfeld and Burger (1997), and Cech and Tudor (2005).

When I tried the same mark and recapture studies with *Speyeria cybele leto* in the far more open habitats of the lower slopes of the Beartooth Mountains in Montana, I never got a recapture among 22 marked individuals.[16] The wanderings of the Montana Fritillaries were, like those in New York, along woodland edges and between copses, but sloping open meadows were far more extensive and less confining, with violets widely scattered, albeit concentrated near woods.

Given that the New York Fritillaries ranged fully over a "leaky" total study area (Figure 8.1), it is even more interesting that some of their circuits were closed (Figures 5.3–5.6).

### 8.6.4 Viceroy

My study site for the Viceroy was of an ideal size for interpreting the territories of males as in some sense optimal (Figure 6.5). It is well within the wide range of scale for the typical habitat, wet meadows and streams with willows, and such scales are likely "natural" in that they are determined as much by topography and drainage as by agricultural history. I have observed all the elements of interactive behavior described in Chapter 6 for unmarked male Viceroys in more extensive linear habitats along streams in central New Jersey, though the boundaries of activity seem more dependent on such interactions than on space per se.[17] So the explicitness of geographic territories, and especially the "leapfrogging" of males between them (Figure 6.2 and Table 6.3) may be peculiar to the size and configuration of my study site.

### 8.6.5 Pearly Eye

The habitat of the Pearly Eye is wet open deciduous woodland, especially with glades and open northern edges and wide-bladed grasses, but the scale of most observable movements of both males and females is within the potential attractive radius of fermenting exudates of woody plants. The only artificiality of my study site is my provisioning of extra ferments (e.g., Figures 7.3–7.4). The simplified graph for the Pearly Eye (Figure 8.7C) assumes that individual ferments could have attractive domains, though their geometry and potential overlap would vary with the condition of each ferment and the prevailing local wind. That and the variation in individual patterns of activity (Section 7.7) may be partly responsible for variations in bait attendance in the experiments of Figure 7.3, but there are still many examples of individual males traversing the attractive domains of others to settle on their "own."

I have observed consistent behavior of Pearly Eyes at all sites where I have found them in New York, northern New Jersey, eastern Massachusetts, southern New Hampshire, and central Illinois. These consistencies include frantic activity in late afternoon where natural ferments are clustered, and some individual males that are unusually active and exploratory. So I am confident that my study records "natural" behavior for the Pearly Eye.

## 8.7 SUMMARY OF RESULTS *AND SPECULATIONS*

The following summarizes results and definitive inferences *with speculations in italics* or in permissive construction. The speculations are *not* to be interpreted as new findings; they are merely suggested as worth further study. This summary is ordered by the logic of behavioral interpretations, whereas the body of the chapter is more by habitat, general rules, and species. So for each statement or paragraph below there is a terminal reference to the appropriate section in the body of this or other chapters.

Mark-and-recapture ranging is a result of interaction between the individual butterfly's behavior, the habitat, the landscape, and my own behavior (Behavior: Section 8.1.2). This set of heterogeneities needs to be kept in mind in all interpretations. The pessimistic view is: . . . that's why I do not trust traditional methods, . . . and why you should be skeptical of my ad hoc machinations. The optimistic view is: . . . that is why it is a pleasant surprise that my ad hoc machinations seem to produce consistent differences between species, . . . and amazing that some of the traditional methods seem to work anyway.

### 8.7.1 Some Fixed Cues and Rules of Behavior Need Not Involve Decisions.

Moving straight ahead helps to efficiently explore new ground, and to traverse an unfavored matrix (Rule & Behavior: Sections 8.2.4 and 3.6.3).

Within the preferred habitat, some form of reflection from the boundary with the matrix is crucial to staying within the habitat (Section 8.2.5). It helps to travel in nearly straight lines in the matrix and to turn at a modest rate in the preferred habitat, but a strict rule of "turn to stay; proceed to explore" in itself is ineffective as a mechanism of habitat preference (Section 8.2.4, especially Figure 8.4). Given its importance to staying within the preferred habitat (Figure 8.7), explicit boundary behavior is woefully understudied, as is behavior in the less preferred matrix. (Cue & Rule & Behavior)

All my studied species seem to "run for daylight" to some extent (Cue & Rule: Section 8.2.1).

The *Eyed Brown*, Fritillary, and Viceroy orient to low spots in the near or distant horizon (Cue & Rule: Section 8.2.2).

*Pearly Eye aggregations can be initiated by tacking up odor plumes and turning when other individuals are detected* (Cue & Behavior: Sections 8.2.4 and 8.3.5).

### 8.7.2 The Ringlet and Eyed Brown May Behave like Robots.

The Ringlet is confined to coherent patches of favorable habitat by reflection from its boundaries (Rule & Behavior & Life History: Sections 8.4.2, 3.2.1, 3.3.1).

The Eyed Brown is also confined to coherent patches of favorable habitat, but the boundary behavior has not been explicitly observed (Section 8.4.2). *What individual localization there is may be due to physical fragmentation of pathways through the vegetation* (Behavior: Section 4.3).

Neither the Ringlet nor the Eyed Brown shows evidence of recognizing particular real estate. Their male-male interactions are short, without qualitatively different behavior from that of "investigating" other fluttering objects, including females (Cue & Behavior: Section 8.4.2).

### 8.7.3 Other Species May "Decide" about Real Estate and Sexual Discrimination.

If different males treat the same location in different but individually consistent ways, I infer that a decision is being made (Section 8.4.2). It is more difficult to infer a decision behind sexually discriminative behavior for at least two reasons (Section 8.5). *There may be a fixed response to a sexually dimorphic signal that I do not perceive, e.g., a pheromone or a flash pattern in ultraviolet reflectance. Sexual differences in interactions may be generated by the interaction itself*, rather than by a decision made by either participant (Section 7.6). *Accordingly I tentatively infer that a decision is needed to terminate an excessively long interaction for which there is a plausible visual signal of the nature of that interaction* (Rule & Behavior).

*Excursions of Fritillaries may be bounded by the configuration of navigable paths between patches of favored habitat*, but individuals follow *different closed circuits that may overlap* (Section 8.4.2). Interactions often take place over long periods and long distances, but one cannot rule out a guiding role for physical cues in the environment (Rule & Behavior: Section 8.5.2).

A male Viceroy executes a direct turnaround when outside its territory or its favored habitat (Section 8.5.3). Different males appear to defend different exclusive territories against other males, engaging in interactions over many seconds and many meters (Section 6.3). The interactions differ qualitatively from those between males and females (Rule & Behavior).

The Pearly Eye is drawn in from a distance by the odor of its favored aggregating sites, but particular males are faithful to an individual selection of those sites (Sections 8.4.2 and 7.1.2, especially Figure 7.3). *Male-male interactions differ from male-female interactions in the visual signals that they afford the butterflies over a period of many seconds, and may differ in olfactory and vibrational signals as well* (Sections 8.5.4 and 7.6). They also look dramatically different to a human observer (Section 7.2, especially Figure 7.6). *Although I believe that Pearly Eyes are likely making decisions, particularly about when to terminate a given interaction*, I can generate most of the variations in behavior that I have recorded with a NetLogo model of fixed autonomous behavior (Cue & Rule & Behavior: Sections 8.5.4 and 7.3).

### 8.7.4. Quasi-Natural Behavior Can Be Observed in a Postagricultural Landscape.

Particular behavior may be fostered or constrained by an interaction between the butterfly and the spatial distribution of its needed resources (Section 8.1.1). The favored habitat is defined by needs (permissive physical parameters, adult and larval food, potential mates, and safety) that differ among species, even in the same landscape (Section 8.1.2). The less preferred matrix is defined by lack of the needed, species-specific resources. The degree to which the postagricultural landscape of my study arena resembles the "natural" landscape of a species' evolutionary history also varies among species (Sections 8.1 and 8.6). Accordingly, landscape issues need to be addressed separately for each species. (Behavior & Life History)

*The landscape of my study site seems natural for the Ringlet and Eyed Brown* (Behavior: Sections 8.6.1–2).

Fritillaries are traditionally known for repetitive behavior, and *may be more regular and periodic in my study area than in other places* (Behavior: Section 8.6.3).

*The Viceroy may be more faithful to particular territories in my study* than elsewhere, but its qualitative behavior is conserved (Behavior: Section 8.6.4).

*Details of local behavior of my Pearly Eyes may be more consistently frenetic in my study* due to my setting out abundant fermenting baits, but the qualitative and short-term quantitative behaviors are similar to those at other sites where I have observed the species under natural conditions (Cue & Behavior: Section 8.6.5).

CHAPTER NINE

# Life History Consequences of Individual Behavior

> An evolutionary tendency is perceived intuitively, and expressed in terms which simplify, and therefore necessarily falsify, the actual biological facts. The only reality which stands behind such abstract theories is, in each case, the aggregate of all the incidents of a particular kind, which can occur from moment to moment to members of a species in the course of their life-histories.
> –Ronald Aylmer Fisher, 1930, *Genetical Theory of Natural Selection*, p. 150

This chapter is *not* in any sense a population biology of the species, which would require a study of the dynamics of all life stages, and of immigration/emigration between neighboring subpopulations. Rather this is an account of observed behavior of adult butterflies over a reproductive season, and of how that behavior plays out as observable effects in the life history of individuals. The aggregated effect of these life histories within the local subpopulation will of course affect the persistence and fate of that subpopulation, but for such patterns I am limited to speculation and to a call for more intensive studies that I wish I had the time and opportunity to do.

I start by characterizing the ranging pattern for each species, using a common analysis, as a supplement and partial confirmation of the qualitatively different stories and maps of Chapters 3 through 7 or the idealized diagrams of Chapter 8. Then the daily rates of local loss to death or emigration are characterized, again using a common analysis. When the analyses of ranging and of local loss are compared, there is an apparent benefit to the local population from individuals that remain in place and perhaps even "learn the lay of the land." Ultimately, differences at the small scale of simple cues and behavioral rules produce differences in ranging that lead to differences in parameters of life history for the local population (cf. Kingsolver 1989 for physiological effects of weather propagating to the scale of the population, and Watt 1991 for biochemical physiology affecting population genetics). Conversely, population structure, in the context of habitat geometry, sets the stage for important aspects of behavior (Southwood 1977).

## 9.1 SHORT-TERM VAGRANCY (OR NOT)

I use the methods introduced in Chapter 3 with the Tactical Vagrancy Model to compare and contrast ranging of the five study species. Minute-to-minute movements are plotted as direct distance between sequential perches against length of flight path between those perches. On a log-log version of such a graph, a Brownian random walk would on average be linear with a slope of 0.5 and variance that increases with flight length—and movement in a straight line would obviously be linear with slope 1.0. A forward-biased random walk would be something in between, depending on the bias, with initial slope close to 1.0, decreasing to about 0.5 for very long movements. Conversely, a horizontal cloud of points would suggest varied movements that either return to the same general area, i.e., a "territory" or a "home range," or are confined by the boundaries of a sufficiently small patch of habitat.

Figure 9.1 plots such movements for the males of all five species. Recall from Chapter 3 that both simulations and field data for the Ringlet show intrinsic variation that blurs both expectation and test (Section 3.3.2 and Figure 3.10). Nevertheless, the predominantly horizontal clouds of points for Fritillary, Viceroy, and Pearly Eye suggest that they tend to stay within a given home range that is a fraction of the available habitat. Conversely, the consistent rise with increasing variance of points for the Ringlet and the Eyed Brown are consistent with a forward-biased random walk, though they are ultimately confined by the boundaries of their preferred habitat. These differences in ranging correspond to the differences in behavior anticipated in Chapter 1 and described in Chapter 8, where the Fritillary, Viceroy, and Pearly Eye show evidence of recognizing individual home ranges, but the Ringlet and Eyed Brown do not (Section 8.4.2). I infer that in overlapping regions of their wanderings, the Fritillary, Viceroy, and Pearly Eye actively learn something about their home ranges, even if only the conventions that bias each individual to a particular piece of real estate.

---

FIGURE 9.1. Perch-to-perch distance versus flight distance for males. A slope of about 1 suggests straight-line movement, about ½ suggests Brownian movement, and a shallower slope suggests either active homing, or, for longer flights, confinement by habitat boundaries. The Ringlet and Eyed Brown show a major component of random walk, albeit forward-biased; the Fritillary, Viceroy, and Pearly Eye all return to very near their starting points after traveling varied distances. There is some continuity between the Eyed Brown and the Viceroy; the former being site-faithful, but not exactly so; the latter in many instances returning exactly to its previous perch.

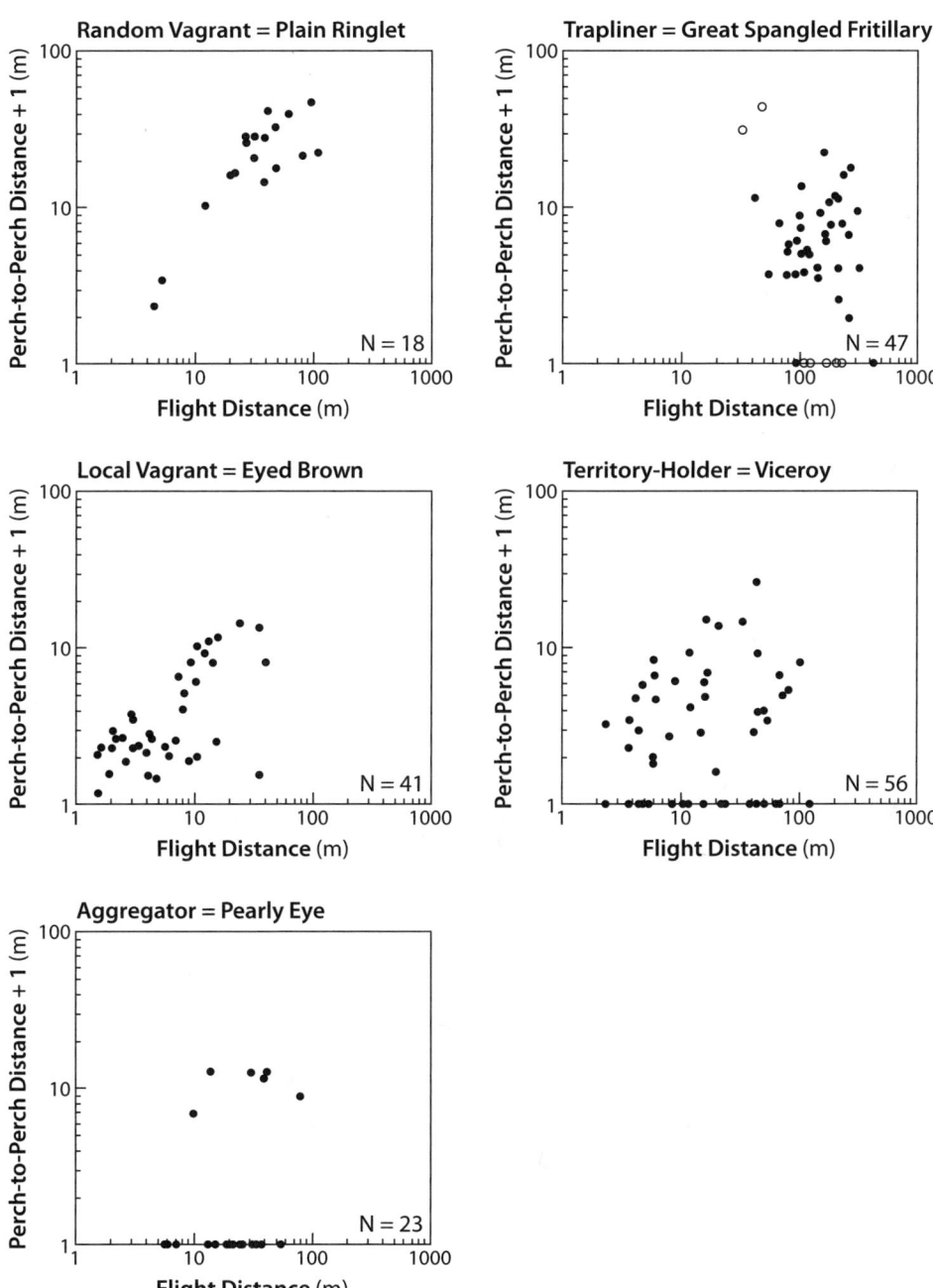

## 9.2 DAY-TO-DAY RANGING IS CONSISTENT WITH SHORT-TERM BEHAVIOR

Measures of individual male movement between day-to-day sightings are subject to cautious interpretation, and may be misleading for the reasons presented in the Preface (Section 0.3) and Chapter 8 (Sections 8.1.2 and 8.6). Nevertheless, Shreeve (1981) presents recipes for simple and elegant parameters of day-to-day movement that put my studied species into the same behavioral categories as my preceding analyses of short-term movements. Shreeve's "velocity" in Table 9.2 was calculated from mapped day-to-day sightings, summing the point-to-point distances between sequential sightings and dividing by the number of days between first and last sightings. Velocity is reported ± its standard error. The root-mean-squared-deviation (RMSD) of location of male sightings is essentially the standard deviation of the spatial coordinates of daily sightings; it measures the average spread of sightings from a central point.

The measures of long-term (day-to-day) male movement in Table 9.1 interact with one another in a way that is interpretable, but not explicitly defined, in terms of the behavioral characterizations of males of each species given in Figure 1.2. The "vagrant" Ringlet wanders rapidly (45 meters/day) and widely (RMSD spread of 57 meters) from its average position in its short lifetime (ca. 7 days), and is often re-sighted (736 times for 711 marked), because its favored habitat (dry upland field) is bounded within the study area. The "traplining" Fritillary also wanders rapidly (38 m/d) and very widely (RMSD spread of 106 m), but its catholic habitat (Figure 8.1 and Table 1.1) is permeable at the edges of the study area, and so it can escape detection over its long lifespan (71 re-sightings for 138 marked; lifespan ca. 13 days). The other species move more slowly day-to-day and cover a smaller range: Eyed Brown (15 m/d; RMSD 49 m; 19 days), Viceroy (16 m/d; RMSD 46 m; 19 days), and Pearly Eye (10 m/d; RMSD 33 m; 19 days).

The Eyed Brown and Viceroy are both limited by their patchy habitat of wet meadows (Figure 8.1), and although those meadows are connected by streams that run beyond the study area, both species are reliably re-sighted during their long lives (200 times for 104 marked Eyed Browns, and 113 times for 67 marked Viceroys). Both Eyed Brown and Viceroy show a pattern of remaining on a small piece of land over several days, moving a relatively short distance and then again staying put for a few days. Their respective difference between "site-fidelity" versus "territory" has more to do with observed behavioral interactions than with patterns of day-to-day ranging (Section 4.2). The Pearly Eye's slow (10 m/d) and local (RMSD 33 m) wanderings are largely an artifact of my providing artificial "singles bars" to supplement the natural supply, and of my monitoring those sites preferentially, but the pattern is sharpened by aggregative behavior, discussed throughout Chapter 7. My providing baits very likely increased the local number

TABLE 9.1. Statistics of Day-to-Day Sightings of Marked Individuals

| Species (sex) and Social System | Ringlet Vagrancy | Eyed Brown Site-fidelity | Fritillary Trapline | Viceroy Territory | Pearly Eye Singles' Bar |
|---|---|---|---|---|---|
| Years of intensive study | 1974, 75 | 1977, 84 | 1977, 82 | 1975, 77, 81–84 | 1974–75, 77–78, 80–84, 93 |
| Number marked (m/f) | 711 / 461 | 104 / 52 | 138 / 69 | 67 / 30 | 660 / 275 |
| Number re-sighted (m/f) | 309 / 183 | 74 / 20 | 37 / 6 | 36 / 3 | 305 / 102 |
| Number re-sightings (m/f) | 736 / 406 | 200 / 36 | 71 / 13 | 113 / 8 | 1,056 / 196 |
| % re-sighted/marked (m/f) | 43 / 40 | 71 / 38 | 27 / 09 | 54 / 10 | 46 / 37 |
| # re-sightings/re-sighted (m/f) | 2.4 / 2.2 | 2.7 / 1.8 | 1.9 / 2.2 | 3.1 / 2.7 | 3.5 / 1.9 |
| Local half-life (days, m/f) | 1 / 1 | 5 / 1 | 1 /– | 3 /– | 1 / 1 |
| Local 95%-life (days, m/f) | 7 / 7 | 19 / 10 | 13 /– | 19 /– | 19 / 11 |
| Most days observed (m/f) | 11 / 11 | 29 / 15 | 18 / 18 | 34 / 13 | 36 / 19 |
| "Velocity" (meters/day, m) | 45 ± 15 | 15 ± 3 | 38 ± 10 | 16 ± 6 | 10.0 ± 1.5 |
| "Velocity" (meters/day, f) | 41 ± 9 | 17 ± 6 | – | – | 8.5 ± 2.0 |
| RMSD of captures sightings (meters, m/f) | 57 /48 | 49 /42 | 106 | 46 /31 | 33 |

Note: Percentage of marked individuals that are re-sighted, and number of re-sightings per re-sighted individual, are related measures of site-fidelity versus wandering, especially comparing male vursus female. Mean "Velocity" ± its standard error is calculated from day-to-day sightings after Shreeve (1981). RMSD of captures = square Root of Mean Squared Deviation of capture locations from their average or centroid. Em-dashes replace data that are too few for meaningful calculations

of Pearly Eyes, and it surely increased my rate of re-sightings (1,056 times for 660 marked).[1]

In the Preface (Section 0.3) I described my discovery that minute-to-minute behavior could not be reliably inferred from day-to-day movements. Nevertheless, the data of Table 9.1 cleanly support Scott's (1986) characterizations of the rapid and widely wandering Ringlet and Fritillary as "patrollers" versus the slower and local Eyed Brown, Viceroy, and Pearly Eye as "perchers" (Sections 1.2 and 1.3, and Figure 1.2).

## 9.3 STATISTICAL SUMMARY OF DAY-TO-DAY SIGHTINGS

### 9.3.1 Overall Local Residency

Table 9.2 summarizes statistics from 5,320 captures or sightings of 2,498 marked individuals. The local half-life is the number of days after marking beyond which

TABLE 9.2. Local Persistence Parameters

| Species | Number Marked and Sex | Vagrant Day X to Day 1 to Day 2 | Resident Presence shifts to Day X+1 (for X>2) | Vagrant Eclosed Resident per day | Freshly Presence when Marked | Day 1 to Day 2 per day if Fresh when Marked | Sighting Rate per individual per day of visit to habitat |
|---|---|---|---|---|---|---|---|
| | A | B | C | D | E | F | G |
| Ringlet | 711 m | 43 ± 2 | 62 ± 2 | 100 | 67 | 48 | 61 |
| | 461 f | 40 ± 2 | 65 ± 3 | 88 | 74 | 44 | 61 |
| Eyed Brown | 104 m | 71 ± 4 | 86 ± 1 | 100 | 32 | 73 | 65 |
| | 52 f | 38 ± 3 | 78 ± 2 | 92 | 31 | — | 66 |
| Fritillary | 138 m | 27 ± 4 | 85 ± 3 | 84 | 22 | — | |
| | 69 f | — | 87 ± 5 | — | 36 | — | |
| Viceroy | 67 m | 54 ± 6 | 85 ± 3 | 100 | 45 | 47 | 92 |
| | 30 f | (8) | (88) | (100) | 29 | — | — |
| Pearly Eye | 660 m | 46 ± 2 | 86 ± 1 | 98 | 26 | 47 | |
| | 275 f | 37 ± 3 | 78 ± 2 | 100 | 31 | 41 | |

*Note:* All entries beyond Column A are proportions expressed as percentages. The "±" sets off a technical binomial "standard error," but it should be interpreted less as sample variation than as an index of relative adequacy of counts to generate meaningful percentages. Em-dashes replace percentages that would be based on numerators <10 and/or denominators <30. Sighting rate is not given for Fritillaries because of the patchiness of my coverage of their habitat, nor for Pearly Eyes because of the patchiness of their coverage of my habitat. Vagrant and Resident Presence (i.e., local persistence) are inverse measures of the slopes in Figure 9.2. See Section 9.3 and Appendix B.1 for more details and discussion.

fewer than half of the marked individuals were sighted. The local 95%-life is the same, but to the more stringent criterion of fewer than 5% being sighted. The most days observed is the championship time from first to last observation for an individual of the species.

All three measures of adult "lifespan" (more accurately, "local persistence") are consistent in characterizing the Ringlet as short-lived (days to a week as an adult) and the other species as long-lived (weeks to a month). Females of the Fritillary and Viceroy are seriously under-characterized, as my teaching obligations interrupted their flying seasons, especially for the females, in all years of the study.

### 9.3.2 Changes in Local Residency

Day-to-day recaptures provide data for analysis of the number of individuals known to be locally present as a function of time since original marking. I interpret this as "local persistence," though it obviously combines both global survival and local emigration. I do not partition these formally, but I do give circumstantial evidence for different interpretations from one species to another. The convention of starting with the time of original marking, rather than restricting to captures of fresh individuals or estimating and using age, is discussed in Section 1.6.2.

I model local persistence with three parameters: rate of local persistence for the first day, presumably as a wanderer searching for a home, rate of conversion from a wanderer to a homebody during that day, and rate of local residency as a homebody after the second day. These parameters are calculated as simple ratios expressed as percentages, summed counts of individuals expressing the given behavior on a given day divided by summed counts of individuals initially present that day. Each day is considered to represent an independent exposure to risk of death or departure for each initial resident; thus the counts used to measure each parameter are statistically independent.[2]

The parameters of local persistence are presented in Table 9.2, along with technical binomial standard errors. Those "errors" should be interpreted less as sample variation than as indices of the adequacy of counts to generate meaningful measures of the parameters. To a first approximation, parameters that look different likely are, and those that look about the same are not different. In particular the numbers in Table 9.2, and the details in Appendix B.2, justify the patterns interpreted below in Figure 9.2.

Figure 9.2 shows the pattern of local residency with time since marking for the five studied species. Dots are the numbers known to be alive, and the lines are fit to the parameters of Table 9.2. Except where counts are fewer than 20, the lines fit the data almost to the tolerance of the minimum dot size for legibility of the graphs. The Ringlets and male Eyed Browns die and/or depart from local residency at a roughly constant rate each day (the difference in loss rate between the first and subsequent days, while statistically significant, is not substantial). All others, and in particular the males of the Fritillary, Viceroy, and Pearly Eye, are lost at a substantially and statistically greater rate in their first day than in subsequent days.

### 9.3.3 Emigration versus Mortality

Potential causes for a high initial loss rate include: altered behavior due to the marking itself, de facto departure from the study area of individuals that do not become resident, and increased mortality before settling locally. An effect of marking per se is unlikely for two reasons. First, I noted those few individuals that showed atypical behavior when released after marking, and left them out of all tallies. And second, the initial loss rate is not substantially higher in the Ringlet, the smallest species and likely the most sensitive to handling.

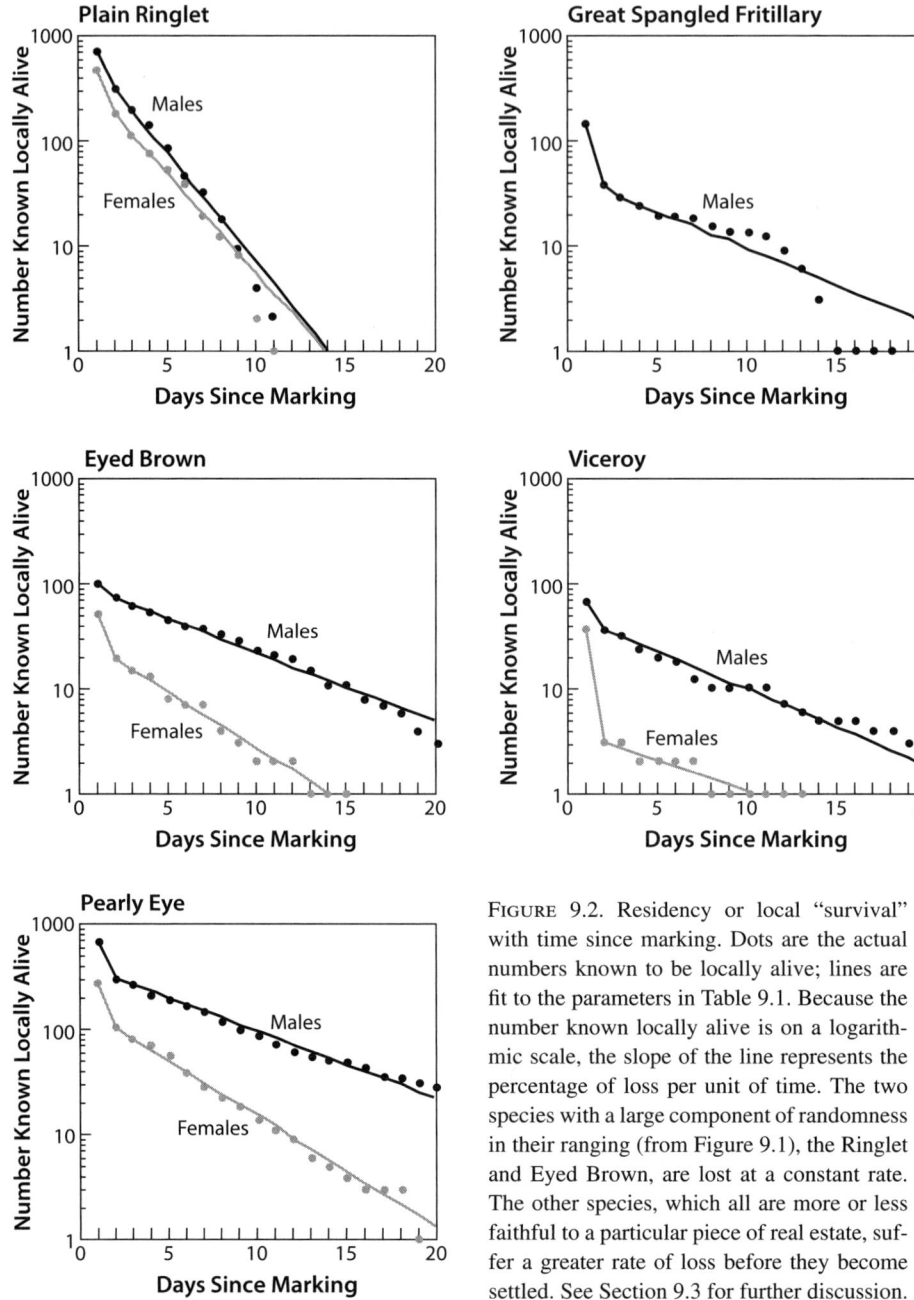

FIGURE 9.2. Residency or local "survival" with time since marking. Dots are the actual numbers known to be locally alive; lines are fit to the parameters in Table 9.1. Because the number known locally alive is on a logarithmic scale, the slope of the line represents the percentage of loss per unit of time. The two species with a large component of randomness in their ranging (from Figure 9.1), the Ringlet and Eyed Brown, are lost at a constant rate. The other species, which all are more or less faithful to a particular piece of real estate, suffer a greater rate of loss before they become settled. See Section 9.3 for further discussion.

Do some newly marked individuals wander beyond the study area to settle elsewhere?

If so, one expects some worn individuals entering the study area from outside prior to marking. Figure 9.3 plots the relative wear and tear on males when first marked.[3] Sure enough, some males are well worn when first marked for the Eyed Brown, Fritillary, and Pearly Eye, and less so for the Viceroy. The effect is less for females all species, as shown in Figure 9.4.[4]

Male and female Ringlets are predominantly fresh when marked (Figures 9.3 and 9.4), suggesting that there is not much interchange among adjacent subpopulations, and that perhaps the losses of Figure 9.2 have a large component of local mortality. The lack of interchange for Ringlets is consistent with their restriction to tightly bounded patches of appropriate habitat that are of a scale similar to that of my study (Section 8.1.2 and Figure 8.1).

For those individuals that are badly worn when first marked and that are indeed immigrants, the wear itself may be a cost of wide movement. In Chapter 2, I argued that the extraordinary efficiency of butterfly flight depends on intact scales (Section 2.2.3), and intact wing edges (Section 2.2.6, especially Figure 2.8). Thus flight efficiency is severely compromised for the higher categories of wear and tear in Figures 9.3 and 9.4, namely Lots to Fray (considerable loss of scales to fraying of wing edges), Tattered (parts of edges of wings missing), and Gross (wings damaged enough to visibly affect flight). Accordingly, damaged immigrant and emigrant males of Eyed Brown, Fritillary, and Pearly Eye (Figure 9.3) may have a greater risk of mortality than resident males.

### 9.3.4 Behavior and Exposure to Predation

An increased mortality prior to settling locally is plausible by several mechanisms involving predation, though the evidence is by no means definitive. Learning the local lay of the land may increase the efficiency of evading or escaping predation or exposure to poor weather. Accordingly, I shall describe for each species of butterfly their predators and factors affecting the risks of predation. However, these factors could simply remove those unlucky individuals that encountered predators in characteristic places, and leave those that had survived, without implying any "learning" of the lay of the land.

Table 9.3 records observed instances of predation or attempted predation. All invertebrate predators were successful, except for one escape each of a Pearly Eye from a Bald-faced Hornet and a Fritillary and a Viceroy from a spider web. Not reported are several instances in which hornets were attracted to a ferment-baited trap (a la Platt 1969) and dispatched all the trapped butterflies within a

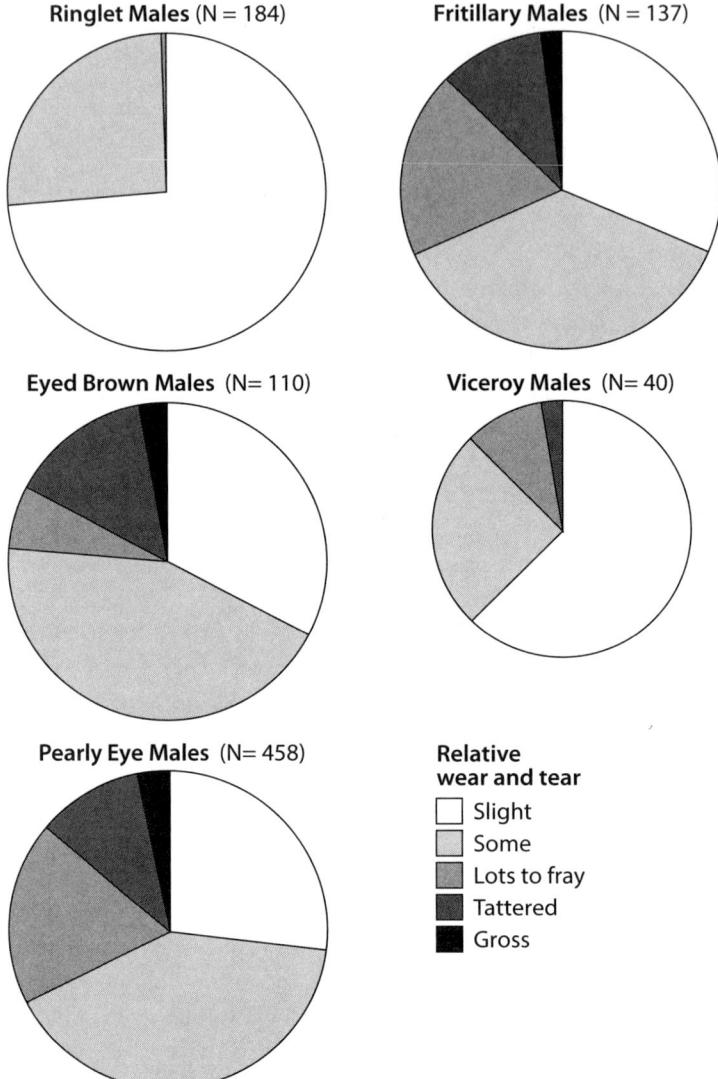

FIGURE 9.3. Wear and tear of males when first captured and marked. Pie diagrams based on fewer than 100 butterflies are scaled down to suggest lower confidence in their interpretation. The degrees of wear and tear are: Slight (no to very slight loss of scales); Some (slight to patchy loss of scales); Lots To Fray (considerable loss of scales to fraying of edges of wings); Tattered (parts of edges of wings missing); Gross (wings damaged enough to visibly affect flight). See Section 9.3.3 for details that suggest that most males were marked within a few days of eclosion, especially the Ringlets and Viceroys. And even for the other species, fewer than one quarter to one third were badly worn when marked.

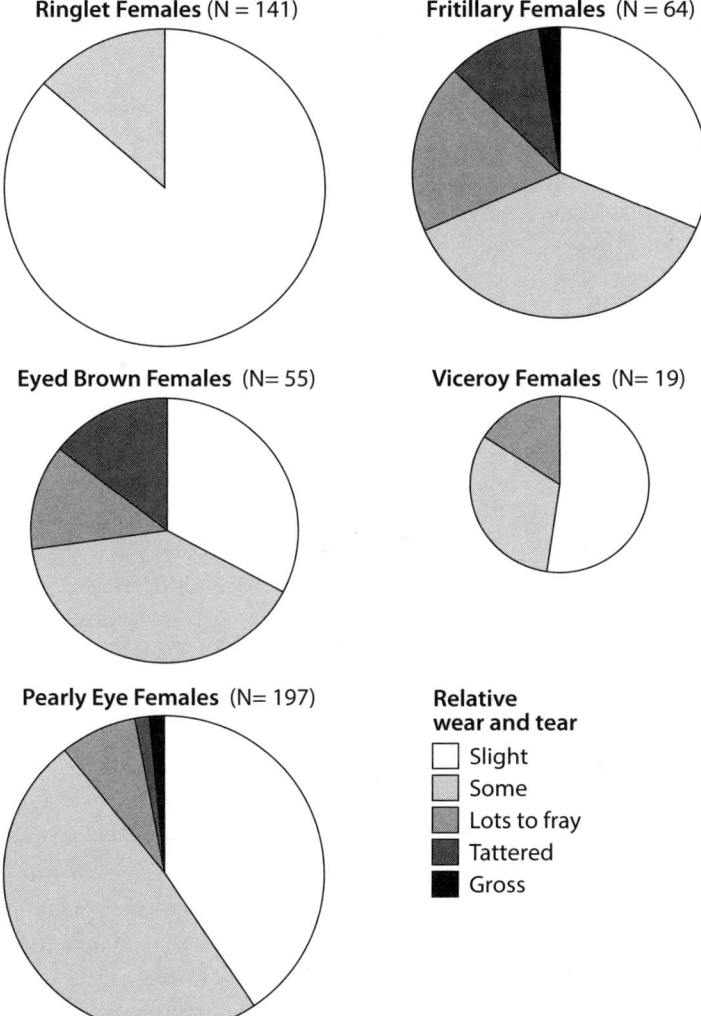

FIGURE 9.4. Wear and tear of females when first captured and marked. Pie diagrams based on fewer than 100 butterflies are scaled down to suggest lower confidence in their interpretation. The degrees of wear and tear are described in legend of Figure 9.3. See Section 9.3.3 for details that suggest that most females were marked within a few days of eclosion, especially the Ringlets, Viceroys, and Pearly Eyes.

TABLE 9.3. Observed Predation (+ Active Escapes in parentheses)

| Predatory Species & Behavior | Micro-habitat | Ringlet | Eyed Brown | Fritillary | Viceroy | Pearly Eye | Other Species |
|---|---|---|---|---|---|---|---|
| *Ambush* | | | | | | | |
| Orb-web Spider (Araneae) | V-notch in vegetation | 3 | 1 | 1 (+1) | 1 | 3 | 5 |
| Ambush Bug (Reduviidae) | Flower | | | 3 | 3 | | 19 (+2) |
| Crab Spider (Thomasidae) | Flower | | | | | | 2 |
| *Short Pursuit* | | | | | | | |
| Dragonfly (Aeshnidae & Libellulidae) | Air | 3 | | | (+1) | | (+1) |
| Baldfaced Hornet (Vespidae) | Fermenting Sap | | | | | 2 (+1) | (+1) |
| *Long Pursuit* | | | | | | | |
| Birds "Beak"-marks = likely Bird attacks | Ad hoc | 3/1219 = 0.2% | 8/157 = 5% | 5/207 = 2% | 5/102 = 5% | 62/969 = 6% | 5* 24/– |

half hour. "Beak" tears are symmetrical tears in the wings of butterflies such as might be made when escaping a bird that had grasped the butterfly while its wings were closed over its back. I was very conservative in attributing wing damage to birds. To qualify, the damage had to be very nearly laterally symmetrical, had to show none of the curling that can accompany damage during emergence from the chrysalis, and had to be severe enough to transect one or more veins of the wing. Interpretation requires even more care. A count of such "beak" tears does not include those butterflies that were attacked successfully and eaten, or the many that may have been damaged asymmetrically while flying. So the percentage of individuals showing such damage is only a very rough index of an underestimate of the relative danger that a species faces from attacks by birds. That the danger is real is shown by my direct observations of successful predations on Pearly Eyes by Wood Pewee *Contopus virens*, Rose-breasted Grosbeak, and Ovenbird *Seiurus aurocapillus*. Sixteen other species of birds were observed lurking around my artificial baits and natural ferments, often lunging unsuccessfully at Pearly Eyes and Viceroys. A Yellow-billed Cuckoo *Coccyzus americanus* was observed stalking a

Fritillary. I occasionally found detached Pearly Eye wings right under my baits, but the predator was more likely a hornet than a bird (birds tend to perch away from the baits and to shake the prey as they chomp, dispersing the wings).

Given the variation in habitats, predators, and behaviors of both butterfly and predator, the interpretation of relative percentages is problematic.[5] Nevertheless each recorded observation represents at least some interaction between that predator and that species of butterfly. Thus the predation observations are conservative underestimates of risk with unknown variance (though the daily risk of predation is bounded by the average daily loss rate). In any case, the observed predations are also too few to expect statistical significance for differences comparing species, sexes, or predators.

Nevertheless, the different behaviors of predators expose different species of butterflies to different risks of predation. Birds find and pursue fluttering butterflies and may loiter where butterflies are aggregated. Hornets (Vespidae) are drawn to fermenting sap and fruits, which may also draw butterflies as potential prey. In open air, dragonflies (especially Aeshnidae and Libellulidae) encounter and pursue smaller prey, but may take butterflies as well. Fine scale paths and V-notches in vegetation provide the anchor-points for orb-weaving spiders (Araneae) to set their traps. And flowers are favored hides for ambush bugs (Reduviidae) and crab spiders (Thomasidae) to ambush perched or nectaring insects.

The Ringlet stops for rest and for nectar on star thistle *Centaurea* flowers, where it is exposed to crab spiders and ambush bugs. Otherwise Ringlets tend to fly just above the vegetation, which fosters visibility to potential mates (Section 3.2.4), and avoids spiderwebs, though it exposes them to dragonflies.[6]

The paths in the vegetation favored by the Eyed Brown (Section 4.1) are ideal places for spiderwebs. Accordingly, the males tend to perch at the edges of paths, and they often fly high between perches. If there is partial reality to the Fortuitous Site-Fidelity Model for the Eyed Brown (Section 4.4), such high flights might shift males between attractive domains, offsetting the potential advantages of site-fidelity.

Running traplines by Fritillaries of both sexes costs energy and requires nectaring. That puts them in harm's way from crab spiders and ambush bugs. Males often crisscross a patch of flowers before landing to nectar, but I cannot tell whether this behavior is related to avoiding predators or to finding females.

Male Viceroys prefer nonterminal perching places in their territories, which makes them less likely to settle on flowers, and so more likely to avoid crab spiders and ambush bugs (see Section 6.3.1 for more speculation about choosing a perch with an awning).

The Pearly Eye is doomed to a high risk of predation by favoring ferments that are also attractive to hornets. In addition, their active aggregations attract the attention of birds. It may be more than coincidental that of the five study species, the Pearly Eye is the most exploratory and skittish, including being sensitive and responsive to vibration and sounds (Sections 2.5 and 7.1.4). When caught in a butterfly net, the Pearly Eye often folds its wings down and attempts to escape with the strong upstroke characteristic of satyrines generally (Section 2.2.4).

Thus each species has some pattern of behavior that reduces the risk from predators to which it is unavoidably exposed by habitat choice and other aspects of behavior. Whether familiarity with a particular bit of real estate fosters survival is an open question. But repeated exposure to specific places where an individual has survived may be safer than continually exploring new territory.

*9.3.5 Oviposition and Parasitism*

Many female butterflies lay eggs singly and walk or fly some distance before laying the next one, in a sequence that lasts most of their adult lifespan. The subsequent dispersion of eggs in space and time has been interpreted as an adaptation to reduce parasitism by wasps and nematodes (Scudder 1889; Southwood and Comins 1976).

Females of three species in my study likewise disperse single eggs: the Ringlet by straight line movement even in preferred habitat (Figure 3.10), the Fritillary by repeating traplines (inferred in Section 5.2, and observed by Clark 1932), and the Viceroy by laying eggs over a long period (Scudder 1889).

The other two species tend to lay eggs in small clusters, 3 eggs each for the Eyed Brown, and 3–7 for the Pearly Eye. While attempting to raise Pearly Eyes I observed a procession of several first instar larvae, chewing their way along the edge of a grass blade. The first was "breaking ground" along the original edge of the blade, and the others were exploiting the presumably softer open edge.

## 9.4 DISCUSSION: RANGING AND LOCAL RESIDENCY

The species with a large component of vagrancy in their ranging, the Ringlet and the Eyed Brown (at least the males), gradually explore new territory, and are lost to death or emigration at a rate that is roughly constant from day one. Conversely, the species that establish and retain a restricted home range, the Fritillary, the Viceroy, and the Pearly Eye, are lost from the local population at a greater rate initially than later.

A pessimist could argue that there is a cost to finding a suitable piece of real estate, a higher local rate of death until one learns the local landscape (cf. Stamps 1995), or wide wanderings beyond the natal region before finding a place to settle. An optimist could argue that there is an advantage once an individual has settled in place. The cognitive burden imposed by settling is unclear. Different individuals of the Fritillary, Viceroy, and Pearly Eye restrict their movements to different pieces of real estate. Chapter 8 suggests that these differences imply individual decisions, or equivalently, learning to treat a common cue differently. For all five species, the cues that bound suitable habitat may be learned or they may simply be responded to robotically.

However, whether the individuals lost from my study are emigrated or deceased, they are lost to the local population, and so there is a cost to the population of the behavior of individuals, particularly when individuals do not establish local residence, or prior to when they do so. Whether local losses through emigration are favorable to the emigrants depends on their subsequent fates. That is a fertile area for future work.

## 9.5 SUMMARY OF RESULTS *AND SPECULATIONS*

The following summarizes results and definitive inferences, for males unless otherwise noted, *with speculations in italics* or in permissive construction.

Contrasts among species in day-to-day movements are largely consistent with differences observed from moment-to-moment (Behavior & Spatiotemporal Scale & Life History: Section 9.2).

The Ringlet and the Eyed Brown *tend to wander into new territory as they accumulate flying time* (Section 9.1), subject to strong confinement within bounded patches of preferred habitat (Section 8.1.2). Accordingly, they are lost from the local population at a nearly constant rate with time since eclosion (Behavior & Spatiotemporal Scale & Life History: Section 9.3.2).

The Fritillary, Viceroy, and Pearly Eye tend to stay within a portion of each continuous patch of their preferred habitat (Sections 9.1 and 8.1.2), and *their occasional departures are usually followed by returns "home," their excursions becoming more local with time since eclosion* (Sections 9.1 and 9.3.2). They are lost from the study area at a higher rate initially than their lower and roughly constant rate once they become resident. *Their higher initial loss rate* is due partly to wandering per se, but *may have a component of increased predation or increased wear and tear* (Behavior & Life History: Section 9.3).

Each species has behavior that exposes it to a different mix of potential predators, and each also has behavior that ameliorates the risk from those predators (Behavior & Life History: Section 9.3.4).

The *ranging behavior of females is crucial to the fate of local subpopulations* and is woefully understudied. All that I can say is that females of all species have behavior that spreads their eggs in time and space, *which may help to hide them from parasites* (Behavior & Life History: Sections 9.3.5 and 9.4).

CHAPTER TEN

# Summary *and Speculations*

> I am part of all that I have met;
> Yet all experience is an arch wherethro'
> Gleams that untravell'd world whose margin fades
> For ever and forever when I move.
> —Alfred, Lord Tennyson, 1842,
> from "Ulysses," in *Poems*

This chapter summarizes the new ideas and new data of this study, in each case citing the closest work of others. The new data are observations of natural history that differ qualitatively from species to species, but all of them have parallels among other studies of butterflies. The new ideas involve handling simple sensory cues with simple rules that generate superficially complex behavior. Again, some elements of these ideas have been argued for other species, but their full application is new, as is the use of overly simplified models as baselines against which to discover greater subtleties of behavior.

From everything that I have discovered, new mysteries arise. Accordingly, this chapter also lists unsolved problems, with suggestions for their exploration. Most of these involve speculations about adaptive patterns of sensory and motor structure and physiology, and questions that arise about the behavior of females.

From my formal study of five species, I can also speculate about the behaviors that I have observed casually in many more local species. This speculation suggests adaptive relations between ranging behavior and social system on the one hand, and overall population density and temporal stability of the habitat on the other. They provide an example of Southwood's (1977) suggestion that localized behavior is adaptive to stable habitats and vagrancy to ephemeral habitats.

A trope for this summary is to present the complexities of butterfly behavior and then to propose plausible cues and rules that rigidly programmed automata could use to generate such behavior. Laying out such cues and rules explicitly is a partial interpretation of the behavior, but more importantly, it highlights additional bits of behavior that, if observed, would contradict the rules and suggest that something more is going on.

# SUMMARY AND SPECULATIONS

## 10.1 MALES OF FIVE SPECIES SPAN SOCIAL BEHAVIOR DESCRIBED FOR BRAINIER ANIMALS

The male Ringlet wanders with no detectable pattern within its prescribed open habitat, *traversing places where it may be "found" by virgin females* (Chapter 3, especially Section 3.5). The male Eyed Brown patrols paths in appropriate habitat *where females may be found*, but I have not observed intersexual behavior; *the paths may form individually attractive domains, resulting in fortuitous site-fidelity* (Chapter 4, especially Sections 4.4 and 4.5). Male Fritillaries repeatedly traverse long, looping paths *that visit sites where freshly eclosed females may be found* (Chapter 5, especially Section 5.1.4). Male Viceroys literally defend individual territories against one another, in places frequented by females, *though the contention that free-flying females are mateable is only indirectly supported* (Chapter 6, especially Section 6.3). Male Pearly Eyes aggregate at fermenting slime fluxes on trees and shrubs, which are also attractive to females (Chapter 7, especially Section 7.1). Accordingly, I stand by the characterizations, given in Figure 1.2, of the Ringlet, Eyed Brown, Fritillary, Viceroy, and Pearly Eye as respectively exemplifying Vagrancy, Site-fidelity (albeit perhaps by accident), Trapline, Territory, and Singles Bar.

The different behaviors are initially documented in different ways for each species (Chapters 3–7). But shared analyses of maps of short-term behavior, and analyses of locations of recapture from day to day, *classify the species into the correspondingly different categories:* in particular, explorers of new territory for Ringlet and Eyed Brown versus eventual local residents for Fritillary, Viceroy, and Pearly Eye (Section 9.1), and patrollers for Ringlet and Fritillary versus perchers for Eyed Brown, Viceroy, and Pearly Eye (Section 9.2).

Differences in wing morphology can be interpreted adaptively for perchers versus patrollers, but the exceptions to general rules are so numerous that the interpretation degenerates into storytelling. The Ringlet, Eyed Brown, Pearly Eye—all satyrines that often fly in vegetation with a strong up-and-down structure (stems of grasses, sedges, shrubs, saplings, and trees)—*spend an appreciable proportion of their wingbeat cycle with the wings closed dorsally, riding the reaction to a ring-vortex of air expelled down and back on the upstroke* (Sections 2.2.2 and 2.2.4).

## 10.2 ADAPTATIONS TO HABITAT ENCODE POTENTIAL CUES AND RULES FOR BEHAVIOR

The eyes of different species, and different parts of the visual field within the eye of each species, differ in the presence of pigments and structures *that can shield, enhance, and/or reflect different wavelengths of light* (Sections 2.3.2 and

2.3.6). Some of these differences *suggest filters that may be adaptive by enhancing differences between habitats, by enhancing perception of cues for behavior like choice of habitat, movement within habitat* (Sections 2.3.4 and 2.3.5), *and mate-finding—and even by dynamically adjusting sensitivity.*

Patterns of changing flicker and of optic flow can encode: degrees of shade and of dappling; habitat boundaries and paths; and horizons and especially V-notches (Section 2.3.4). Orientation to V-notches in the horizon is experimentally confirmed for the male Viceroy (Section 6.3.1; Figure 6.3; Table 6.2), and *strongly suggested for the female Viceroy and the Fritillary* (Figures 6.2 and 5.4).

Experiments with the Butterfly Rotisserie (Appendix C, Figure C.6) confirm that all five species can respond to flickering images of conspecifics when those images span the field of view of only a few of their more than 10,000 ommatidia (Section 2.3.1; Figure 2.1; Table 2.1). Keeping such a flicker in the frontal visual field is sufficient to move toward and/or follow the butterfly generating the flicker. Keeping it in the lateral field of view will result in circling it.

## 10.3 SIMPLE CUES MAY SIMULATE COMPLEX BEHAVIOR: A FORWARD VAGRANCY MODEL FOR THE RINGLET; A FIGHT-OR-FOLLOW MODEL FOR THE PEARLY EYE

A forward-directed random walk can be tuned to mimic a Lévy walk, and so it can achieve the varied compromises between exploration and local coverage that have endeared Lévy walks to theoreticians of animal movement (Section 3.3.3). *Cues to generate a forward-directed random walk can be encoded by instantaneous lateral asymmetrical flow of the visual field.*

A Forward Vagrancy Model mimics local wanderings of the Ringlet in appropriate habitat. In particular, a directed random walk with a forward bias of $N(0°, 55°)$ represents the short-term twistiness of the real paths of male Ringlets, and allows an *estimation of their long-term exploration of their habitat for receptive females* (Sections 3.3.2 and 3.4.2). Patrolling under all permissible conditions for its entire adult life *a male still has a probability of only 0.4 of encountering a virgin female.*

A Fight-or-Follow Model is proposed to interpret the complex behavior of interacting Pearly Eyes (Sections 7.3 and 7.4). A rule that "males approach a fluttering object, but females avoid or ignore it" produces a whirling "conflict" between males, a "chase" of a female by a male, and a "conga line" of males "pursuing" a female. A rule that "butterflies in motion remain in motion, but butterflies initially at rest return to rest" is sufficient to generate "home court advantage" between "competing" males. And if male-male interactions are concatenated by aerial reencounters, two males that tend to rest in proximity to each other will appear to "escalate" their "conflicts."

Behavior very like videos of interacting Pearly Eyes *can be generated* by the following NetLogo model: Males may initially be either at rest or flying. When within detection distance of another butterfly, they follow the rules of the Fight-or-Follow Model. Otherwise males and females move according to the Forward Vagrancy Model (Section 7.4 and Figure 7.9).[1]

## 10.4 A FORTUITOUS SITE-FIDELITY MODEL IS A NULL MODEL FOR THE EYED BROWN, AND SHOWS SITE RECOGNITION IN THREE OTHER SPECIES

A Fortuitous Site-Fidelity Model shows that *cues in the configuration of vegetation, even if randomly oriented, can produce separate attractive domains for individual butterflies* (Section 4.4; Figure 4.5). Systematic orientation of cues would tend to further separate such attractive domains. Whether this model underlies the partial site-fidelity of the Eyed Brown is an open question. But the model predicts that individual butterflies whose wanderings overlap will settle in places determined by the vegetation, rather than returning to individually distinctive "home" sites (Sections 4.5 and 4.6).

*So overlapping wanderings with returns to individually characteristic domains strongly suggests either "learning" those domains or individual differences in response to the same environmental cues in the region of overlapping wanderings, either of which implies, at least weakly, "decisions."* Three species show evidence of such behavior. Individual Fritillaries run different but overlapping traplines (Section 5.1.4). Male Viceroys defend an outright territory, coursing into other defended territories, and even out of the favored habitat, but each individual returning to its "own" territory (Sections 6.2–6.4). Individual male Pearly Eyes favor particular fermenting sites for nutrients and females, but episodically visit sites favored by other males (Section 7.1.2; Figure 7.3).

## 10.5 STAYING WITHIN A CIRCUMSCRIBED AREA MAY BENEFIT THE LOCAL POPULATION

The males of species that settle in a circumscribed area, the Fritillary, Viceroy, and Pearly Eye, have a lower rate of loss from that area once settled (Section 9.3.2). That is of course partly a tautology given that my study area is itself circumscribed, and is smaller than some observed wanderings of these species. *However, there is some indirect evidence that remaining in a circumscribed area can reduce the risks of predation and of general wear and tear, either through "learning the lay of the land," or simply by repeatedly traversing areas within which each individual has a history of having avoided predation and wear* (Sections 9.3.4–9.3.5).

Each species has behavior that exposes it to specific predators, and each has some partially compensating evasive behavior (Section 9.3.4). *Even escape from predators is potentially debilitating,* through decreased flight efficiency due to loss of scales, breaking of wing veins, and/or fraying of wing edges (Section 2.2.6).

How much the persistence of local residents is statistical versus a result of adaptive behavior is unclear, and likely varies among the species. Nevertheless, whatever the mechanism of loss, those individuals that do not persist are lost to the local subpopulation (Section 9.4). The fate of emigrants, especially gravid females, may be crucial for the population at large but is unknown in my study.

## 10.6 NEW WONDERS

I began lying on my belly just enjoying butterflies 65 years ago, I started chasing them for a collection 60 years ago, I started working with them in field and lab 45 years ago, and I have sat working episodically on the analysis and writing for 30 years. You would think that by now my curiosity had been satisfied. Not so! For each discovery or answered question, a new question arises. So an important part of what I have accomplished is setting the stage to explore these new questions.

### 10.6.1 Boundary Behavior and Movement in the Unfavored Matrix

Butterflies are easiest to study where they are common and visibly active. So my study has concentrated on their favored habitats. And I have been additionally limited by my academic and family schedules. The crucial fate of emigrants from my study populations depends on how they traverse the unfavored matrix, and indeed on their behavior at the boundary of the favored habitat, which determines whether they enter the matrix (Section 9.4). The techniques of Chapter 1—marking, following, and mapping, which I have employed with males in their favored habitats—could usefully be extended to females, especially ovipositing females, and to the unfavored matrix.

### 10.6.2 Speculative Insights into Vision and Flight

My experiments with vision and flight, described in Chapter 2, have led to highly speculative insights that are worth exploration by someone with more training in bioengineering and/or physiological ecology.

*Butterflies do not have an eye for the continuous detail that we see, but they are well equipped to be very sensitive to patterns of flicker and optical flow that we barely notice* (Section 2.3). *And the mosaic distribution of color sensitivity among ommatidia should enhance the flicker for color patterns.* Robotic vision has begun to take advantage of spatial patterns of flow and flicker (Floreano et al. 2013, Pericet-Camara et al. 2015), and there is much for the sensitive naturalist to observe that could lead to new research. For example, one can identify a Silver-Spotted Skipper *Epargyreus clarus* from a distance of 15 meters by first noticing its flutter, but then seeing the dramatic flashes of its silver spot with each wingbeat. Other species show subtler defining patterns at closer distances once you look for them. When I look for parallax flicker at the landscape scale, I begin to notice potential cues for butterflies to navigate. On many field trips I have amazed my companions by watching various species disappear into shrubbery, predicting where they will emerge, and easily catching them for close observation and release.

My analyses of flight, using strobe photos and wingbeat-phase-analysis of rudimentary videos, have led to insights that bode well for future analyses of high-speed digital videos of butterfly flight (Figures 2.2, 2.3, 2.5, 2.6, and 2.8). In addition, such videos could elucidate the role of wing flexure in postulated "slithering" propulsion. And, if it is observed, alternate up-down bowing between veins, respectively on downstroke and upstroke, *could contribute to postulated vortex management by the veins* (Section 2.2.3).

The efficiency of butterfly flight may be significantly enhanced by inhibition of vortex shedding by rows of scales, and by confinement of vortices between veins (Section 2.2.3).

*10.6.3 Females Are "Calling the Shots" and Are Woefully Understudied.*

I started this project with the idea of documenting and interpreting the ranging and interactive behaviors of male butterflies as they actively searched for females. But time and again I discovered that it is the woefully understudied behavior of females that "calls the shots."

An adaptation of the Forward Vagrancy Model shows that only one sex need move for mate-finding to be efficient (Figure 3.12). If the male wanders, the female may "search" for a mate by sitting still and watching, avoiding the risk of mortality from movement (Sections 3.5 and 9.3.4).

If females initially sit still and eventually mate with only one male, they force males to pay the cost of moving. To expose themselves to the maximum number of virgin females, males of monandrous species must pre-emerge most females, which the Ringlet and the Fritillary do by an interval almost equal to their average adult lifespan (Figures 3.4 and 5.1). The male Ringlet potentially pays the additional cost of continuing to explore new territory over its whole adult lifespan (Section 3.4.2).

The seemingly complex interactive behavior of male Pearly Eyes—male-male contests, male-female chases, and male-male-female conga lines—can be interpreted as fixed investigative behavior on the part of the male, with the differences in interaction being due to the degree of avoidance by the female (Section 7.3). Those behaviors are mimicked by a Follow-or-Fight Model and directly induced with the Butterfly Rotisserie (Section 7.4, Figure 7.9). High resolution videos of such interactions would allow testing a particular prediction of the model interpretation of real behavior. The pathway of a male should in general point to the pathway of a female or preceding male at a roughly constant distance, but the path of the leading individual will often not aim directly away from a follower. This is like the observation that the rear wheel of a bicycle will always aim toward the track of the front, but not vice versa.[2]

### *10.6.4 Do Butterflies Make Decisions?*

A major theme of this study has been documentation of complex patterns of butterfly behavior, followed by proposing simple rules of response to simple environmental or biotic cues that can achieve those complex behaviors without outright cognition and decisions. There is a possible residual role for decisions to explain differences in behavior between individuals exposed to the same set of cues. If these differences remain consistent with repeated visits to the same patch of the landscape, *the suggestion of decisions, or at least of learning the lay of the land, becomes stronger, albeit still not compelling.* The departures shown by the Fritillary, Viceroy, and Pearly Eye from the Fortuitous Site-Fidelity Model, reviewed above in Section 10.4, suggest that *individuals at least learn cues that tend to keep them respectively in their own traplines, territories, and aggregation sites.*

In each case what might need to be learned is simplified by the geographical configuration of the habitat. This is less clear for the Fritillary than the other species, for while navigation through V-channels in vegetation would align the travels of independent individuals, the closure of these channels into repeatable circuits of convenient size (specifically, few enough potential points of decision to avoid the "traveling salesman problem") might be an unnatural artifact of the postagricultural landscape (Sections 5.3–5.4). For the male Viceroy, a favorable territory is defined by wet meadow encircled by trees, with perches from which to observe the movements of females. And active driving of interlopers from the territory of a resident is a sufficient mechanism to maintain individual territories (Sections 4.2 and 6.4). The ferments that are both nutritional resource and a potential site for mating of

## SUMMARY AND SPECULATIONS 209

Pearly Eyes are attractive from a distance by olfactory cues that are easiest to follow in the preferred habitat of woodland edge or glade (Section 7.1.1).

And there is a similar potential role for cognition, or at least for memory, in differences in behavior among individuals with different histories of prior behavior. I have observed such individual differences among male Viceroys and male Pearly Eyes.

Interactions between male Ringlets, Eyed Browns, and Fritillaries were infrequently observed, and generally took less than 5 seconds, consistent with their ending when casual movements took them out of the 1–3 meter visual detection distance of each other.[3] So for these species, there is no need to postulate a "decision" to terminate the interaction.

I frequently observed interactions between males of Viceroy and Pearly Eye that twisted and spun over tens of meters and tens of seconds (Figures 6.6 and 7.14). The patterns of optic flow generated during such interactions were surely of sufficient character, strength, and duration to provide cues of the nature of the activity, at least as it differed from other activities (Section 2.3.4). I know a lot about initiation of those interactions, from both observation and experiment, but little about what terminates them, which may be where any adaptive interpretations lie. High-speed videos might help to interpret those interactions, especially for the Viceroy, which engages in qualitative changes in behavior over distances of a meter or more in the course of a given interaction. Pearly Eye interactions typically persist at distances of less than a meter, a scale at which seemingly complex behavior could be generated by automatic responses to simple physical stimuli: dynamic visual cues, vortex entrainment or buffeting, or odors that the vortex trail retains (Section 2.2.2). *So for the Pearly Eye, high-speed videos might be difficult to interpret.*

It is important to know what terminates the long interactions of Viceroy and Pearly Eye, in particular whether the termination is consistently initiated by one of the individuals, and if so, whether that individual was the initiator of the interaction. I could not tell for the Viceroy because long interactions so often terminated out of sight of their starting point, and videos might not help. The Pearly Eye would be an ideal subject for videos, given that baits can be used to manipulate the "studio" so that many interactions would be initiated and terminated within a single focal plane.[4] However it is possible that the long interactions of the Viceroy and the Pearly Eye, like short interactions, also end when casual movements take individuals out of the visual range and field of view of one another. These species might simply be more tenacious in maintaining mutual proximity than are the Ringlet, Eyed Brown, and Fritillary.

So individual males of three species (Fritillary, Viceroy, and Pearly Eye) act as though they recognize particular pieces of real estate, and males of two of the species (Viceroy and Pearly Eye) show individualistic behavior in interactions with others. *These behaviors are consistent with learning and/or making decisions.* But it is possible to suggest plausible cues and rules that automatically generate all the variants of behavior that I have observed so far.

### *10.6.5 Varied Parameters May Account for Individualistic Behavior.*

Minor differences in response to cues can have self-reinforcing effects on behavior. For example, an individual flying at a slightly greater height than average above the vegetation can be exposed to a dramatically different configuration of visual horizons, which lure it into a characteristically different path from its fellows. This might partially account for the individual differences in traplines of Fritillaries and territories of male Viceroys. Being drawn to a greater height by interaction with another individual might allow shifts between attractive domains in the Fortuitous Site-Fidelity Model, but I do not have observations that test this notion.

Individual variations in behavior with respect to habitat and/or other individuals could also be generated by variation in visual acuity or mosaic color sensitivity, and by how they map onto the visual field (e.g., Section 2.3.2).

An individual attracted to fluttering objects at a slightly greater than average distance would experience more frequent and longer interactions, and would stand out among its fellows by doing so consistently throughout its life. Longer latency in response to cues to change behavior would result in greater variance and hysteresis in behavior. Different values of such parameters may be all that is needed to interpret individual variation in the duration of interactions between male Pearly Eyes, and even the diurnal changes in their behavior that shift in their timing among individual males (Sections 7.7 and 7.5). The differences in behavior are sufficiently consistent for each particular male to correspond to what other authors have called a "personality" (Section 7.1.4).

So dramatically different "personalities" can be determined by mere persistence of behavior that results from natural variation in perception and response to cues. Once interpreted in these terms, one can make predictions about the suites of behavior that will be associated in given individuals. Such predictions are readily testable with detailed videos of marked individuals, looking for consistency, or exposing potential nonmechanical complexities.

### *10.6.6 There Are Many More Species Out There!*

The field techniques and analyses that I have used can be applied to other species as well. The following is a speculative summary of information presented in

Appendix B1, analogizing my five intensively studied species to others that I have observed at the same site (this section is essentially a "bookend" to Section 1.3.1).

Rough analogs to the vagrancy of the Ringlet can be seen in the Wood Nymph, Little Wood Satyr, Whites, Sulphurs, and Swallowtails, though the latter three groups also aggregate on hilltops and at mud puddles. Most of the nymphalines and several lycaenids show behavior between the site-fidelity of the Eyed Brown and the outright territoriality of the Viceroy. The distinction between site-fidelity and territoriality is ripe for studies with videos of interacting males. All of my "fritillaries," both large and small, show some evidence of repetitive circuits, like the Great Spangled Fritillary. The aggregative behavior of the Pearly Eye is uniquely extreme in my experience, but it is worthwhile to watch for it wherever nymphalines, especially those that are attracted to ferments, reach high local densities. And by adding artificial ferments, it may be possible to increase those densities and to engineer informative interactions.

Even species that wander with no detectable pattern may be susceptible to tracking and modeling movement, but it should be applied in less favored as well as favored habitats, emphasizing gravid females. To predict metapopulation dynamics requires dynamic versions of my static maps of habitat. And of course the maps of suitable habitats and permissive corridors between them will vary from species to species.

*Among my species, those that wander widely without a detectable pattern typically inhabit the upland fields that are ephemeral in a postagricultural landscape. In such a landscape the natal area may or may not be suitable for the next generation, and depositing eggs both near and far is adaptive. The species that settle in a particular piece of real estate typically inhabit openings of modest size that tend to persist, due to stable patterns of drainage or to the relatively slow turnover of tree populations. Once an appropriate opening is found, it is likely to remain favorable for several generations, and staying put is likely adaptive. An apparent exception that may prove the rule is the American Copper, which appears to be combative and site-faithful. It inhabits upland fields that are ephemeral overall, but favors edaphically dry microsites that are likely persistent.*

## 10.7 GENERAL CONCLUSION

The bulk of my new definitive work is outright natural history, essentially semiquantitative description and modeling of behavior, in the introduction that is Chapter 1, and in the accounts of species in Chapters 3–7. At the level of physiological ecology, the inquiry into underlying roles of flight, vision, olfaction, and hearing in Chapter 2 is partly speculative, but firmly inspired by novel observations. At the level of population ecology, the measures of habitat size in Chapter 8 and residency in Chapter 9 are direct and novel, and add to the more detailed work done by others on different species. The discussions in Chapters 8 and 9 are highly speculative, but lead to the testable insights here in Chapter 10, which return the call for new observations of natural history.

The cues and rules that I have tested with models and data have direct adaptive value in mate-finding. This could be viewed by a mechanist as explaining away any need for adaptive strategic interpretation. Alternatively, the cues and rules could be viewed as a simple mechanism for achieving a result that is practically indistinguishable from the strategic one. Equivalently, one could ask whether butterflies have been underrated, or whether vertebrates have been overrated. I think the answer is a little of both. I can safely conclude that mechanistic arguments, cleverly applied, can mimic an astonishing array of behavior that the recent tradition of behavioral ecology has explained with strategic arguments, equally cleverly applied. I honestly think that much of the strategic thinking in behavioral ecology is overextended and underchallenged because of the beauty of its pure theory, the charm of its descriptive language, and the number, fame, and ubiquity of the people using that language. It has been enlightening, hard work, and great fun trying to become one of those people. But so far, my success has been ambiguous, and that may be a *good thing*.

APPENDIX A

# Taxonomy

This study has taken long enough that the consensus names of the major players have changed. During this period, the official guardians of taxonomy have argued over whether the common names or the "scientific" Latin binomials were the more stable and less confusing (Miller and Brown 1981, 1983, Ehrlich and Murphy 1982, 1983a, 1983b, Murphy and Ehrlich 1983, 1984, Pyle 1984). So, to be on the safe side, I present names and aliases of my most intensively studied butterflies in Table A.1. Other species are listed in Table A.2 with common and Latin names according to Glassberg 2017 and others (for updates, please see http://www.naba.org/images/index.html ).

Basically, I follow Cech and Tudor 2005, Opler and Krizek 1984, and Glassberg 2017, shortening the common name for euphony. I do not mean to imply that other species or subspecies that share the shortened name will share the behavior that I have described. Indeed, for the Fritillary, Viceroy, and Pearly Eye, there is considerable behavioral variation within my study among individuals of the same species. I prefer Shapiro's 1974 discussion of the status of *Coenonympha tullia inornata* to Opler and Krizek's. My population is clearly a single bivoltine species, rather than two temporally separated univoltine species (Wiernasz 1989 agrees). Indeed in artificial culture I have raised the August brood from the June.

Here are some synonyms for European readers. The Ringlet is a different species from that of the same name in Europe, where what I call the Ringlet is called the Large Heath (*Coenonympha tullia*). *Speyeria* would probably be called *Argynnis* by someone who believes in Holarctic genera (Hovanitz 1962 classically, but see Simonsen et al. 2006 for a modern justification supported by a molecular phylogeny). *Enodia*, the Pearly Eye, shares morphological, behavioral, and ecological similarities with *Pararge aegeria*, the Speckled Wood (personal observation in southern England, northern Spain, and southern France).

Among the large Fritillaries, the Great Spangled *Speyeria cybele* was common during the study, the Aphrodite *S. aphrodite* was occasional, and the Atlantis *S. atlantis* was very rare. All data cited for marked individuals come only from the Great Spangled (e.g., Figures 5.1–5.3, 5.7, and 9.1–9.4). Compiled observations of free-flying large Fritillaries (e.g., Figures 5.4 and 5.8) include unmarked

TABLE A.1. Taxonomy of Intensively Studied Species.

| Authority | (Common or Inornate) Ringlet | (Northern) Eyed Brown | (Great Spangled) Fritillary | Viceroy | (Northern) Pearly Eye |
|---|---|---|---|---|---|
| Miller 1992 alternative common names | Large Heath; Plain Yellow Quaker | Grass Nymph; Ten-spotted Quaker | Cybele | Mimic | |
| Glassberg 2017 | *Coenonympha tullia* | *Satyrodes eurydice* | *Speyeria cybele* | *Limenitis archippus* | *Enodia portlandia* |
| Cech & Tudor 2005 | *Coenonympha inornata* | *Satyrodes eurydice* | *Speyeria cybele* | *Limenitis archippus* | *Enodia anthedon* |
| Miller 1992 | *Coenonympha tullia inornata* | *Satyrodes eurydice* | *Speyeria cybele* | *Basilarchia archippus* | *Enodia anthedon* |
| Opler & Krizek 1984 | *Coenonympha tullia* | *Satyrodes eurydice* | *Speyeria cybele* | *Limenitis archippus* | *Enodia anthedon* |
| Shapiro 1974 | *Coenonympha tullia inornata* | *Lethe eurydice* | *Speyeria cybele* | *Limenitis archippus* | *Lethe anthedon* |
| Klots 1951 | *Coenonympha tullia inornata* | *Lethe eurydice Lethe canthus* | *Speyeria cybele* | *Limenitis archippus* | *Lethe portlandia* |

individuals and marked individuals whose numbers could not be read; accordingly, most of them are of the Great Spangled, but a few may be of the Aphrodite. While this detracts from the ritual purity of such compilations, the substantive effect is likely small since the behavior of marked Aphrodites seemed not to differ from that of marked Great Spangleds.

In addition to the five intensively studied species, many others were studied less systematically. Data and observations from them are cited where they gave general insights (especially flight behavior and vision in Chapter 2), or where they extended the insights gained from the five main subjects (e.g., Chapter 10). The species observed during the study are listed in Table A.2, and most of them are pictured in Figure 1.1. All Tiger Swallowtails closely observed were Canadian *Papilio canadensis*, rather than Eastern *P. glaucus*, fidé Hagen et al. (1991). Pearl Crescents *Phyciodes tharos* were common, but some individuals were clearly identifiable as Northern Crescent *P. selenis*, and a few were intermediate; the study site, Whitehall, NY, is at the tentative border between the two forms (Williams 2009).

TABLE A.2. Butterflies (exclusive of Skippers) of the Study Area.

| Common Name | Latin Name | Presence during study |
|---|---|---|
| Black Swallowtail | *Papilio polyxenes* | Variable presence |
| Tiger Swallowtail | *Papilio canadensis* | Common but fleeting |
| Cabbage (Small) White | *Pieris rapae* | Variable abundance |
| Clouded Sulphur | *Colias philodice* | Common |
| Orange (Alfalfa) Sulphur | *Colias eurytheme* | Episodic |
| American Copper | *Lycaena phlaeas* | Sparse but frequent |
| Bronze Copper | *Lycaena hyllus* | Patchy, with variable abundance |
| Harvester | *Feniseca tarquinius* | Rarely encountered |
| Coral Hairstreak | *Satyrium titus* | Episodic |
| Banded Hairstreak | *Satyrium calanus* | Common but fleeting |
| Hickory Hairstreak | *Satyrium caryaevorum* | Episodic |
| Edwards' Hairstreak | *Satyrium edwardsii* | Episodic |
| Striped Hairstreak | *Satyrium liparops* | Common but fleeting |
| Acadian Hairstreak | *Satyrium acadica* | Rare |
| Gray Hairstreak | *Strymon melinus* | Rare |
| Eastern Tailed Blue | *Everes comyntas* | Common |
| Spring Azure | *Celastrina ladon* | Common |
| Great Spangled Fritillary | *Speyeria cybele* | Common |
| Aphrodite Fritillary | *Speyeria aphrodite* | Sparse |
| Atlantis Fritillary | *Speyeria atlantis* | Rare |
| Meadow Fritillary | *Boloria bellona* | Episodic |
| Pearl Crescent | *Phyciodes tharos* | Common, also *P. selenis* |
| Baltimore | *Euphydryas phaeton* | Highly Variable |
| Question Mark | *Polygonia interrogationis* | Sparse but dependable |
| Comma (Hop Merchant) | *Polygonia comma* | Rare |
| Compton Tortoise Shell | *Nymphalis vau-album* | Sparse |
| Mourning Cloak | *Nymphalis antiopa* | Sparse but dependable |
| Milbert's Tortoise Shell | *Nymphalis milbertii* | Rare |
| Red Admiral | *Vanessa atalanta* | Common but scattered |
| American Painted Lady | *Vanessa virginiensis* | Sparse but frequent |
| Painted Lady (Cosmopolite) | *Vanessa cardui* | Sparse but frequent |
| White Admiral | *Limenitis arthemis* | RsP = most common morph; but |
| Red-spotted Purple | *Limenitis arthemis astyanax* | WA = frequent; hybrids are found. |
| Viceroy | *Limenitis archippus* | Local |
| Eyed Brown | *Satyrodes eurydice* | Episodic, in general decline |
| Appalachian Eyed Brown | *Satyrodes appalachia* | Rare |
| Northern Pearly Eye | *Enodia anthedon* | Common but patchy |
| Common Wood Nymph | *Cercyonis pegala* | Common |
| Little Wood Satyr | *Megisto cymela* | Common |
| Ringlet | *Coenonympha tullia* | Common |
| Monarch | *Danaus plexippus* | Variable |

APPENDIX B

# More Natural History

Here are details of natural history that are peripheral to the central discussion of Chapters 3–9. Notes on my study versus other places and other species give some idea of how general my results are. Following the prose are conventional presentations of data gathered in like fashion for all species: local persistence data (Table B.1) and ovipositions by females (Table B.2).

### B.1 MY STUDY VERSUS OTHER PLACES AND OTHER SPECIES

My observations for this study were mostly of five species in one locality, a post-agricultural landscape in Whitehall, NY, on the Vermont border just southwest of the southern end of Lake Champlain. Here are notes that compare my study with observations of other species and in other places.

#### *B.1.1 Ringlet*

My observations of the Ringlet in my study area are similar to those I have made elsewhere: large fields in New England and the Eastern Midwest, shortgrass prairies in the Western Midwest, montane meadows of the Rockies and Sierra Nevada, as well as in open meadows in Asturias and Provence. I see similar behavior, suggestive of random wandering and casual encounters, in the Cabbage White, Sulphurs, Wood Nymph, and Little Wood Satyr in like environments, though they transgress field boundaries more readily than the Ringlet, and the Little Wood Satyr explicitly favors woodland edges and glades.

#### *B.1.2 Eyed Brown*

The signature behavior of the Eyed Brown in my study was wandering that seemed to remain local by following depressed paths in the vegetation, and by occasional returns to fixed perches from short distances. I have seen like behavior in Coppers and Hairstreaks, respectively, centered in wet or dry meadows versus forest glades and clearings, where the "depressed paths" are invaginations in any of the three dimensions of the wooded landscape. Swallowtails are the early canonical hilltoppers, but on those hilltops they follow depressed pathways in the canopy, as I have observed from towers and cliff-tops in New Jersey, New York, Illinois, and Minnesota.

### *B.1.3 Fritillary*

The local and interactive behavior that I have observed in the Great Spangled Fritillary is similar to what I have observed in other Fritillaries (all genera of both large and small species, in the United States and Western Europe). Repetitive circuits are characteristic of the traditional lore of large Fritillaries, but I could not have observed them in patchy New York meadows without marking individuals. However, when I marked Great Spangled Fritillaries in the more open montane meadows of western Montana, I did not see evidence of closed circuits, albeit in a short study. It is possible that closed circuits are favored by the scale of patchiness in human agriculture and gardening.

### *B.1.4 Viceroy*

I routinely see the rudiments of Viceroy behavior, specifically seemingly territorial behavior between males, and brief courtship of females, by a wide variety of nymphalines throughout the United States. This is most obvious along mowed paths in lush meadows or along streams in both dry and wet settings for the Viceroy and its congeners, the other Admirals, Buckeyes, Emperors, Painted Ladies, and Red Admiral. I observe likewise in and between woodland glades and forest paths for Angle Wings, Tortoiseshells, and the Mourning Cloak.

### *B.1.5 Pearly Eye*

The interactive behavior of male Pearly Eyes is qualitatively like that of the Viceroy and others above, but characterized by greater sensitivity, activity, intensity, and persistence—which might all be automatically enhanced by their spatial and temporal aggregation, and vice versa.

### B.1.6 Other Species

I wish I knew more about the Baltimore, Blues, Pearl Crescent, Skippers, and interactive behavior of everyone at mud puddles and on hilltops.

## B.2 LOCAL PERSISTENCE (TABLE B.1)

Table B1 presents the raw data on which calculations and discussion of local persistence are based (Section 9.3.3, Table 9.2, and Figure 9.2). Table B.1 records the number of individuals marked on Day #1 as $S_0$, and the number of marked individuals known to be alive on Day #X as $S_X$. On the first day everyone is assumed to be a vagrant, and an estimate of daily vagrant "survival" (actually of local persistence) is $V = S_1/S_0$. "Survival" of the second day is estimated as $D = S_2/S_1$. Figure 9.2 suggests that most individuals transition to at least partial residency within the first 2 days. So an estimate of resident "survival" (which includes some losses to persistent vagrancy) is $R = \Sigma S_3$ to $S_{last}/\Sigma S_2$ to $S_{next\text{-}to\text{-}last}$. Assuming that losses of individuals are independent, the three parameters V, D, and R are statistically independent, and allow estimation of the daily rate of conversion between initial and eventual loss, i.e., $C = (V-D) / (V-R)$. Results are presented in Table 9.2.

## B.3 OVIPOSITION (TABLE B.2)

Table B.2 records observed instances of oviposition, each entry being the number of instances of a cluster of eggs of the given size being attached to the given substrate.

The observations for the Ringlet were mostly from confined females in a 1 × 1 × 0.2 m outdoor arena roofed with cheesecloth, plus a few fortuitous observations of unconfined females. To count as a "cluster," eggs had to be attached within about an egg's width of the nearest neighbor, as was the case for the clusters laid at one time by the unconfined females. The most common thin-bladed grass was poverty grass *Danthonia spicata*, on which I was able to rear Ringlets to adulthood.

The two observations for the Eyed Brown were of unconfined females. Both chose a wide-bladed yellow-green sedge *Scirpus*, and the three eggs were laid together beneath the blade at the peak where the blade bent over to hang down.

The observations for the Pearly Eye were of unconfined females, and the physical setting was in all cases like that of the Eyed Brown, on the underside of a wide blade at the peak where it turned down. The chosen species included

TABLE B.1. Locally Persisting (= "Surviving") Individuals versus Time since Marking

| Day # | 1 | 2 | 3 | 4 | 5 | 6 | 7 | 8 | 9 | 10 | 11 | 12 | 13 | 14 | 15 | 16 | 17 | 18 | 19 | 20 |
|---|---|---|---|---|---|---|---|---|---|---|---|---|---|---|---|---|---|---|---|---|
| Survivors xxxxx | $S_0$ | $S_1$ | $S_2$ | $S_3$ | $S_4$ | $S_5$ | $S_6$ | $S_7$ | $S_8$ | $S_8$ | $S_{10}$ | $S_{11}$ | $S_{12}$ | $S_{13}$ | $S_{14}$ | $S_{15}$ | $S_{16}$ | $S_{17}$ | $S_{18}$ | $S_{19}$ |
| Ringlet m | 711 | 309 | 205 | 138 | 84 | 46 | 32 | 18 | 9 | 4 | 2 | | | | | | | | | |
| Ringlet f | 461 | 183 | 114 | 74 | 53 | 38 | 19 | 12 | 8 | 2 | 1 | | | | | | | | | |
| Eyed Brown m | 104 | 74 | 64 | 55 | 46 | 40 | 38 | 34 | 30 | 24 | 22 | 20 | 15 | 11 | 11 | 8 | 7 | 6 | 4 | 3 |
| Eyed Brown f | 52 | 20 | 15 | 13 | 8 | 7 | 7 | 4 | 3 | 2 | 2 | 2 | 1 | 1 | 1 | | | | | |
| Fritillary m | 138 | 37 | 28 | 24 | 19 | 19 | 18 | 15 | 13 | 13 | 12 | 9 | 6 | 3 | 1 | 1 | 1 | 1 | | |
| Fritillary f | 67 | 6 | 4 | 4 | 3 | 3 | 3 | 3 | 2 | 2 | 2 | 1 | 1 | 1 | 1 | 1 | 1 | 1 | | |
| Viceroy m | 67 | 36 | 31 | 24 | 20 | 18 | 12 | 10 | 10 | 10 | 10 | 7 | 6 | 5 | 5 | 5 | 4 | 4 | 3 | 3 |
| Viceroy f | 34 | 2 | 2 | 1 | 1 | 1 | 1 | | | | | | | | | | | | | |
| Pearly Eye m | 660 | 305 | 261 | 214 | 189 | 170 | 147 | 117 | 99 | 85 | 72 | 61 | 56 | 50 | 48 | 43 | 36 | 34 | 31 | 28 |
| Pearly Eye f | 275 | 102 | 80 | 68 | 54 | 38 | 28 | 23 | 18 | 14 | 11 | 9 | 6 | 5 | 4 | 3 | 3 | 3 | 1 | |

V = Vagrant Survival = $S_1/S_0$
D = Day 2 Survival = $S_2/S_1$
R = Resident Survival = $\sum S_3$ to $S_{last} / \sum S_2$ to $S_{next\text{-}to\text{-}last}$
C = Probability of conversion from Vagrant to Resident Day 1 to Day2 = $(V - D)/(V - R)$.

TABLE B.2. Oviposition.

| Species | Substrate | Eggs/Cluster→ 1 | 2 | 3 | 4 | 5 | 6 | 7 | 8 |
|---|---|---|---|---|---|---|---|---|---|
| Ringlet | Green thin-bladed grass | 13 | 2 | 2 | | | | | |
| Ringlet | Dry thin-bladed grass | 11 | 6 | 2 | 1 | 1 | | | 1 |
| Ringlet | Cloth | 3 | | | | | | | |
| Eyed Brown | *Scirpus* | | | 2 | | | | | |
| Pearly Eye | Wide-bladed grass | 1 | | 2 | 3 | 2 | 5 | 3 | |

the grasses *Phleum*, *Agrostis*, *Muhlenbergia*, and an unidentified spring-flowering species, plus an unidentified sedge. In 10 instances, I observed an ovipositing female landing on the peak of a wide-bladed species, "testing" the underside with her abdomen, and flying on without laying an egg. I was able to raise larvae to the fourth instar on *Phleum* alone. Shapiro (1974) lists the grasses *Brachyelytrum* and *Muhlenbergia* as host species.

I have observed unconfined female Viceroys "inspecting" willows (*Salix nigra* and *Salix amygdaloides*) and aspens (*Populus tremuloides* and *Populus grandidentata*), fluttering near leaves, sometimes landing and tapping with their abdomens, but not laying eggs.

I have many observations of female Fritillaries fluttering down into and through vegetation that includes the widely documented host plant, violets (*Viola* spp.), but only one observation of closer inspection. A female *Speyeria aphrodite* fluttered near, but did not land on several plants: a mustard, bladder campion, blackberry, ash, sumac. She landed on *Hieracium* but immediately launched again. She landed on, and tapped with her abdomen, several dark, round leaved plants: plantain, basil, *Rumex acetosella*, and roseate goldenrod, but then moved on. Once on the ground, she gave the impression of an energetically walking insect, unencumbered by wings. These observations are consistent with the following hypothetical rules: visual signals induce approach to dark leaves, and landing on small round ones; hairy leaves are rejected by gross texture; and the rest are inspected for chemical cues.

APPENDIX C

# Machinery

Following are designs and specifications for idiosyncratic machinery used in this study: a clamp to hold butterflies while marking them, a timer to provide metronomic signals, a portable stroboscope, a simulator for compound-eye vision, an eye-portrait studio, a "butterfly rotisserie," an artificial horizon, and a baited olfactory lure.

## C.1 MARKING AND MAPPING GEAR

The clamp for marking butterflies is shown in Figure 1.6, diagrammed in Figure C.1A, and described in more detail by Horn (1976b). The clamp allows speedy and efficient marking at the site of capture by a single person. Marking is done by light taps of a permanent fiber-tipped pen (Sanford Sharpie at the suggestion of L. E. Gilbert). The transparent clip allows accurate linear measurements to be made on the butterfly. The clamp can even be used to capture torpid or placid butterflies directly, with less risk of damage than netting.

A great aid in mapping movements of a butterfly is the mini-table diagrammed in Figure C.1B. A fine-scale map is put on a standard clipboard, attached around the waist by a cord with an S-hook on one end, and suspended from the neck by another cord. When moving between mapping sites, the neck cord can be pulled through the clipboard and tied in a bow so that the clipboard remains comfortably next to the chest. The clipboard could also be a portable table for a light laptop computer or tablet.

## C.2 BEHAVIORAL TIMER

The older timer in Figure C.2A is of interest mainly as a historical artifact. Component values are: $B = 9V$; $S = SPST$; $E = 2K$ piezoelectric earphone; $Q1 = Q3 = 2N170$; $Q2 = Q4 = 2N1097$; $C1 =$ chosen for timing intervals; $C2 = 30$ mfd electrolytic; $C3 = 0.01$ mfd; $R1 = 100K$; $R2 = 2.2$ meg; $R3 = 10$ meg linear pot;

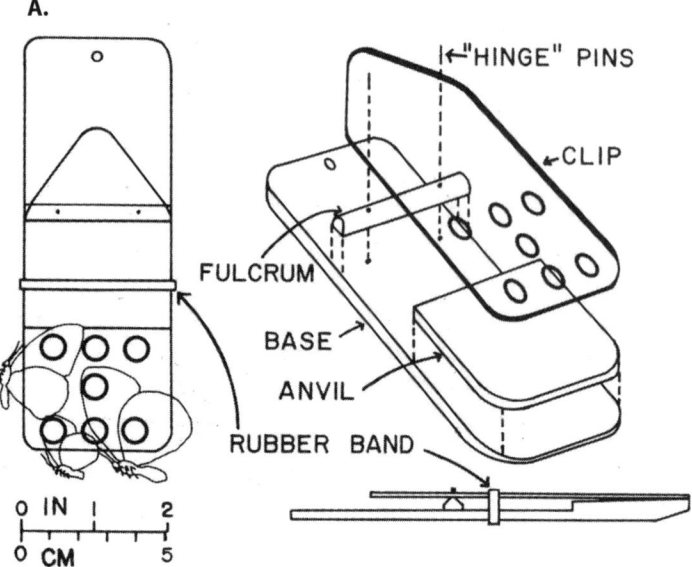

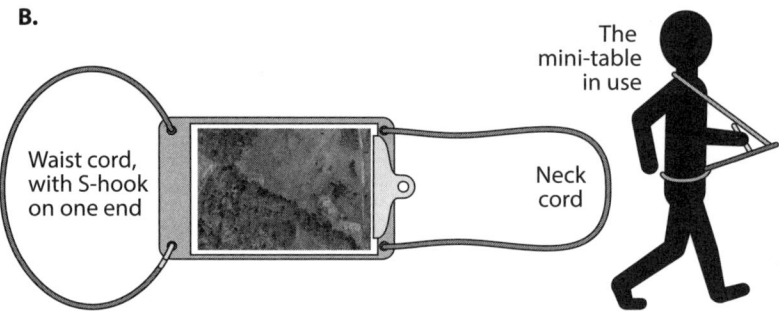

FIGURE C.1. A. Butterfly-marking clamp. The base and anvil are 1/8 inch Masonite. The clip is 1/16 inch clear acrylic plastic with 1/16 inch holes to clear the hinge-pins (#18 brads with the heads cut off), and is supported on a softwood fulcrum. Marking is done through one of seven 5/16 inch holes, drilled in the clip and chamfered from the top with a countersink. The rubber band can be adjusted to give sufficient pressure to hold a butterfly, but little enough pressure not to injure it. A hole in the top of the base allows the clamp to be carried on a belt-hook or neck-strap. The figure is from Horn (1976b). B. Mini-table. A clipboard holds a copy of a map with fine details of vegetation, on which movements of a butterfly are traced while following it in the field.

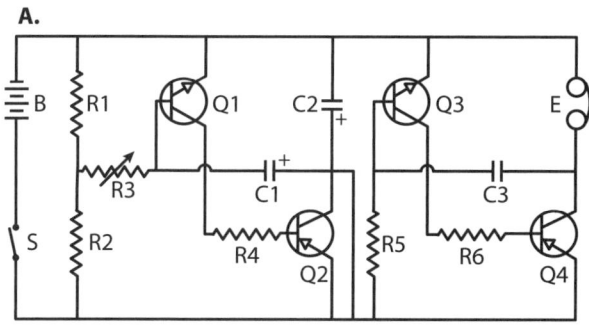

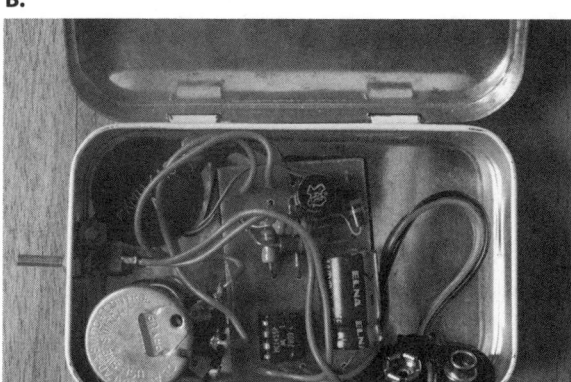

FIGURE C.2. A. Older version of the Behavioral Timer. Q3 and Q4 constitute a blocking oscillator, an amplifier with positive feedback through C3, that provides a tone to the earphone and serves as the load for the short-cycle oscillation of the timing oscillator driven by Q1 and Q2. See text for choice of component values. Figure from Horn (1966). B. Intermediate version of the Behavioral Timer. A single transistor is overdriven to oscillate as an audio signal through a piezoelectric transducer, and it is the load for the short-cycle oscillation of an IC 455 timer (the CMOS version of the IC 555). An obvious update would use an IC 556 dual timer or an integrated piezoelectric tone generator.

R4 = 1K; R5 = chosen to stabilize oscillation; R6 = 1K. For R5, try values between 300K and 1 meg; if no tone is heard, replace C3 with another of the same nominal value and vary R5 again. Available timing intervals vary from about 0.4 C1 (in mfd) to about 4.7 C1 as R3 is adjusted from 0 to 10 meg.

The intermediate timer in Figure C.2B uses a transistor to generate its tone, and an integrated circuit as the timer. It would be easy to use an IC 556 dual timer, inspired by Lancaster (1974), but the timer shown has been so robust, and holds its settings so accurately, that I have used it for decades, and never had to make another. Of course, were I starting this project now, I would simply find or write a timer app for a mobile device.

## C.3 PORTABLE STROBE FLASH

My inspiration for strobe photography of butterflies was Dalton (1975), but his rig was laboratory-bound. So rather than rapid firing of a single strobe flash, which requires high power and elaborate cooling for the short recycling times, I designed circuitry to fire five small inexpensive flashguns in sequence (Figure C.3).

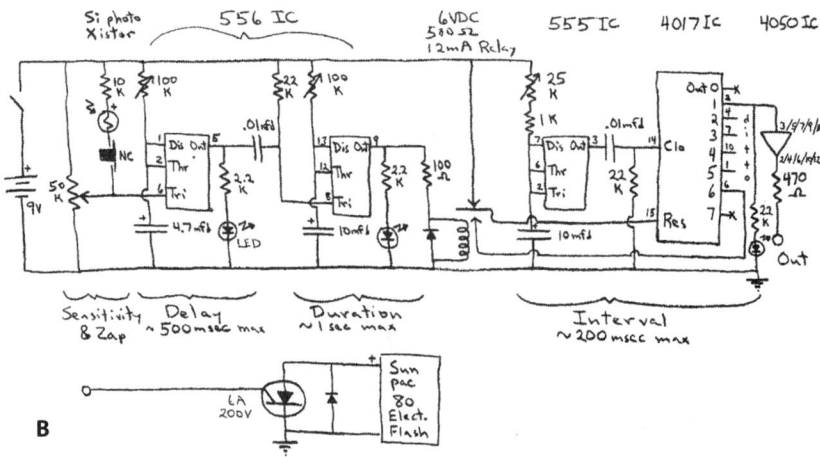

FIGURE C.3. A. Sequenced strobe flash, camera, and wooden pika. B. Circuit of the sequencer. See text for an overview of the design.

Technical help came from Lancaster (1974). The input trigger comes either from the camera, or from a photocell that is darkened when the butterfly breaks a rasterbeam of light as it flies between two parallel mirrors (after Dalton 1975). The trigger is conditioned to be bounceless by a Schmitt trigger made with half an IC 556 dual timer; the other half of the timer provides an adjustable delay (to allow the butterfly to fly clear of the mirrors). An IC 555 timer provides pulses that step an IC 4017 decade counter that ultimately triggers five flashguns in sequence.

That's the theory. In practice, capacitive coupling of internal signals, and voltage spikes as the flashguns fired, caused spurious behavior. Trial and error led to a patchwork that solved these problems. Each flashgun is triggered by a GE C122D SCR thyristor at its base. An IC 4050 Hex noninverting buffer in the sequencing box makes all circuitry less sensitive to capacitive coupling in the cable connecting the box to the flashguns. A physical relay between the delay and the pulsing circuits improved reliability, and did so surprisingly better than another IC Schmitt trigger or IC buffer. Obvious improvements would be to mount the thyristors and flashguns directly on the box holding the main circuitry and to separate the short, doubly shielded cables connecting them to the signal-output buffer. Wireless synchronization with the camera would eliminate the need for any other long cable.

## C.4 SIMULATING A COMPOUND EYE VIEW

Figure 2.11B was generated from Figure 2.11A by using *Adobe Photoshop Elements 6* Stained Glass Artistic Filter and blackening the surrounding matrix. A regular array of hexagonal ommatidial fields would be more elegant, and would add more difficulty without any more interpretive power. I had previously made more elegant pictures by rear-projecting poorly focused slides onto a screen made by sandwiching vellum tracing paper between two precisely oriented fluorescent light diffusers, each a hexagonal array of plano-convex lenses with the convex side to the paper. This was inspired by Gould's (1979) ingenious similar technique, making a screen of two pieces of thin translucent white plastic, separated by a hexagonal array of short pieces of soda straws. Photoshop is much easier. Figure 2.11 highlights the accuracy with which static details of contrasting edges and objects can be discerned and located within the visual field. However, it does not convey the potentially large amount of information about texture, pattern, and/or object distance and movement that could be encoded in flickers as the butterfly or object moves or as the butterfly performs small scanning movements of its head.

Figure C.4 shows the design of a simple viewer that gives a crude but effective idea of the environmental and social information that can be encoded in both static patterns and flickers. I have not given detailed measurements and parameters

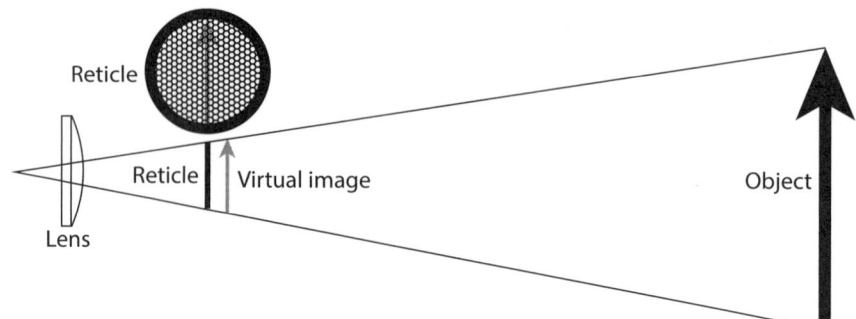

FIGURE C.4. Optical design to simulate a partial field of view of a compound eye (a solid field of about 30° with each ommatidium viewing about 1.5° and overlapping its next neighbor). The focal length of the lens is about 3 centimeters. The reticle is a hexagonal array of about 350 pinholes in heavy aluminum foil that has been lamp-blackened; it is placed somewhat short of the focal length from the lens. The image is seen at about a focal length from the lens. The intent is not to show what a butterfly sees, but rather to remove the details that we readily see and/or interpolate, and thus to emphasize potential that we might not notice, but that a butterfly might. See text (Sections 2.3.3 and C.4) for details and alternative constructions, and for many more cautions about interpretation.

because the result can vary with the numerical aperture of the lens, the viewing distance from the eye, and details of the viewer's own eye, including pupil size in response to lighting conditions. Accordingly, I have made several such apparatus with different parameters that work well, and many more with like parameters that work poorly. One has to tinker and try, measuring acceptance angles and overlaps using a pair of single LED flashlights at an appreciable distance as targets; when properly adjusted to the viewer's eye, views of the natural environment are literally enlightening.

The reticle is made by punching a hexagonal array of pinholes in a piece of heavy aluminum foil, backed up by a piece of firm cardboard, and then carefully blackening both sides of the aluminum by holding it over a candle. The angle between the central axes of adjacent ommatidial views is relatively easy to adjust by the ratio between the distance between holes in the reticle and the focal length of the lens (though you will see that it is also affected by the comfortable viewing distance). The size of each hole and the pattern of protrusion of the aluminum displaced in making it, as well as the focal length and viewing distance of the lens, all affect the acceptance angle of each simulated ommatidium, and hence the degree of overlap in the fields of view of adjacent ommatidia. As fussy and tedious as the process is, I have had better success with it than with experiments that scale up the apparatus and drill holes in thicker aluminum, or that use patterns printed on acetate sheets.

FIGURE C.5. Eye-portrait studio. The camera is a Canon EOS Rebel T3 with a 40 mm lens and 65 mm of extension tubes, giving an image 1.6 times the size of the object. The butterfly is held in clamp like that of Figure C.1, which is binder-clipped to an L-shaped bracket, binder-clipped in turn to the base. Critical focus is achieved by gently nudging the bracket. Main light is a Bescor 96 LED light bank, with an adjustable flat aluminum shadow relief, optionally highlighted with a Mighty Bright LED reading light. An 18% reflective gray background facilitates post-processing for optimal white balance and exposure. Inset is the initial test with a female Cabbage White.

## C.5 EYE-PORTRAIT STUDIO

Figure C.5 shows the Eye-Portrait Studio, which holds and lights a butterfly for macrophotography. Details are best explained by referring back and forth between Figure C.5 and its caption.

## C.6 BUTTERFLY ROTISSERIE

The Butterfly Rotisserie is used to present a rotating visual lure to butterflies in the hope of eliciting an approach or retreat (after Magnus 1958). Details are best explained by referring back and forth between Figure C.6 and its caption.

## C.7 ARTIFICIAL HORIZON

Artificial low spots in the horizon were engineered by placing 3×4-foot mirrors on tripods so that they could be tilted to reflect the sky into an otherwise vegetated

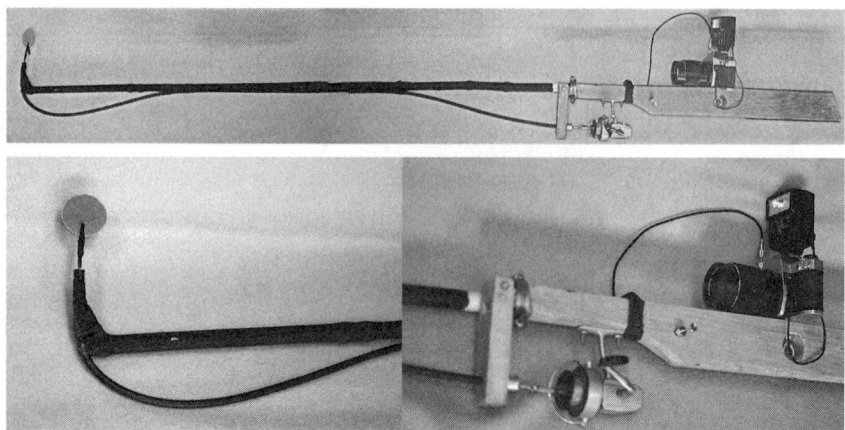

FIGURE C.6. Butterfly Rotisserie. The total apparatus is about 2.2 meters long. The lure, which can be varied in size, shape, and color, is held by an electrical alligator clip, crimped to one end of an archaic automobile speedometer cable. The other end of the cable is attached to the bail of a fishing rod spinning reel with another alligator clip. The camera is tripped by a cable release, mounted to be pressed by the trigger finger of the right hand, which cradles the stock while the left hand rotates the lure (see Figure 7.8). Flat black paint moderates flare from the flashgun.

region of the field of view from a butterfly's perch (see Figure C.7). The artificial horizon could be "turned off" by tilting the mirrors forward to reflect vegetation. Confirmation that the mirrors worked was immediate; as I set up the first one, there were multiple satisfying "plocks" as biting deer flies (Tabanidae) flew into them at full speed and fell stunned to the ground. The artificial horizon was used experimentally to confirm orientation cues used by male Viceroys (Chapter 6). I also observed instances in which other species bumped into the mirrors as though they were flying toward an opening in the vegetation—specifically, the Ringlet, the Eyed Brown, and the Fritillary.

## C.8 BAITED OLFACTORY LURE

Figure C.8 is shown just for fun. It documents a successful initial experiment to attract Pearly Eyes to an artificial ferment. Much more effective was a traditional brew of brown sugar in water, with baker's yeast, a pinch of flour, and crushed raspberries *Rubus occidentalis*, being maximally attractive between days 5 and 8 after initial mixing, i.e., when a slight vinegary smell was first detected, but the mixture still foamed readily. The attractive life of the ferment could be extended for several days by adding more brown sugar.

FIGURE C.7. Artificial horizon. The mirrors are aluminized Mylar "space-blankets," attached to 3×4-foot Masonite panels with wallpaper paste, edged with duct tape. Tripods are poles and lashings remembered from youth as a Boy Scout (e.g., Kephart 1921). The side mirrors reflect sky as an artificial visual horizon; the middle mirror is tilted to reflect vegetation.

FIGURE C.8. Baited olfactory lure. This early experiment, filtering beer through a wick of basswood (*Tilia americana*) bark, successfully attracted Pearly Eyes. The home-brew described in the text was far more effective, renewed once or twice a day on cellulose sponges (see Figure 7.1B).

# APPENDIX D
# NetLogo Programs

### D.1 RINGLETFANCYHOTCOLDDEC18.NLOGO

This NetLogo program simulates random wanderings with an adjustable bias in the forward direction. On a 31 × 31 grid, with boundary movement either reflecting or wrapping to appear at the opposite side, a track is recorded on the 961 patches, showing which are visited, and how many times they are visited.

This program was used to generate Figures 3.11 and 8.2, along with the data for Figures 3.10A–B. A simpler version was used to generate Figures 3.9B–C and 3.13. A more complex version was used to generate Figures 8.3 and 8.5, along with the data for Figures 8.4 and 8.6.

Figure D.1 shows the program's screen for input of parameters, control of the run, and output of graphs and numbers. See the legend of Figure D.1 for description of its features.

*The program itself:*

```
globals [steps visits distmove origx origy visits0
visits1 visits2 visits3
  {sl11}visits4 visits5 visits6 visits7 visits8 visits9
  visits10]

to Setup
   clear-all
   ask patches [set pcolor 9.9]
   create-turtles 1
     ask turtles [set shape "butterfly"
        setxy 0 0
        set color 0
        set size 6
        set heading 0
        pen-down
        set pen-size 1]
```

# NETLOGO PROGRAMS

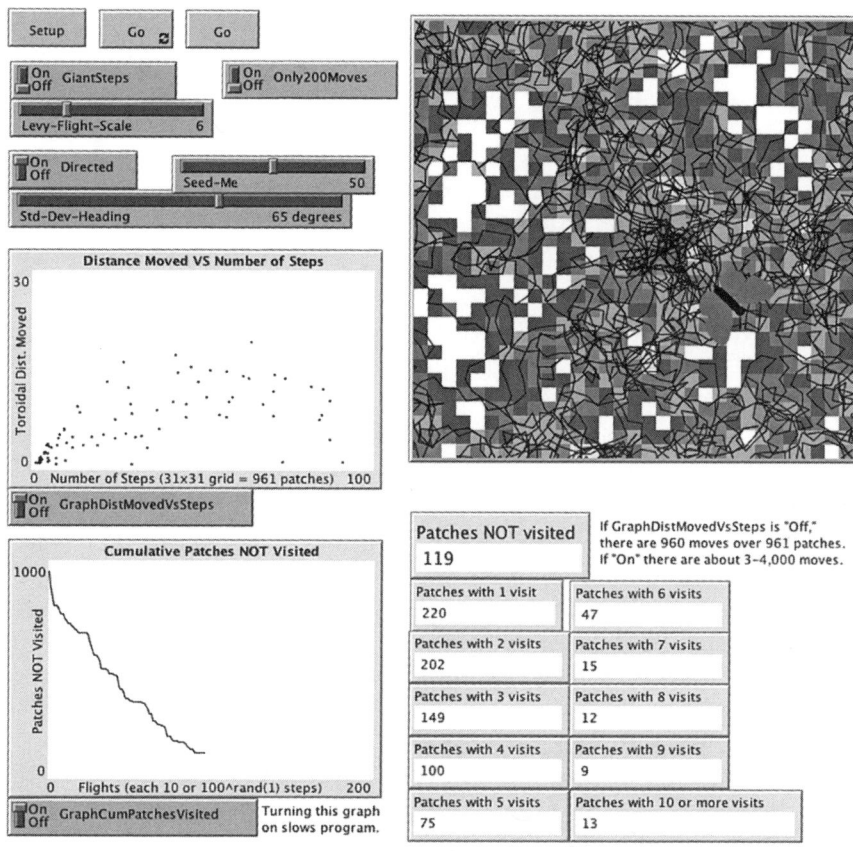

FIGURE D.1 Interface screen for NetLogo program FancyHotColdDec18.nlogo. Buttons allow Lévy flight (GiantSteps) or forward bias for a random walk (Directed) to be turned on or off, and their parameters to be adjusted. The Seed-Me slider provides the opportunity to set a particular seed to the random number generator, which is useful for repeating a particular run to optimize the rhetoric of an example. The numeric output should be self-explanatory, and can be exported as comma-separated value files for analysis in other programs.

```
set steps 0
    set visits0 0
    set visits1 0
    set visits2 0
    set visits3 0
    set visits4 0
    set visits5 0
```

```
          set visits6 0
          set visits7 0
          set visits8 0
          set visits9 0
            set visits10 0
      setup-plot
      random-seed Seed-Me
      reset-ticks
   end

   to Go
      set steps 10
      if Only200Moves [set steps 1]
      if GraphDistMovedVsSteps [set steps 100 ^ random-
      float 1]
      ask turtles [set origx xcor set origy ycor]
      repeat steps [move]
      ask turtles [set distmove distancexy origx origy]
      if GraphDistMovedVsSteps [set-current-plot
      "Distance Moved VS Number of
         Steps"
         plotxy steps distmove]
      if GraphCumPatchesVisited [ask patches [set
      visits count patches with
         [pcolor = 9.9]]
            set-current-plot "Cumulative Patches NOT
            Visited"
            plotxy ticks visits]
      tick
      if ticks = 97 [stop]
      set visits0 count patches with [pcolor = 9.9]
      set visits1 count patches with [pcolor = 105]
      set visits2 count patches with [pcolor = 95]
      set visits3 count patches with [pcolor = 85]
      set visits4 count patches with [pcolor = 75]
      set visits5 count patches with [pcolor = 65]
      set visits6 count patches with [pcolor = 55]
      set visits7 count patches with [pcolor = 45]
      set visits8 count patches with [pcolor = 35]
```

```
        set visits9 count patches with [pcolor = 25]
        set visits10 count patches with [pcolor = 15]
end

to move
     ask turtle 0 [
     ifelse Directed [set heading heading + random-
     normal 0 Std-Dev-Heading]
          [set heading heading + random 360]
     ifelse GiantSteps [fd Levy-Flight-Scale * random-
     float 1 ^ 2]
          [fd 1]
     ask patch-here
          [ifelse pcolor = 9.9 [set pcolor 105]
          [set pcolor pcolor—10]
          if pcolor = 5 [set pcolor 15]]]
end

to setup-plot
     set-current-plot "Distance Moved VS Number of
     Steps"
     set-plot-x-range 0 100
     set-plot-y-range 0 30
     set-plot-pen-mode 2
     set-current-plot "Cumulative Patches NOT Visited"
     set-plot-x-range 0 200
     set-plot-y-range 0 1000
     set-plot-pen-mode 0
end

;; So here is a way to make pseudo-Levy flights and
still avoid edges.
;; Set up a separate on-the-edge procedure. Then
repeat to a random
;; integer (with a slider-set upper limit, say 1+4):
;; If on-the-edge: exit to general move and do edge.
;; OR maybe don't exit; just run out the pseudo-Levy
distance with the
;; edge corrections.
```

## D.2. FANCYNEXTKISS.NLOGO

This NetLogo Program starts with two butterflies in random positions and then moves them with a forward-biased random walk, and counts the number of moves before they meet. Alternatively, one can move while the other stays put.

This program generated the data for Figure 3.13. I also modified the program slightly and changed the graphics to represent an ornithologist in a canoe and an Ivory-billed Woodpecker *Campephilus principalis*. Given the additional parameters of the bird-human interaction, the ornithologist would do best to stay put.

## D.3. PEARLYEYEMOVIETRAIL.NLOGO

This NetLogo program starts with a model male Pearly Eye on the trunk of a tree, and a female (or another male) at a random position in the air. Both then begin a forward-biased random walk with an adjustable bias. The formerly perched male returns more directly to its perch if it travels a set distance from that perch. If that male comes within a set distance of the other butterfly, it turns toward it before continuing with its forward-biased random walk. If the formerly moving butterfly is a male, it continues with the same behavior as the other male, namely turning toward it before continuing its movement. If the formerly moving butterfly is a female, it either ignores the male and continues moving, or it avoids by turning away and continuing to move, maintaining its forward-biased random walk.

This program was used to generate Figures 7.9E–F.

Figure D.2 shows the program's screen for input of parameters, control of the run, and output of graphs and numbers. See the legend of Figure D.2 for description of its features.

*The program itself:*

```
;; PearlyEyeMovie.nlogo =
;; Two males OR a male and a female, who wander in a correlated random walk
;; with forward bias from a normal distribution. When they encounter within
;; a given distance, they turn toward (males) or away from (female) the
;; other,and then continue to choose the correlated random direction of the
;; next step.
```

# NETLOGO PROGRAMS

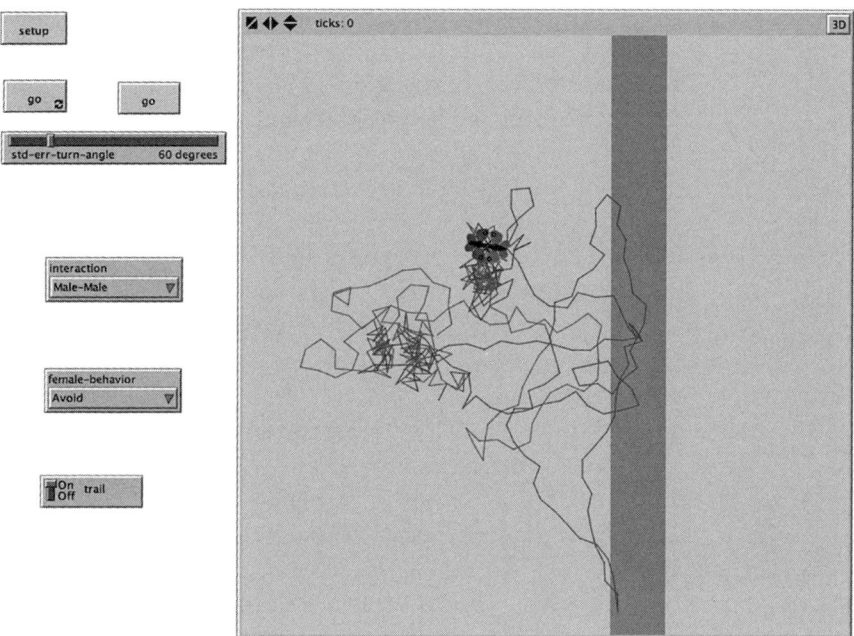

FIGURE D.2 Interface screen for Netlogo program PearlyEyeMovieTrail.nlogo. The action of the model butterflies is described in the text (Section D.3), and interpreted in Section 7.4. A slider (std-err-turn-angle) allows setting the forward bias in otherwise random movements. The weaker the forward bias, the twistier the path apart from close interaction, and the more readily the interactions are broken. Interactions can be toggled between Male-Male and Male-Female. Female behavior can be either to ignore a nearby male or to actively avoid it by initially turning away. The trail of the individuals can be recorded, as in the figure, or that recording can be turned off to simulate a live video. NetLogo provides an adjustment for the speed of the simulation, which can be adjusted to give a remarkably realistic cartoon of the behavior of real Pearly Eyes.

```
breed [butterflys butterfly]

to setup
    _ _ clear-all-and-reset-ticks
    ask patches [set pcolor 57]
    ask patches with [3 < pxcor and pxcor < 7] [set
    pcolor gray]
    set-default-shape butterflys "butterfly"
    create-butterflys 3
        ask butterfly 0 [set color 34 setxy -6 0]
```

```
          ask butterfly 1 [set color 33 setxy 6 0]
          ask butterfly 2 [setxy 5 0 hide-turtle]
          ask butterflys [set size 2]
          ask butterfly 1 [move-to butterfly 2]

   end

   to go
       ask turtles [ifelse trail [pd] [pu] ]
       ask butterfly 0 [set heading heading + random-
       normal 0 std-err-turn-angle
           fd 1]
       ask butterfly 1 [set heading heading + random-
       normal 0 std-err-turn-angle
           fd 1]
       ask butterfly 1 [let dist-from-perch distance
       butterfly 2
           if dist-from-perch > 15 [repeat 20 [ask
           butterfly 1 [face butterfly 2]
           set heading heading + random-normal 0 30
           fd 1]wait .2 + random .6]]
       ask butterfly 1 [let separation distance
       butterfly 0
       if separation < 5 [ask butterfly 1 [face
       butterfly 0]
           if interaction = "Male-Male" [ask butterfly 0
           [face butterfly 1]]
           if female-behavior = "Avoid" [avoid]]]
   end

   to avoid
           ask butterfly 0 [face butterfly 1]
           if interaction = "Male-Female" [ask butterfly
           0 [set heading heading + {sl6}
               180]]
   end
```

# Notes

PREFACE

1. I was devastated when we moved in 1953 from Augusta, where natural land was just down the street, to Cambridge, where natural land was at least a bus or subway ride away. Nevertheless, there were still more insects and greater variety in the vacant lots and unkempt borders of the Cambridge of the 1950s than there were in the more superficially rural Princeton, New Jersey, of 1966, which was dominated by chemically maintained "industrial" lawns (a term coined by Bormann et al. 1993). (0.1)

2. I use the names given in Glassberg (2017) or Cech and Tudor (2005), but with the common name shortened for euphony. See Appendix A for details and synonymy, including European analogs. (0.2)

3. Two such species are the American Copper *Lycaena phloeas americana* and the Bronze Copper *Lycaena thoe*, both of which seem interestingly territorial. The former is too small to withstand handling, and the latter's flight muscles are so powerful relative to the strength of its wings that it is often injured in the marking clip. (0.3)

4. Speaking of libraries reminds me that some books had such a great effect on my thinking that they and their authors deserve special acknowledgment: Scudder 1895 *Frail Children of the Air*, Ford 1945 *The New Naturalist: Butterflies*, Dalton 1975 *Borne on the Wind*, Heinrich 1979 *Bumblebee Economics*, Opler and Krizek 1984 *Butterflies East of the Great Plains*, Vane-Wright and Ackery, eds. 1984 *The Biology of Butterflies*, Douglas 1986 *The Lives of Butterflies*, Scott 1986 *The Butterflies of North America*, Brackenbury 1992 *Insects in Flight*, Ehrlich and Hanski, eds. 2004 *On the Wings of Checkerspots*, Prete, ed. 2004 *Complex Worlds from Simpler Nervous Systems*, Cech and Tudor 2005 *Butterflies of the East Coast*, Seeley 2010 *Honeybee Democracy*, Cronin et al. 2014 *Visual Ecology*, Agrawal 2017 *Monarchs and Milkweed*. (0.5)

CHAPTER 1: INTRODUCTION

1. Scott's (1974, 1986) formal terms for the behavior of my species were not current when I began my intensive fieldwork in 1974, but I had in hand earlier descriptions for my species as summarized by Klots (1951), and an advance manuscript of Shapiro (1974). (1.2)

2. "I'm not complaining, but There It Is."—Eeyore in Chapter 6 of Milne 1926. (1.2)

3. A charming, but untenable, alternative is that Hairstreaks have discovered a macroscopic version of quantum mechanical tunneling (Devault 1980). (1.3.1)

4. Scott (1986) characterizes the males of my species as: Ringlet patrols all day; Eyed Brown patrols with occasional perching; Fritillary patrols; Viceroy perches and patrols; and Pearly Eye perches. Quantitative data on day-to-day ranging can be sorted unambiguously

into categories that match Scott's broad characterizations, and my added details (Section 9.2). (1.4)

5. This description of the study site applies to the years of my intensive fieldwork (1973–1993). The former "dry upland fields" are now second-growth woodland, "woodland edges" of the former "wet meadows" are now densely shrubby, and central areas for the former meadows have been converted to standing water by Beaver *Castor canadensis* dams. The only remaining "woodland edges" are along bordering country roads. The lovely cabins of Figure 0.2 have succumbed to vandalism and weather. (1.6.1)

6. Cf. Winnie-the-Pooh and his plans for a Heffalump trap, Milne 1926. (1.6.2)

7. My current favorite is the *Papilio* by Pentax. Early in my fieldwork, close-focusing binoculars were not available, so I butchered a monocular, installed a shim to extend its optical length, and put it back together with duct tape. (1.6.2)

8. The grid was initially drawn to match the landscape of the aerial photo; the 36 meter spacing was measured later. So my ultimate large-scale mapping was procedurally idiosyncratic. The original aerial photograph was a contact print on resin-coated paper from negative # EHI-3JJ-141, taken on 9-15-68, and obtained from the USDA's Agricultural Stabilization and Conservation Service. Overlapping photos along the flight line and from adjacent flights provided stereo coverage. Although the terrain is hilly, parallax-induced distortion is well within the ±6 meter resolution of recorded locations of coarse-scale observations. (1.6.2)

9. I recall, with no fondness at all, field-processing film with ambient temperature compensation to find out whether the day's photography was successful, and if so with what exposure parameters. (1.6.3)

10. Even GPS positions may need such "ground-truthing," as New York State Police troopers have told me that they experience radio anomalies nearby, and I have found magnetic anomalies while surveying long, straight lines on the boundary of the study area. (1.6.3)

11. Fisher's "lady tasting tea" is the title and motivating example for Salsburg's (2001) engaging history of the statistical revolution in science. I encountered Fisher's discussion while first imprinting on statistics, and so I did some experiments with milk and tea. It turns out that one can indeed learn to distinguish whether milk is added to tea or tea to milk, as long as the tea is sufficiently hot, i.e., near the boiling point, the temperature needed to steam-distill the essential oils—which in turn means that the tea should be sipped from a cup, rather than being swilled from a mug. The milk lowers the temperature of the mixture from that of the tea. When tea is added to milk, the milk reaches this mixed temperature from below. But the first bits of milk added to tea reach the mixed temperature from above, and a bit more of its protein is denatured, scalded, curdled. So the milk added to tea has a very faintly "cheesier" smell than the milk to which tea has been added. Try it. Get good at it. Earn money from wagers in tearooms. (1.6.4)

12. In common parlance the latter is called a "statistical smokescreen." Folks who know more about statistics than I do accuse me of being a Bayesian in Fisherian garb. For help in deciding whether they are right, see Berger and Berry (1998) and Gotelli and Ellison (2013). I once wrote a song about this (Horn and Farnsworth, 1986), to which I recently added the missing couplet: "and if anyone says you're a Fisherian liar, you can rescue yourself with a Bayesian prior." (1.6.4)

13. The best way to learn NetLogo is via its online tutorials (Wilensky 1999) and by modifying the many online examples to adapt to your own project. Ideally, you should do

this with a NetLogo-literate friend because the syntax of the language is highly idiosyncratic, and navigating the online documentation is frustratingly inefficient. An alternative is to use Grimm and Railsback (2005), which is a combination of primer on the language and application to interesting problems in ecology. (1.6.5)

## CHAPTER 2. MORPHOLOGICAL AND PERCEPTUAL ADAPTATIONS TO HABITAT AND SOCIETY

1. In particular I repeat from Chapter 1 my bias as a visual creature. I admit to missing potential cues from other sensory modalities. However, this deficiency on my part must not be taken as evidence that butterflies do NOT use visual cues. (Ch. 2 Intro.)

2. Experimental orthodoxy would require me to record the number of failed presentations of a lure. However, that number depends not only on the state of the tested butterfly, but also on my degree of persistence in instances that are ultimately futile. My usual protocol was to move the lure slowly toward the subject and to record a response distance only if a clear response occurred. (2.1.2)

3. One additional cue is worth noting. Hairstreaks *Satyrium* spp. tended to respond to an approaching (looming) dark object that subtended an appropriate angle. The stimulus was variously: another Hairstreak nearby, other larger butterflies at greater distances, and even a Catbird *Dumetella carolinesis* at several meters, which the Hairstreak adaptively fled as it loomed larger. (2.1.2)

4. Saunders (1932, p.17) anticipated this analysis: "Rounded or convex winged butterflies are slower, weaker fliers, while straight-edged are faster, and concave-edged fastest of all." (2.2.1)

5. Table 9.3 in Chapter 9 records several predations by dragonflies, and also records that between 3% and 10% of captured butterflies carry tears on the wings that are conservatively attributed to attempted predation by birds. Of course, such tears record escape from predation, and the low number of such observations for the Ringlet (0.1–0.4%) may be due to its small size; a "tear" for another species would be large enough to engulf a Ringlet. (2.2.1)

6. In Europe *Pieris rapae* would be called the Small Cabbage White. It is a slightly smaller relative of the Large Cabbage White *Pieris brassicae*, a canonical subject for European studies of flight by Ellington and others. (2.2.2)

7. The dynamics of forward flapping flight without down and up claps is likely more complex than the simple caricature of Figure 2.2F. If the wingbeat is shallow, the shed vortices will be a parallel pair of roughly sinusoidal counter-rotating vortices. Fu and Liu (2015) have generated such vortices in computational and physical models that agree in showing that counter-rotation allows the vortex pair to interact so as to generate more efficient propulsion than would be expected from the analysis of either vortex alone. (2.2.2)

8. Nachtigall (1965, p. 216): "Schmetterlinge ohne Schuppen können schlecht oder gar nicht fliegen." (2.2.3)

9. In the analysis of complex dynamics, the propagation of a tiny shift in initial conditions to a dramatic change in the outcome at a larger scale has been called the "Butterfly Effect," with the canonical hypothetical example being the flapping of a butterfly's wings in Brazil somehow affecting increasing scales of nonlinear meteorology to cross

the intertropical convergence zone and set off a tornado in Texas. The term is traditionally attributed to Lorenz 1963, but Palmer 2009 gives a more nuanced history of the metaphor, still giving Lorenz his deserved credit. My postulated effect of microscopic scales on the formation and nature of shed vortices, and the effect of such vortices on flight and on the behavior of other butterflies is thus a literal as well as metaphorical Butterfly Effect. All ambiguous puns in this endnote are intended. (2.2.3)

10. Nachtigall (1967) showed that scales improve the aerodynamic properties of the wing, but hairs (which are highly modified scales) impair them. In many butterflies, hairs are more abundant near the base of the wing. They are clearly of value in thermoregulation there (Kingsolver 1985a), but that is also the region of the lowest spanwise velocities of air in the more powerful parts of the wingbeat cycle. So they may serve to make the aerodynamic drag more uniform over the wing surface. This speculation brings Nachtigall back into the discussion half a century later. (2.2.3)

11. Scudder (1889, p. 186) notes for the Pearly Eye: "Its general flight seems to me to be similar to that of *Cercyonis alope*, but less languid, the wings closing back to back with something of a snap." Indeed, I have heard this snap in the field, but I do not know whether it is perceived by other Pearly Eyes (Section 2.5). (2.2.4)

12. Contra most of the rest of the nymphalids, which typically hold wings dorsally when captured and escape with a forceful downstroke. (2.2.4)

13. The videotapes were made in 1983, and allow unambiguous measurement of wing angle only to the nearest 22.5°. Far better machinery is now available, but even the old rough measurements are sufficient to show characteristic differences in wing phase cycles between species. (2.2.4)

14. I developed my own adaptive response to this difference when capturing Monarchs versus Viceroys. I swing the net from in front of a Monarch or a V-sailing Viceroy, but from behind for a flat-sailing Viceroy. (2.2.5)

15. I would ignore this as biased wishful thinking except for several comments in the field by our daughter Jennifer, "No, Daddy, that's a dragonfly!" sometimes a correction, sometimes a mistake. (2.2.5)

16. In a series of early and brilliant experiments, Magnus (1958) tested the responses of male Fritillaries (*Argynnis paphia*, a close European analog of *Speyeria cybele*) to stimuli presented with a machine that was my inspiration for the Butterfly Rotisserie. To elicit a strong following response, the stimulus had to be closely Fritillary-orange colored, and the more rapidly it flickered, the more effective it was. The pattern of dark dots was irrelevant, except as it affected flicker frequency, but Magnus' experiments did not reproduce the conditions that could test my hypothesis. (2.3.3)

17. After Cronin et al. 2014, p. 234. (2.3.4)

18. When I move my analog compound eye viewer, subtle differences in regularity of patterning in light and shade are visible as different degrees of flickering; I have no evidence whether butterflies perceive this, but a potential signal is available. Analogous, but less subtle, potential discriminations are discussed for a Fritillary's potential mate (Section 2.3.3 and Figure 2.12) and for a Pearly Eye's view of its environment (Section 2.3.5 and Figure 2.17). And here is an idea that is so wildly speculative that I am hiding it in this endnote: The role of interommatidial bristles is not known, though suggestions have been made about possible measurement of airflow. Suppose the bristles reflect just enough background light into nearby ommatidia (less from further away) to affect adjustments in

sensitivity by migratory distal pigments. This would reduce sensitivity to the distant direct axial light, in proportion to a regional average of the illumination of nearby ommatidia, in imitation of "unsharp mask" in photo and digital image processing, decreasing sensitivity, but increasing acutance. I'm only saying . . . (2.3.4)

19. In case you are wondering, left and right-handed helices are roughly equally common both among individuals and between bouts of a single individual male Pearly Eye. (2.3.4)

20. In addition to responding to substrate-borne vibrations, male Pearly Eyes may even be responsive to airborne sound. More on this is in Section 2.5. (2.3.4)

21. This is only a suggestion. The observed pattern of pseudopupils, especially the presence and intensity of secondary and higher order pseudopupils is highly dependent on lighting conditions. Where secondary pseudopupils are possible, axial lighting favors a small primary pseudopupil; oblique lighting enlarges the primary pseudopupil, and favors the appearance of secondary ones (Stavenga 1979). The butterflies in Figures 2.16 and 2.17 are all lit obliquely and anterolaterally, favoring the observation of secondary pseudopupils if they are visible, but clouding their detailed interpretation. (2.3.6)

22. Clark's excellent 1926 review of "Fragrant Butterflies" is detailed and extensive, and though he did not mention the Pearly Eye, he did cite males with "a delicious vanilla-like scent" in an Asian species of *Lethe*, a genus to which the Pearly Eye was once assigned (Appendix A). (2.4.1)

23. Esters are a plausible attractant. Landolt and Hammond 2001 review the efficacy in attracting moths of a mixture of acetic acid and amyl alcohol (3-methyl-1-butanol). Acetic acid and amyl alcohol under acidic or enzymatic catalysis produce the ester isoamyl acetate, a strongly scented component of the odors of bananas and pears, as well as of a variety of synthetic fruit flavors (Azudin et al. 2013). The odor of isoamyl acetate is an active component of the sting pheromone of Honey Bees *Apis mellifera* (Boch et al. 1962), it has been shown to enhance the activity of the tephritid fruit fly *Bactrocdera tyoni* (Dalby-Ball and Meats 2000). Tang et al. 2013 tested the effectiveness of various fermented fruit juices, including pear juice, for inducing nymphalid Dead-leaf Butterflies *Kallima inachus* to feed at artificial flowers. They also tested volatile components of the juices separately at 5%-by-volume concentrations, and found the top four to be effective in the order: ethanol > amyl alcohol > isoamyl acetate ~ ethyl acetate. (2.4.2)

## CHAPTER 3. TACTICAL FORWARD VAGRANCY

1. On a 21st-century visit, in August 2013, that "field" was a young forest, with white ash, sugar maple, and black birch crowding out the previous invaders, staghorn sumac, apple, and fire cherry. (3.2.1)

2. The dominant thin-bladed species is poverty grass *Danthonia spicata*, on which I raised eggs of the first brood in 1975 to adulthood in time for the flight of the second brood. (3.2.1)

3. A normal, colloquially bell-shaped, distribution with central mean $\mu$ and standard deviation $\sigma$, the latter a measure related to width of the distribution. (3.3)

4. That is not universally true, but on the rare occasions when a male or female Ringlet does encounter dense woody vegetation and cannot, or does not, turn back into the open

field, it tends to rise to about a meter above that vegetation and fly in a straight line for much longer distances than it does in the open field. Where the canopy of the vegetation is irregular, the Ringlet may drop low enough to enter and follow channels between trees or shrubs (cf. the Eyed Brown of Chapter 4 and the Fritillary of Chapter 5). As stated in Section 3.2.1, Ringlets tend to penetrate a hedgerow only from a position at which a bright opening or distant horizon is visible (Figure 3.2B). (3.3.1)

5. This assumption is irrelevant for the Ignore and Reverse Models, and fails for points closer to a corner than the straight-line approximation of a typical path for the other Models, but it is close enough for the comparisons with and within Table 3.2. (3.3.1)

6. My mapping of such tracks was more difficult than that of tracks in the open field, as I switched from mapping turns in the butterflies' paths to mapping movements relative to the edges, and then back to turns in the butterflies' paths. Butterflies likely faced a similar problem with the switch back and forth between egocentric and geocentric navigation (cf. the elegant work of Peleg and Mahadevan 2016 with dung beetles). (3.3.1)

7. One could count the patches passed in the moves of a Lévy flight as having been searched, and then for a fair comparison, normalize flight length, rather than number of steps, for the Lévy flight and the compared models. "Optic flow" arguments (Section 2.3.4) suggest that simultaneous search is more difficult during rapid movements than slow movement, let alone than stops. This provides an additional physiological reason to Bartumeus and Levins' (2008) argument for separating search from reorientation, in models and discussions of spatial searching behavior. (3.3.3)

8. I have also made explicit models for the effect of boundary behavior and for competitive search by two or more males. Details of boundary behavior matter a lot, but so far, the results reduce to what I have already said qualitatively and heuristically. The Competing Males Model produces an inefficiently gerrymandered map of first visits, and would need an energetic minimization rule and complex bookkeeping to produce site-restricted territories. (3.3.3)

9. However, females were not rare late in the second brood of 1975, as shown in Figure 3.4. (3.6.3)

## CHAPTER 4. FORTUITOUS SITE-FIDELITY?

1. Shapiro (1974) lists several species of *Carex* as host plants. My two observations of oviposition were of clusters of three eggs each on *Scirpus* (Appendix B.3). I have noticed that Eyed Browns tend to be found where wide-bladed yellow-green sedges dominate, but tend to be absent where most sedges are blue-green. The locales of these observations include vicinities of Whitehall and Ithaca, NY, Cambridge and Waltham, MA, Rindge and Bradford, NH, and northern New Jersey. Some of the New Jersey observations may include *S. appalachia*, a sibling species of the Eyed Brown (Cardé et al. 1970). The few Appalachian Eyed Browns that I have encountered at Whitehall, NY have been among sedges beneath wet woodland, not in open wet meadows. (4.1.1)

2. Of 15 males involved in the "Cyclotron Experiment" of Section 4.3, three made the same long journey, but I had initially captured them at the pond site and released them at the meadow site. See Section 4.3 for notes on whether their behavior constitutes long-distance "homing." (4.1.1)

3. Several colleagues have commented that my Fortuitous Site-Fidelity Model shares elements with spin glasses and Ising models in solid state physics. I have yet to find further insights from this metaphor, but I shall keep looking. (4.4)

## CHAPTER 5. SETTING AND RUNNING A TRAPLINE

1. During the 1977 field season, Jeff initially marked 83 Great Spangled Fritillaries and 10 Aphrodites; my respective totals were 38 and 7. Varsity lacrosse letter-winner trumps former badminton amateur! And Jeff wrote an excellent undergraduate senior thesis (Georgia 1978). (Ch.5 Intro.)

2. Recent emergence was inferred from the fact that they had lost no scales either from the wings or from the end of the abdomen. After confinement for two days, they were marked and released; they were not dissected to discover whether they had already received a spermatophore. (5.1.4)

3. My replacement of the traditional "salesman" with "salesfolk" is dedicated to Penna Rose, director of the Princeton Chapel Choir, a tireless and inventive champion of inclusive language. I shall continue to use "saleman" in reference to the traditional literature or to male Fritillaries. (5.3)

4. This idea goes back at least as far as Scudder (1889, pp. 261–262, referring to *Basilarchia* = *Limeniti*s, the genus of the Viceroy). Southwood and Comins (1976, p. 954) describe the mechanism convincingly and memorably as "the immature stages are in some ways . . . 'playing hide and seek' in space and time with their natural enemies." And Douglas (1986, p. 125) bemoans the lack of studies on the likely important effect of egg parasitism on population survival in butterflies. (5.3)

5. Clark (1932, pp. 109–112) describes this behavior of female Great Spangled Fritillaries in detail. In particular their flights late in the season are direct, with drops into the vegetation only when they encounter patches of violets. He does not mention repeated circuits, for either the female or the male. (5.3)

## CHAPTER 6. DEFINING AND DEFENDING A TERRITORY

1. All my observations of Viceroy interactive behavior are parsimoniously interpretable as responses to visual stimuli. Odors and pheromones may be involved, as Clark (1932) reports that males have a slight sugary odor, and females a pronounced and disagreeable odor; but I could not detect either. So I can only claim that I have nothing to say about such odors, and I do not claim that they are unimportant. (6.1)

2. My explicit NetLogo models for male Viceroys have been uninformative since those that generate realistic individual dynamics and patterns of exclusive territories need so much empirical detail and internal bookkeeping that they do no more than generate a motion picture of their extensive input. (6.3.3)

3. Ross, studying two species of *Hamadryas*, taxonomic next-neighbors of *Limenitis*, actually describes a "lack of territoriality," in which a male defends a given perch until another more "pugnacious" male comes along. There is an element of this in my observed behavior of the Viceroy, but additionally, a male defends a neighborhood of perches for an extended period, and that group of perches is what I call a "territory." (6.4)

## CHAPTER 7. SOCIOLOGY AT A SINGLES BAR

1. Gosse refers to *Hipparchia Andromacha*, later called *Lethe portlandia*, which included what are now the Southern Pearly Eye *Enodia portlandia* and the Northern Pearly Eye *Enodia anthedon* (Heitzman and Dos Passos 1974, Glassberg 2017). Isn't taxonomy fun (Appendix A)! Gosse (1840, pp. 246–247) had previously published the same description of Pearly Eye behavior, with only slightly less florid exposition. (Ch. 7 Ref. for Epigraph)

2. Shapiro (1974) cites *Brachyelytrum erectum* as the typical larval food plant, but notes that the Pearly Eye can be raised on *Muhlenbergia* spp. Opler and Krizek (1984) and Cech and Tudor (2005) add species of *Erianthus, Leersia, Chasmanthium, Hystrix, Festuca, Schizachne,* and *Panicum*. I have observed oviposition on *Agrostis* sp. and frequently on *Phleum pratense* (timothy grass, a common pasture grass imported from Europe), and several other wide-bladed species of woodland glades that I did not identify as they were not in flower. I raised Pearly Eyes on *Phleum* to third instar, when they shifted from eating to what seemed like exploratory movements, and I incurred family obligations that prevented experiments with other grasses. Robbins (fide Swearingen and Adams 1977) reported an oviposition on Japanese stiltgrass *Microstegium vimineum*. I have found *Microstegium* at many southern sites that have Pearly Eyes, but I have also found the grass at many places where the butterfly is absent. In all my observations, eggs are laid on the underside of the peak in a bend of a wide-bladed grass; a prospective blade is rejected immediately if it does not bear the female's weight, and if it fails a further test that involves palpation by the tip of the female's abdomen. (7.1.1)

3. Extensive woodland edges are of course a relatively recent artifact of abandonment from agriculture, and are unlikely to be the ancestral habitat of the Pearly Eye. I have found Pearly Eyes in small clearings and glades with a sparse canopy and wide-bladed grassy understory in northern New Jersey, Pennsylvania, and eastern Massachusetts, eastern New York, southern Vermont, southeastern Minnesota, and Illinois. Spatial dispersion observed in my woodland edge populations may not be universal or even widely typical, but individual behaviors and interactions are similar in all these locales. (7.1.1)

4. A few of these episodes of inactivity are likely due to outright alcohol intoxication, and males occasionally fall off the bait and lie or flop on the ground below. (7.1.4)

5. In Wales in June 1985, I observed behavior of the Speckled Wood butterfly similar to that of my Pearly Eyes. (7.2)

6. Davies (1978) had reported this individual behavior, and hinted at the alternative interpretation as well. My male Pearly Eyes can also be classified at any given moment as patrollers (= fliers) or perchers, with diurnal shifts in prevailing status that are discussed in Section 7.5. (7.2)

7. My inspiration for explicitly citing Isaac Newton was Marden (1967). Newton's (1687, p. 2) First Law of Motion is: "Corpus omne perseverare in statu quiescendi vel movendi uniformiter in directum, nisi quatenus a viribus impressis cogitur statum illum mutare" = "Every body perseveres in its state of rest or of uniform motion straight ahead, unless it is compelled to change its state by impressed force." (7.3.3)

8. An observational asymmetry limits quantitative representation of long-term shifts of status. Most of my individually identified males were initially caught and marked as perchers, and their identifying numbers can be read only when they are perching. So the

observation of a marked patroller is likely evidence of a switch from perching—but it is hard to confirm the individual identity of a marked patroller, and a doubly rare event to encounter it later as a percher. (7.5)

9. Male Pearly Eyes can follow females (and other males following females) very closely through seemingly complex maneuvers, as long as the separation does not exceed about 20 centimeters. That distance is on the scale of the likely "footprint" of vortices shed in flapping flight (Section 2.2.2). Males can follow approximately at distances of as much as 75 centimeters, the scale of reliable detection and orientation toward a butterfly-sized object in the anterolateral field of view (Section 2.1.2). (7.6)

## CHAPTER 8. DO BUTTERFLIES MAKE DECISIONS?

1. For a nuanced definition and valuable discussion of "habitat," see Dennis (2010) (especially pp. 9 ff.). However, I am using "habitat" to connote a landscape element in a loosely conventional sense, with a meaning that depends on context. Thus "habitat" may signify a consistent vegetative composition and geometrical configuration (e.g., field, meadow, woodland, or forest), corresponding to Dennis' "biotope." But the "habitat" of a butterfly species may signify the ecological or geographical union of vegetative types that provide all the resources that an individual needs in order to carry out its life cycle, together with other vegetative types and edges that the individual must traverse to garner those resources, Dennis' "habitat" sensu stricto, which would differ from species to species. For me to use a more nuanced terminology would require a note as long as this endnote every time "habitat" is mentioned. (Ch. 8 Intro.)

2. There is direct evidence of this in a long right-hand tail of the distribution of individual dispersion distances for the Pearly Eye. Even in 1974 and 1975, before my intensive experiments with artificial baits, of a total 320 marked males, 75 were observed over several days; of these, 61 stayed well within 50 meters of their marking sites, but 14 relocated over distances of 100 to 300 meters. (8.1.2)

3. I was tempted to characterize this rule as "Dance if you are having fun; run if not," but wisely decided to banish the thought from the text to this endnote. (8.2)

4. It is possible that disorientation without a horizon is related more to size than to openness of habitat. Among the species photographed more than 10 times in the strobe apparatus of Chapter 2, the proportion of abnormal flights (loop, roll, yaw, or outright crash) to total photos increased as 1/15 for Pearl Crescents (typical wingspan = 3 cm), 4/21 for Clouded Sulphurs (5 cm), 5/17 for undamaged Wood Nymphs (6 cm), 10/24 for Great Spangled Fritillaries (8 cm), and 7/14 for Monarchs (10 cm). (8.2.1)

5. The only one of the local species that occasionally flies in significant wind is the Monarch. (8.2.1)

6. Eastern Tiger Swallowtails *Papilio glaucus* aggregate atop Bowman's Hill, Washington Crossing State Park, Pennsylvania. From a memorial tower above the forest canopy there, I have watched flights of the butterflies as they followed depressed pathways between the crowns of adjacent trees. So Tiger Swallowtails appear to "hilltop" at large scales and to "valley-bottom" at small scales. (8.2.2)

7. Whether this is a direct response to topography or to another environmental cue is unclear, as is how much it may be adaptive or even effective for mating (Baughman

et al. 1990, Boggs and Nieminen 2004). In any case, it shows that significant differences in population density can be associated with very subtle environmental gradients. (8.2.2)

8. Even some now-classical models place a high cognitive burden on individual agents in terms of some combination of: information, memory, computation, and game-theoretic strategizing (e.g., Root and Kareiva 1984, Turchin et al. 1991, Weiss and Murphy 1988, and Wiens et al. 1993). (8.2.4)

9. Cody's (1971) forward-biased exploratory model achieved maximal and most uniform coverage when Pr(ahead) = 0.6–0.7, Pr(right) ~ Pr(left) = 0.15–0.20, and Pr(back) = 0.05. A normally forward-biased model with standard deviation equal to Cody's is about $N(0°, 67°)$, only a bit more twisty than the $N(0°, 55°)$ that I have used for the Ringlet. (8.2.4)

10. That is to say, such a sloppy forward bias that it might as well be unbiased. For a uniform distribution between a and b, the mean = $(a + b)/2$, and the variance = $(a-b)^2/12$. For turns varying uniformly between $-180°$ and $180°$, the mean = $0°$, the variance = $360°^2/12$, and the standard deviation = $104°$. So the forward bias in the central region of Figures 8.3E–F is $N(0°, 104°)$; well, actually, for convenience, $N(0°, 105°)$. (8.2.4)

11. Jarome Ali has suggested (pers. comm., 6 Apr. 2018) that this bias against the favored habitat near the boundary might be alleviated by a one or two-step "memory," or equivalently by delaying response to a change in habitat for one to several moves. Some preliminary modeling suggests that the effectiveness of such a mechanism is sensitive to parametric details, and it is not yet clear (to me) whether the realization of effective behavior is less or more complex than reflecting from the boundary, which is considered in Section 8.2.5. (8.2.4)

12. Trust me on this one, even though I have shown you neither animations nor videos. The match is impressive. (8.2.5)

13. The Fortuitous Site-Fidelity Model of Figure 4.5 ideally predicts several attractive domains; Figure 8.7B is a more realistic caricature of the behavior of a male Eyed Brown, and might represent movement within an individual attractive domain. (8.3.2)

14. Given that, for the Ringlet, virgin females may "pursue" potentially eligible males (Section 3.5), a female rule of "approach if virgin; avoid if mated," i.e., status-dependent fixed behavior, would predict the observed lack of response to fluttering lures by free-flying females. But given the rarity and crypticity (to me) of stationary females, there is no easy way to test this proposed rule. (8.5.1)

15. Tinbergen (1951, pp. 47 ff & 102 ff) argued that a "reaction chain" of simple stimuli and fixed responses could lead to complex overall behavior. Janetos and Cole (1981) argued that simple choices can come close to optimality that would require complex memory and integration. Real (1991) extended Tinbergen's idea to show that a series of simple choices can mimic the results of cognition and eidetic images. (8.6)

16. That failure rate for a three-day study, though disappointing, is not statistically different from expectation based on my New York data from Table 9.2. The probability of failure to recapture one or more of 22 marked = $(1.00-.12 \text{ local persistence to day 2})^{22}$ = 0.06. (8.6.3)

17. Ditto along streams and woodland edges for the Viceroy's congeners, *Limenitis arthemis* in New York, New Jersey, Massachusetts, and Vermont, and *L. weidemeyerii* in Utah and northern Arizona. However, the interactions of these other species do not include, at least in my experience, the elaborate male-male "dogfights" of the Viceroy (Figure 6.6). (8.6.4)

## CHAPTER 9. LIFE HISTORY CONSEQUENCES OF INDIVIDUAL BEHAVIOR

1. Colleagues have routinely been amazed at the number of Pearly Eyes I have captured and kept under observation. I share their impression of the species' patchy rarity from my encounters with it elsewhere. My engineered ferments added to those already present on an abundance of volunteer apple trees, and my Pearly Eyes' oviposition on abundant timothy grass *Phleum* (Ch. 7 endnote 2), may have augmented their local abundance. This bodes well for artificially promoting high local abundance in future studies of this species. (9.2)

2. There is a statistical nicety here. I am using the ratio described as a maximum likelihood measure of each parameter, not as an estimator of the mean of a range of values that the parameter might take, for which Bayesian estimators would be more appropriate (Pascual and Kareiva 1996). (9.3.2)

3. The categories of wear and tear and their field-note codes and descriptions are as follows: "Slight" = vf (very fresh); f (fresh); vsw (very slightly worn) = no to very slight loss of scales. "Some" = sw (slightly worn); w (worn) = slight to patchy loss of scales. "LotsToFray" = vw (very worn); varied intermediate categories that fit here = considerable loss of scales to fraying of edges of wing. "Tattered" = wt (worn and tattered); vwt (very worn and tattered) = parts of edges of wings missing. Gross = gwt (grossly worn and tattered) = wings damaged enough to visibly affect flight. These categories were compiled from field codes to produce the best overall cross-species unimodal distributions for all individuals when initially marked. (9.3.3)

4. Emigration could be documented directly only for the Ringlet, and Wiernasz (1983) did so. During the 1980 season she heroically extended the surveyed Ringlet habitat to four times what I covered, captured emigrants from my study area, and estimated the daily mortality of males as 21% and of females as 27%. Thus of the daily losses of Ringlets, roughly one third to half were likely mortality. (9.3.3)

5. For example, the "beak"-marks on most species are sufficiently large to encompass the body of the Ringlet, making that species less likely to escape a bird and be recorded as suffering a "beak"-mark. So the low percentage of such marks on Ringlets (Table 9.3) vastly underestimates its risk of predation by birds. (9.3.4)

6. Several times in my field notes I have recorded that Ringlets seem to be less apparent when dragonflies are abroad. I doubt that this is an adaptive response of the Ringlets to the dragonflies, I suspect that it is partly due to Ringlets being less active and dragonflies more active when the humidity is very low, and I hope that this pattern is not due to dragonflies having eaten Ringlets that I would otherwise have seen. (9.3.4)

## CHAPTER 10. SUMMARY AND SPECULATIONS

1. It is interesting that a forward-directed random walk with a bias of $N(0°, 55°)$ fits simulations for both the Pearly Eye and the Ringlet in entirely different contexts. I am tempted to compare that ±55° with the forward-directed acute region that Rutowski (et al. 2009) has measured in several species of butterfly. However, my standard deviation is an arbitrary measure, idealized not for physiology but for statistics that I don't even use—and Rutowski's physiological criterion could be either the region of maximum acuity or the

region bounded by the maximum decrease in acuity. With so much freedom of choice it would be too easy to claim a spurious match. (10.3)

2. As physicists are wont to point out, Sherlock Holmes (Doyle 1905, The Adventure of the Priory School) missed this crucial observation when he deduced from the track of a bicycle which direction its rider had been going (Konhauser et al., 1997). (10.6.2)

3. I observed a few instances of one Fritillary following another at 2–3 meters separation for 10–15 seconds (Section 5.1.4), but both were also paralleling channels in the vegetation, and so the longer period of association may or may not have been an explicit interaction between butterflies. (10.6.4)

4. The sole disadvantage for videography is that the Pearly Eye is most active at places and times when environmental lighting is suboptimal (Figure 7.11). Low-light machinery would be needed as supplementary artificial light would almost certainly affect behavior (Section 2.2.6). (10.6.4)

# Bibliography

> Bookes must follow Sciences, and not Sciences, Bookes.
> –Sir Francis Bacon, 1657 (posthum.), "Proposition, . . .
> Touching, . . . Amendment, of the Lawes,
> of England," p. 277 in *Resuscitatio*.

N. B. Each reference is followed by a parenthetical set of citations of the places where it is mentioned. Unadorned numbers denote sections of the main text. A, B, C, D, and E denote their respective appendices. N denotes endnotes, cited by Chapter#.Note#. F denotes figures, and T, tables. Other abbreviations are self-explanatory.

Agrawal, A., 2017. *Monarchs and Milkweed: A migrating butterfly, a poisonous plant, and their remarkable story of co-evolution*. Princeton Univ. Press, Princeton, NJ. xi + 283 pp. (N.0.4)

Alcock, J., C. E. Jones, and S. L. Buckmann, 1976. Location before emergence of female bee, *Centris pallida*, by its male (Hymenoptera: Anthophoridae). J. Zool. 179(Jun.): 189–199. (2.3.3)

Alcock, J., C. E. Jones, and S. L. Buckmann, 1977. Male nesting strategies in the bee *Centris pallida* Fox (Anthophoridae: Hymenoptera). Am. Nat. 111(977): 145–155. (2.3.3, 5.1.4)

Amstrup, S. C., T. L. McDonald, and B. F. J. Manly, eds., 2005. *Handbook of Capture-Recapture Analysis*. xviii + 313 pp. Princeton Univ. Press, Princeton, NJ. (1.6.4)

Anderson, D. J., 1983. Optimal foraging and the traveling salesman. Theor. Popul. Biol. 24: 145–159. (1.7, 5.3, 8.4.2)

Angevine, M. W., and P. F. Brussard, 1979. Population structure and gene frequency analysis of sibling species of Lethe. J. Lepidopt. Soc. 33(1): 29–36. (Ch. 4 Introd.)

Arikawa, K., M. Kinoshita, and D. G. Stavenga, 2004. Color vision and retinal organization in butterflies. Pp. 193–219 and Plates 8–10 in F. R. Prete, ed., *Complex Worlds from Simpler Nervous Systems*. M. I. T. Press, Cambridge, MA. (2.3.2)

Austad, S. N., W. T. Jones, and P.M. Waser, 1979. Territorial defence in speckled wood butterflies: Why does the resident always win? Anim. Behav. 27: 960–961. (7.2)

Averbakh, I., V. Lebedev, and V. Tsurkov, 2008. Nash equilibria solutions in the competitive salesmen problem on a network. Appl. Comput. Math. 7: 54–65. (5.3)

Azudin, N. Y., M. D. Mashitah, and S. R. Abd Shukor, 2013. Optimization of isoamyl acetate production in a solvent-free system. J. Food Qual. 36: 441–446. (N.1.23)

Bacon, Sir Francis, 1657 (posthum.). *Resuscitatio*. Ed. W. Rawley, S. Griffin/ W. Lee, London, UK. Early English Books Online: Text Creation Partnership. Univ. Michigan Library. Accessed 26 Jul 2018 http://name.umdl.umich.edu/A28378.0001.001. (Bibliog. Epigraph)

Bailey, N. T. J., 1995. *Statistical Methods in Biology*, 3rd ed., Cambridge Univ. Press, Cambridge. 255 pp. (1.6.4)

Baker, R. R., 1972. Territorial behaviour of the nymphalid butterflies, *Aglais urticae* (L.) and *Inachis io* (L.). J. Anim. Ecol. 41: 453–469. (6.4)

Baker, R. R., 1983. Insect territoriality. Ann. Rev. Entomol. 28: 65–89. (7.2)

Bartumeus, F., and S. Levin, 2008. Fractal reorientation clocks: Linking animal behavior to statistical patterns of search. Proc. Natl. Acad. Sci. USA 105(49): 19072–19077. (3.3, N.3.7)

Baughman, J. F. and D. D. Murphy, 1988. What constitutes a hill to a hilltopping butterfly? Am. Midl. Nat., 120: 441–443. (8.2.2)

Baughman, J. F., D. D. Murphy, and P. R. Ehrlich, 1990. A reexamination of hilltopping in *Euphydryas editha*. Oecologia 83: 259–260. (6.4, N.8.7)

Bechert, D. W., G. Hoppe, and W.-E. Reif, 1985. On the drag reduction of the shark skin. AIAA Shear Control Conference, Boulder CO: Paper # AIAA-85–0546. (2.2.3)

Bekoff, M., 1994. Should scientists bond with the animals whom they use? Why not? Int. J. Comp. Psychol. 7: 78–86. (0.3)

Bekoff, M., 2007. *Animals Matter: A biologist explains why we should treat animals with compassion and respect*. xxi + 202 pp. Shambhala, Boston. (0.3)

Bekoff, M., L. Gruen, S. E. Townsend, and B. E. Rollin, 1992. Animals in science: some areas revisited. Anim. Behav. 44: 473–484. (0.3)

Benjamin, A., G. Chartrand, and P. Zhang, 2015. *The Fascinating World of Graph Theory*. xiii + 322 pp. Princeton Univ. Press, Princeton, NJ. (8.3)

Berger, J. O., and D. A. Berry, 1988. Statistical analysis and the illusion of objectivity. Am. Sci. 76: 159–165. (N.1.12)

Bernard, G. D., 1971. Evidence for visual function of corneal interference filters. J. Insect Physiol. 17: 2287–2300. (2.3.2)

Bernard, G. D., and C. L. Remington, 1991. Color vision in *Lycaena* butterflies: spectral tuning of receptor arrays in relation to behavioral ecology. Proc. Natl. Acad. Sci. USA 88: 2783–2787. (2.3.2)

Bernhard, C. G. 1965. Opening address. Pp. 1–11 in C. G. Bernhard, ed., *The Functional Organization of the Compound Eye*. Pergamon Press, Oxford. (2.3.3)

Betts, C. R., and R. J. Wootton, 1988. Wing shape and flight behaviour in butterflies (Lepidoptera: Papilionoidea and Hesperoidea): a preliminary analysis. J. Exp. Biol. 138: 271–288. (2.2.1)

Bitzer, R. J., and K. C. Shaw 1979(80). Territorial behavior of the red admiral, *Vanessa atalanta* (Lepidoptera: Nymphalidae). J. Res. Lepidopt. 18: 36- 49. (6.1)

Bitzer, R. J., and K. C. Shaw, 1983. Territorial behavior of *Nymphalis antiopa* and *Polygonia comma* (Nymphalidae). J. Lepidopt. Soc. 37: 1–13. (6.4)

Bitzer, R. J., and K. C. Shaw, 1995. Territorial behavior of the Red Admiral, *Vanessa atalanta* (Lepidoptera: Nymphalidae) I. The role of climatic factors and early interaction frequency on territorial start time. J. Insect Behav. 8: 47–66. (6.4)

Bixler, G. D., and B. Bushan, 2013. Fluid drag reduction and efficient self-cleaning with rice leaf and butterfly wing bioinspired surfaces. Nanoscale 5: 7685–7710. (2.2.3)

Boch, R., D. A. Shearer, and B. C. Stone, 1962. Identification of iso-amyl acetate as an active component in the sting pheromone of the honey bee. Nature 195: 1018–1020. (N.2.23)

Boggs, C. L., and M. Nieminen, 2004. Checkerspot reproductive biology, pp. 92–111 in P. R. Ehrlich and I. Hanski, eds., *On the Wings of Checkerspots: A model system for population biology*. Oxford Univ. Press, Oxford. (N.8.7)

Bond, A. B., 1980. Optimal foraging in a uniform habitat: the search mechanism of the green lacewing. Anim. Behav. 28: 10–19. (3.1)

Bormann, F. H., D. Balmori, and G. T. Gaballe, 1993. *Redesigning the American Lawn: A search for environmental harmony*. 166 pp. Yale Univ. Press, New Haven, CT. (N.0.1)

Bouferrouk, A., 2014. On the applicability of trapped vortices to ground vehicles. International Vehicle Aerodynamics Conference 2014, Loughborough, UK, pp. 101–111. Woodhead Publishing (Elsevier, Amsterdam) . (2.2.2, 2.2.3)

Brackenbury, J., 1992. *Insects in Flight*. 192 pp. Blandford, London. (2.2.2, N.0.4, F.2.2)

Bradbury, J. W. 1985. Contrasts between insects and vertebrates in the evolution of male display, female choice, and lek mating. Pp. 273–289 in B. Hölldobler and M. Lindauer, eds., *Experimental Behavioral Ecology and Sociobiology*. G. Fisher Verlag, Stuttgart. (7.1.1)

Bradbury, J. W., and S. L. Vehrencamp, 1977. Social organization and foraging in emballonurid bats. III. Mating systems. Behav. Ecol. Sociobiol. 2: 1–17. (0.2)

Brown, J. L., 1964. The evolution of diversity in avian territorial systems. Wilson Bull. 76: 160–169. (0.2)

Burns, J. M., 1968. Mating frequency in natural populations of skippers and butterflies as determined by spermatophore counts. Proc. Natl. Acad. Sci. USA 61: 852–859. (0.3, 1.3.4, T.1.3)

Calabrese, J. M., C. H. Fleming, and E. Gurarie, 2016. CTMM: an R package for analyzing animal relocation data as a continuous-time stochastic process. Methods Ecol. Evol. 7: 1124–1132. (8.2.4)

Cardé, R. T., and R. E. Charlton, 1984. Olfactory sexual communication in Lepidoptera: strategy, sensitivity and selectivity. Pp. 242–265 in T. Lewis, ed., *Insect Communication*, R. Entomol. Soc. Lond., Academic Press, London. (2.4.2)

Cardé, R. T., Shapiro, A. M., and Clench, H. K., 1970. Sibling species in the *eurydice* group of *Lethe*. (Lepidoptera: Satyridae). Psyche 77: 70–103. (N.4.1)

Cartwright, B. A., and T. S. Collett, 1982. How honey bees use landmarks to guide their return to a food source. Nature 295: 560–564. (8.4.2)

Cartwright, B. A., and T. S. Collett, 1987. Landmark maps for honeybees. Biol. Cybern. 57: 85–93. (5.1.4, 8.4.2)

Catchpole, C. K. 1989. Pseudoreplication and external validity: playback experiments in avian bioacoustics. Trends Ecol. Evol. 4: 286–287. (1.6.4)

Cech, R, and G. Tudor, 2005. *Butterflies of the East Coast: An observer's guide*. xii, 360 pp. Princeton Univ. Press. (1.3.2, Ch. 4 Introd., Ch.5 Introd., 6.2.2, 8.6.3, N.0.2, N.0.4, N.7.2, A, T.A.1)

Chai, P., and Srygley, R. B., 1990. Predation and the flight, morphology and temperature of Neotropical rain-forest butterflies. Am. Nat. 135(6): 748–765. (2.2.1)

Chandrasekhar, S., 1943. Stochastic problems in physics and astronomy. Rev. Mod. Physics 15: 1–87. (7.1.2)

Clark, A. H. 1932. The butterflies of the District of Columbia and vicinity. U. S. Natl. Mus. Bull. 157; 337 pp. (2.2.5, Ch. 4 Introd., Ch. 5 Introd., 5.1.1, 5.2, 6.2.2, 6.2.3, 9.3.5, N5.5, N.6.1)

Clark, A., 1926. Fragrant Butterflies. Annual Report of the Smithsonian Institution, 1926: 421–446 + 13 Plates. (2.4.1, N.1.22)

Cody, M. L., 1971. Finch flocks in the Mohave Desert. Theor. Popul. Biol. 2: 142–158. (3.3, 8.2.4, N.8.9)

Collett, T. S. 1992. Landmark learning and guidance in insects. Phil. Trans. R. Soc. Lond. B 337: 295–303. (Ch. 8)

Comstock, A. B., 1911. *Handbook of Nature Study*. Comstock Publishing, Ithaca, NY, 887 pp. (Ch. 2 Epigraph)

Cook, W. J., 2012. *In Pursuit of the Traveling Salesman Problem: Mathematics at the limits of computation*. xiii + 228 pp. Princeton Univ. Press, Princeton, NJ. (1.7, 5.3)

Cordero, C., 2002. Matings without spermatophore transfer and with transfer of two spermatophores in *Callophrys xami* (Lycaenidae). J. Lepidopt. Soc. 56: 106–108. (1.3.4, 1.7)

Corner, E. J. H., 1964. *The Life of Plants*. World Publ. Co., Cleveland, OH. xii + 315 pp. Republ. 2002 with new Foreword by K. J. Niklas, Univ. Chicago Press, Chicago. (3.5)

Crane, J., 1955. Imaginal behavior of a Trinidad butterfly, *Heliconius erato hydara* Hewitson, with special reference to the social use of color. Zoologica (NY) 40: 167–196. (1.6.2, 2.1.2)

Cronin, T. W., S. Johnsen, N. J. Marshall, and E. J. Warrant, 2014. *Visual Ecology*. xx, 405 pp. Princeton Univ. Press, Princeton, NJ. (2.3.1, 2.3.3, 2.3.4, N.0.4, N.2.17, F.2.13)

Daily, G. C., P. R. Ehrlich, and D. Wheye, 1991. Determinants of spatial distribution in a population of the subalpine butterfly *Oeneis chryxus*. Oecologia 88: 587–596. (6.4)

Dalby-Ball, G., and A. Meats, 2000. Influence of the odour of fruit, yeast and cue-lure on the flight activity of the Queensland fruit fly, *Bactrocera tryoni* (Froggatt) (Diptera: Tephritidae). Aust. J. Entomol. 39: 195–200. (N.2.23)

Dalton, S., 1975. *Borne on the Wind: The extraordinary world of insects in flight*. 160 pp. Reader's Digest Press & E. P. Dutton, New York. (1.6.2, 2.2.2, N.0.4, C.3)

Dalton, S., 1977. *The Miracle of Flight*. 168 pp. McGraw-Hill, New York. (1.6.2, 2.2.5)

Davies, N. B., 1978. Territorial defence in the speckled wood butterfly *(Pararge aegeria)*, the resident always wins. Anim. Behav. 26: 138–147. (1.7, 7.2, 7.3.3, N.7.6)

Davies, N. B., 1979. Game theory and territorial behavior in speckled wood butterflies. Anim. Behav. 27: 961–962. (7.2)

Dennis, R. L. H., 2010. *A Resource-based Habitat View for Conservation: Butterflies in the British landscape*. Wiley-Blackwell, Chichester, West Sussex, UK. xii + 406. (8.1.1, N.8.1)

Devault, D., 1980. Quantum mechanical tunnelling in biological systems. Q. Rev. Biophysics 13(4): 387–564. (N.1.3)

Dickinson, Emily, 1862, *fide* Matchett 1962. (Epigraph Ch. 3)

Douglas, M. M., 1979. Hot butterflies. Nat. Hist. 88(9): 56–65. (6.3.1)

Douglas, M. M., 1986. *The Lives of Butterflies*. Univ. of Michigan Press, Ann Arbor. xvii + 241 pp. (6.3.1, N.0.4, N.5.4)

Doyle, A. Conan, 1892, The Boscombe Valley Mystery, in *The Adventures of Sherlock Holmes*. Reprinted 1993, R. L. Green, ed., Introd., annot., xlix + 389 pp. Oxford Univ. Press, Oxford. (Ch.1 Epigraph)

Doyle, A. Conan, 1905, The Adventure of the Priory School, in *The Return of Sherlock Holmes*. Reprinted 1994, R. L. Green, ed., Introd., annot., xlvii + 408 pp. Oxford Univ. Press, Oxford. (N.10.2)

Dudley, R., 2000. *The Biomechanics of Insect Flight: Form, function, evolution*. Princeton Univ. Press, Princeton, NJ, 476 pp. (2.2.3)

Ehrlich, A. H., and P. H. Ehrlich, 1978. Reproductive strategies in the butterflies: I. Mating frequency, plugging, and egg number. J. Kansas Entomol. Soc. 51: 666–697. (T.1.3)

Ehrlich, P. R., and L. E. Gilbert, 1973. Population structure and dynamics of the tropical butterfly *Heliconius ethilla*. Biotropica 5: 69–82. (2.3.3, 5.1.4)

Ehrlich, P. R., and I. Hanski, eds., 2004. *On the Wings of Checkerspots: A model system for population biology*. xx + 371 pp. Oxford Univ. Press, Oxford. (1.2, 1.3.1, 1.6.2, N.0.4)

Ehrlich, P. R., and D. D. Murphy 1982. Butterfly nomenclature: a critique. J. Res. Lepidopt. 20: 1–11. (A)

Ehrlich, P. R., and D. D. Murphy, 1983a. Nomenclature, taxonomy, and evolution. J. Res. Lepidopt. 20: 199–204. (A)

Ehrlich, P. R., and D. D. Murphy, 1983b. Butterfly nomenclature, stability., and the rule of obligatory categories. Syst. Zool. 32(4): 451–453. (A)

Ehrlich, P. R., and D. Wheye, 1986. "Nonadaptive" hilltopping behavior in male checkerspot butterflies (*Euphydryas editha*). Am. Nat. 127: 477–483. (6.4)

Ellington, C. P., 1980. Vortices and hovering flight. Pp. (782–819) 64–101 in Werner Nachtigall (Hrsg.) *Instationäre Effekte an schwingenden Tierflügeln*. Franz Steiner Verlag, Wiesbaden. (F.2.2)

Ellington, C. P., 1984a. The aerodynamics of flapping animal flight. Am. Zool. 24: 95–105. (2.2.2)

Ellington, C. P., 1984b. The aerodynamics of hovering insect flight (I-VI). Phil. Trans. R. Soc. Lond. 305: 1–181 and 7 plates. (2.2.2)

Ellington, C. P., 1995. Unsteady aerodynamics of insect flight. Symp. Soc. Exp. Biol. 49: 109–129. (2.2.2)

Eltringham, H., 1919. *Butterfly Vision*. Trans. (R.) Entomol. Soc. Lond. 1–49 + 5 plates. (2.3.3)

Estrada, C., S. Yildizhan, S. Schulz, and L. E. Gilbert, 2010. Sex-specific chemical cues from immatures facilitate the evolution of mate guarding in Heliconius butterflies. Proc. R. Soc. Lond. B 277: 407–413. (2.4.1)

Exner, S., 1891. *Die physiologie der fascettierten Augen von Krebsen und Insecten*. English translation and annotation by R. C. Hardie, 1989. *The Physiology of the Compound Eyes of Insects and Crustaceans*. Springer-Verlag, Berlin. 177 pp. (2.3.3)

Fagerström, T., and C. Wiklund, 1982. Why do males emerge before females? Protandry as mating strategy in male and female butterflies. Oecologia (Berl.) 52: 164–166. (3.2.3)

Farnsworth, E. J., 1995. Oikos and Ethos: setting our house in order. Trends Ecol. Evol. 10: 56–57. (0.3)

Farnsworth, E. J., and J. Rosovsky, 1993. The ethics of ecological field experimentation. Conserv. Biol. 7: 463–472. (0.3)

Fei, Y.-H. J., and J.-T. Yang, 2015. Enhanced thrust and speed revealed in the forward flight of a butterfly with transient body translation. Physical Rev. E 92, 033004: 1–10. (2.2.3)

Fekete, S. P., R. Fleischer, A. Fraenkel, and M. Schmitt, 2004. Traveling salesmen in the presence of competition. Theor. Comput. Sci. 313: 317–392. (1.7, 5.3)

Fenk, L. M., A. Poehlmann, and A. D. Straw, 2014. Asymmetric processing of visual motion for simultaneous object and background responses. Curr. Biol. 24: 2913–2919. (2.3.4)

Ferris, C. D., 1970. Occurrence of *Coenonympha inornata* (Satyridae) in Maine. J. Lepidopt. Soc. 24: 202. (0.2)

Finney, D. J., and J. L. Harper, 1993. Editorial code for presentation of statistical analyses. Proc. R. Soc. Lond. B 254: 287–288. (1.6.4)

Fisher, R. A., 1930. *The Genetical Theory of Natural Selection*. Facsimile variorum edition of 1999, with introduction and notes by Henry Bennett, Oxford Univ. Press, Oxford. xiv + 318. (Ch. 9 Epigraph)

Fisher, R. A., 1935. *The Design of Experiments*. Oliver and Boyd, Edinburgh, UK. (1.6.4)

Fisher, R. A., 1958. *Statistical Methods for Research Workers*, 13th ed. xv + 356 pp. Hafner, New York. (1.6.4)

Fleming, C. H., J. M. Calabrese, T. Mueller, K. A. Olson, P. Leimgruber, and W. F. Fagan, 2014. From fine-scale foraging to home ranges: a semivariance approach to identifying movement modes across spatiotemporal scales. Am. Nat. 183: E154–E167. (8.2.4)

Fleming, C. H., Y. Subasi, and J. M. Calabrese, 2015. Maximum-entropy description of animal movement. Physical Rev. E 91: 032107 (7 pp.). (8.2.4)

Floreano, D., R. Pericet-Camara, S. Viollet, F. Ruffier, A. Brückner, R. Leitel, W. Buss, M. Menouni, F. Expert, R. Juston, M. K. Dobrzynski, G. L'Eplattenier, F. Recktenwald, H. A. Mallot, and N. Franceschini, 2013. Miniature curved artificial compound eyes. Proc. Natl. Acad. Sci. USA 110(23): 9267–9272. (2.3.3, 10.6.2)

Ford, E. B., 1945. *The New Naturalist: Butterflies*. Collins, London. xiv + 368 pp. (N.0.4)

Forsberg, J., and C. Wiklund, 1989. Mating in the afternoon: time-saving in courtship and remating by females of a polyandrous butterfly, *Pieris napi* L. Behav. Ecol. Sociobiol. 25: 349–356. (6.4)

Fraenkel, L. E., 1972. Examples of steady vortex rings of small cross-section in an ideal fluid. J. Fluid Mech. 51(part 1): 119–135. (2.2.2)

Fu, Z., and H. Liu, 2015. Transient force augmentation due to counter-rotating vortex ring pairs. J. Fluid Mech. 785: 324–348. (N.2.7)

Georgia, J. D., 1978. Behavioral Study of Six Common Northeastern Butterfly Species. Undergraduate Senior Thesis, Department of Biology, Princeton University, 100 pp. Archived at Mudd Manuscript Library, Princeton Univ., Princeton, NJ. (0.5, Ch. 4 Introd., N.5.1)

Gilbert, L. E., 1976. Postmating female odor in *Heliconius* butterflies: a male-contributed aphrodisiac? Science 193: 419–420. (5.1.4)

Glassberg, J., 2017. *A Swift Guide to Butterflies of North America, Second Ed*. Princeton Univ. Press, Princeton, NJ, 420 pp. (First Ed. 2012, Sunstreak Books, Morristown, NJ, 416 pp.) (1.3.2, N.0.2, N.7.1, A, T.A.1)

Gochfeld, M., and J. Burger, 1997. *Butterflies of New Jersey: A guide to their status, distribution, conservation, and appreciation*. Rutgers Univ. Press, New Brunswick, NJ. 327 pp. (Ch. 5 Introd., 8.6.3)

Goldsmith, T. H., and G. D. Bernard, 1974. The visual system of insects. Pp. 165–272 in M. Rockstein, ed., *The Physiology of Insecta II*. Academic Press, New York. (2.3.2)

Gosse, P. H., 1840. *The Canadian Naturalist: A series of conversations on the natural history of lower Canada*. John Van Voorst, London. xii + 372 pp. (N.7.1)

Gosse, P. H., 1859. *Letters from Alabama, (U.S.) chiefly relating to natural history*. Morgan and Chase, London. xii + 306 pp. (Ch.7 Epigraph)

Gotelli, N. J., and A. M. Ellison, 2013. *A Primer of Ecological Statistics*, 2nd ed. xxi + 614 pp. Sinauer, Sunderland, MA. (1.6.4, N.1.12)

Gould, J. L. 1979. Do honeybees know what they are doing? Nat. Hist. 88(6): 66–75. (2.3.3, Ch. 4)

Grimm, S. F. and V. Railsback, 2005. *Individual-based Modeling and Ecology*. Princeton Univ. Press, Princeton, NJ. 448 pp. (3.3.2, N.1.13)

Hagen, R. H., R. C. Lederhouse, J. L. Bossart, and J. M. Scriber, 1991. *Papilio canadensis* and *P. glaucus* (Papilionidae) are distinct species. J. Lepidopt. Soc. 45(4): 245–258. (A)

Hammond, P. C., D. V. McCorkle, and W. Bergman, 2013. Hybridization studies of genomic compatibility and phenotypic expression in the greater fritillary butterflies (Nymphalidae: Argynnini). J. Lepidopt. Soc. 67(4): 263–273. (2.4.1)

Hargrove, W. W., and J. Pickering, 1992. Pseudoreplication: a *sine qua non* for regional ecology. Landscape Ecol. 6: 251–258. (1.6.4)

Harris, M., 1766. *The Aurelian or Natural History of English Insects; Namely, Moths and Butterflies*. 104 pp. Reprinted 1986 with introduction and commentary by R. H. Mays, Country Life/Newnes Books, Twickenham, GB. (Ch. 0 Epigraph, Ch. 5 Epigraph)

Harrison, S., J. F. Quinn, J. F. Baughman, D. D. Murphy, and P. R. Ehrlich, 1991. Estimating the effects of scientific study on two butterfly populations. Am. Nat. 137: 227–243. (0.3, 1.6.2)

Heinrich B., 1979. *Bumblebee Economics*. viii + 245 pp. Harvard Univ. Press, Cambridge, MA. (N.0.4)

Heinrich, B., 1986a. Thermoregulation and flight activity of a satyrine, *Coenonympha inornata* (Lepidoptera: Satyridae). Ecology 67: 593–597. (Ch. 3 Introd., 3.1, 3.2.3, 6.3.1)

Heinrich, B., 1986b. Comparative thermoregulation of four montane butterflies of different mass. Physiol. Zool. 59: 616–626. (6.3.1)

Heitzman, J. R., and C. F. Dos Passos, 1974. *Lethe portlandia* (Fabricius) and *L. anthedon* (Clark), sibling species, with descriptions of new subspecies of the former (Lepidoptera: Satyridae). Trans. Am. Entomol. Soc. 100: 52–99. (N.7.1)

Hoban, R., 1974. How Tom Beat Captain Najork and his Hired Sportsmen. Illustrated by Q. Blake. 30 pp. Atheneum, New York. (0.2)

Höglund, J., and R. V. Alatalo, 1995. *Leks*. Princeton Univ. Press, Princeton, NJ. (7.1.1)

Horn, D. J., H. S. Horn, C. M. Horn, and W. M. Horn, 1976. Dragonfly territories in Massachusetts marsh. P. 235 in D. J. Horn, *Biology of Insects*. 439 pp. Saunders, Philadelphia. (F.0.1)

Horn, H. S., 1966. Colonial nesting in the Brewer's blackbird (*Euphagus cyanocephalus*) and its adaptive significance. 86 pp. Ph.D. Thesis, Zoology, University of Washington, Seattle. (F.C.2)

Horn, H. S., 1971. *The Adaptive Geometry of Trees*. Monogr. Popul. Biol. 3. xi + 144 pp. Princeton Univ. Press, Princeton, NJ. (0.1, Ch. 8 Epigraph)

Horn, H. S., 1975. Markovian properties of forest succession. Pp. 196–211 in M. L. Cody and J. M. Diamond, eds., *Ecology and Evolution of Communities*. Harvard Univ. Press, Cambridge, MA. (0.1)

Horn, H. S., 1976a. Succession. Pp. 187–204 in R. M. May, ed., *Theoretical Ecology*. Blackwell, Oxford. (0.1)

Horn, H. S., 1976b. A clamp for marking butterflies in capture-recapture studies. J. Lepidopt. Soc. 30: 145–146. (0.2, 1.6.2, 7.1.3, C1, F.1.6, F.C.1)

Horn, H. S., 1981. Some causes of variety in patterns of secondary succession. Pp. 24–35 in D. C. West, H. H. Shugart, and D. B. Botkin, eds., *Forest Succession: Concepts and application*. Springer-Verlag, New York. (0.1)

Horn, H.S., 2013. From the butterfly point of view. *Science First Hand* No. 2(50): 121–122 (Novosibirsk; in Russian). (2.3.3)

Horn, H. S., and J. C. Farnsworth, 1986. Notes on empirical ecology. Am. Sci. 74: 572–573. (0.2, N.1.12)

Horn, H. S., H. H. Shugart, and D. L. Urban, 1989. Simulators as models of forest dynamics. Pp. 256–267 in J. Roughgarden, R. M. May, and S. A. Levin, eds., *Perspectives in Ecological Theory*. Princeton Univ. Press, Princeton, NJ. (0.1)

Horridge, G. A. 1986. A theory of insect vision: velocity parallax. Proc. R. Soc. Lond. B 229: 13–27. (2.2.3, 2.3.4)

Horridge, G. A., 1992. What can engineers learn from insect vision? Phil. Trans. R. Soc. Lond. B 337: 271–282. (2.3.4)

Horridge, G. A., and L. Marcelja, 1992. On the existence of "fast" and "slow" directionally sensitive motion detector neurons in insects. Proc. R. Soc. Lond. B 248: 47–54. (2.3.4)

Horridge, G. A., S.-W. Zhang, and D. O'Carroll, 1992. Insect perception of illusory contours. Phil. Trans. R. Soc. Lond. B 337: 59–64. (2.3.4)

Hovanitz, W., 1962. *Argynnis* and *Speyeria*. J. Res. Lepidopt. 1: 95–96. (A)

Hu, H., and M. Tamai, 2008. Bioinspired corrugated airfoil at low Reynolds numbers. Journal of Aircraft 45(6): 2068–2077. (2.2.3)

Hurlbert, S. H. 1984. Pseudoreplication and the design of ecological field experiments. Ecol. Monogr. 54: 187–211. (1.6.4)

Iftner, D. C., 1997. Southward range extension of the common ringlet, *Coenonympha tullia inornata* (Lepidoptera: Satyridae). Entomol. News 108: 201–202. (0.2)

Jacobs, M. D., and W. B. Watt, 1994. Seasonal adaptation vs physiological constraint: photoperiod, thermoregulation and flight in *Colias* butterflies. Funct. Ecol. 8: 366–376. (7.2)

James, F. C., and C. E. McCulloch, 1990. Multivariate analysis in ecology and systematics: panacea or Pandora's box? Ann. Rev. Ecol. Syst. 21: 129–166. (1.6.4)

Janetos, A. C., and B. J. Cole, 1981. Imperfectly optimal animals. Behav. Ecol. Sociobiol. 9: 203–209. (N.8.15)

Johnson, A. R., J. A. Wiens, B. T. Milne, and T. O. Crist, 1992. Animal movements and population dynamics in heterogeneous landscapes. Landscape Ecol. 7: 63–75. (3.2.1, 8.2.4)

Kareiva, P. M., and N. Shigesada, 1983. Analyzing insect movement as a correlated random walk. Oecologia (Berl.) 56: 234–238. (3.2.5)

Keji, J. A., 1963. *Coenonympha* in Essex Co., New York. J. Lepidopt. Soc. 17: 158. (0.2)

Kephart, H., 1921. *Camping and Woodcraft: A handbook for vacation campers and for travelers in the wilderness*. 405 + 479 pp. MacMillan, New York. (F.C.7)

Kingsolver, J. G. 1985a. Butterfly engineering. Sci. Am. 253(2): 106–114. (6.3.1, N.2.10)

Kingsolver, J. G. 1985b. Butterfly thermoregulation: organismic mechanisms and population consequences. J. Res. Lepidopt. 24: 1–20. (2.2.2, 2.2.6, 6.3.1)

Kingsolver, J. G., 1989. Weather and the population dynamics of insects: integrating physiological and population ecology. Physiol. Zool. 62: 314–334. (Ch. 9 Introd.)

Kitamura, T., and M. Imafuku, 2015. Behavioural mimicry in flight path of Batesian intraspecific polymorphic butterfly *Papilio polytes*. Proc. R. Soc. Lond. B 282: 20150483: 1–7. (2.2.5)

Klots, A. B., 1951. *A Field Guide to the Butterflies of North America, East of the Great Plains*. xvi + 340 pp. Houghton Mifflin, Boston. (N.1.1, T.A.1)

Konhauser, J. D. E., D. Velleman, and S. Wagon, 1997. *Which Way Did the Bicycle Go?: ... and other intriguing mathematical mysteries*. xv + 235 pp. Mathematical Assoc. of America, Washington, DC. (N.10.2)

Lancaster, D., 1974. *TTL Cookbook*. 335 pp. Sams (Prentice Hall), Carmel, IN. (Ch. 2, Ch. 3)

Land, M. F., and D.-E. Nilsson, 2012 (1st Ed. 2002). *Animal Eyes*. Oxford Univ. Press, Oxford. 221 pp. (2.3.1)

Landolt, P. J., and P. C. Hammond, 2001. Species' composition of moths captured in traps baited with acetic acid and 3-methyl-1-butanol, in Yakima County, Washington. J. Lepidopt. Soc. 55: 53–58. (N.2.23)

Lederer, G., 1960. Verhaltensweisen der Imagines und der Entwicklungsstadien von *Limenitis camilla camilla* L. (Lep. Nymphalidae). Zeit. Fur Tierpsychologie 17: 521–546. (T.1.3)

Lederhouse, R. C., 1993. Territoriality along flyways as mate-locating behavior in male *Limenitis arthemis* (Nymphalidae). J. Res. Lepidopt. 47(1): 22–31. (6.2.3)

Levin, S. A. 1986. Random Walk models of movement and their implications. Pp. 149–154 in T. G. Hallam and S. A. Levin, eds., *Biomathematics*, Vol. 17, *Mathematical Ecology*. Springer-Verlag, Berlin. (Ch. 3 Introd., 3.2.5, 8.2.4)

Lewis, H. R., and C. H. Papadimitriou, 1978. The efficiency of algorithms. Sci. Am. 238(1): 96–109. (5.3)

Lewis, O. T., and S. R. Bryant, 2002. Butterflies on the move. Trends Ecol. Evol. 17: 361–362. (0.5)

Lian, Y., and W. Shyy, 2007. Laminar-turbulent transition of a low Reynolds number rigid or flexible airfoil. AIAA Journal 45: 1501–1513. (2.2.7)

Lombardi, V., and W. C. Heinz 1963. *Run to Daylight: Vince Lombardi's diary of one week with the Green Bay Packers*. Simon & Schuster, New York. (2.1.1, 3.2.1)

Lorenz, E. N., 1963. The predictability of hydrodynamic flow. Trans. NY Acad Sci. 25: 409–432. (N.2.9)

MacGregor, J. N., and T. Ormerod, 1996. Human performance on the traveling salesman problem. Percept. Psychophys. 58: 527–539. (5.3, 5.4)

Magnus, D. B., 1950. Beobachtungen zur Balz und Eiablage des Kaisermantels *Argynnis paphia* L. (Lepidoptera: Nymphalidae). Z. Tierpsychol. 7: 435–449. (2.4.1)

Magnus, D. B., 1958. Experimentelle Untersuchungen zur Bionomie und Ethologie des Kaisermantels *Argynnis paphia* L. (Lep. Nymph.). Z. Tierpsychol. 15: 397–426. (1.6.2, 2.1.2, N.2.16, C.6)

Marden, J. H., 1992. Newton's second law of butterflies. Nat. Hist. 101(1): 54–61. (N.7.7)

Marden, J. H., and P. Chai, 1991. Aerial predation and butterfly design: how palatability, mimicry, and the need for evasive flight constrain mass allocation. Am. Nat. 138: 15–36. (2.2.1)

Matchett, W. H., 1962. Dickinson's revision of "Two butterflies went out at noon". PMLA, 77(4): 436–441. (Introd. Ch. 3)

Maynard Smith, J., 1978. The evolution of behavior. Sci. Am. 239(3): 176–192. (1.7, 7.1.2, 7.3.3)
Maynard Smith, J., and G. A. Parker, 1976. The logic of asymmetric contests. Anim. Behav. 24: 159–175. (7.1.2)
Mentis, M . T., 1988. Hypothetico-deductive and inductive approaches in ecology. Funct. Ecol. 2: 5–14. (1.6.4)
Miller, J. Y., ed., 1992. *The Common Names of North American Butterflies*. Smithsonian Institution Press, Blue Ridge Summit, PA. (T.A.1)
Miller, L. D., and F. M. Brown 1981. A catalogue/checklist of the butterflies of America north of Mexico. Mem. Lepidopt. Soc. (2). vii + 280 pp. (A)
Miller, L. D., and F. M. Brown 1983. Butterfly taxonomy: a reply. J. Res. Lepidopt. 20: 193–198. (A)
Miller, W. H., and G. D. Bernard, 1968. Butterfly glow. J. Ultrastruct. Res. 24: 286–294. (2.3.2)
Milne, A. A., 1926. *Winnie-the-Pooh*. E. P. Dutton, New York. (N.1.2, N.1.6)
Mischiati, M., H.-T. Lin, P. Herold, E. Imler, R. Olberg, and A. Leonardo, 2015. Internal models direct dragonfly interception steering. Nature 517: 333–338. (2.3.4)
Morton, A. C., 1982. The effects of marking and capture on recapture frequencies of butterflies. Oecologia (Berl) 53: 105–110. (1.6.2)
Murlis, J., J. H. Elkinton, and R. T. Cardé, 1992. Odor plumes and how insects use them. Ann. Rev. Entomol. 37: 505–532. (2.4.2)
Murphy, D. D., and P. R. Ehrlich, 1983. Crows, bobs, tits, elfs, and pixies: the phoney "common name" phenomenon. J. Res. Lepidopt. 22: 154–158. (A)
Murphy, D. D., and P. R. Ehrlich, 1984. On butterfly taxonomy. J. Res. Lepidopt. 23: 89–93. (A)
Nachtigall, W., 1965. Die aerodynamische Funktion der Schmetterlingsschuppen. Naturwissenschaften 52: 216–217. (2.2.3, N.2.8)
Nachtigall, W., 1967. Aerodynamische Messungen am tragflugel system segelinde Schmettenlinge. Z. vergl. Physiol. 54: 210–231. (N.2.10)
Nathan, R., W. M. Getz, E. Revilla, M. Holyoak, R. Kadmon, D. Saltz, and P. E. Solomon, 2008. A movement ecology paradigm for unifying organismal movement research. Proc. Natl. Acad. Sci. USA 105: 19052–19059. (1.1)
Newton, I., 1687. *Philosophiae Naturalis Principia Mathematica*. vii + 511 pp. Royal Society (with connivance of E. Halley and S. Pepys), London. 1954 Facsimile, Wm. Dawson & Sons, London. (7.3.3, N.7.7)
Odendaal, F. J., Y. Iwasa, and P. R. Ehrlich, 1985. Duration of female availability and its effect on butterfly mating systems. Am. Nat. 125: 673–678. (3.2.3)
Odendaal, F. J., P. Turchin, and F. R. Stermitz, 1988. An incidental-effect hypothesis explaining aggregation of males in a population of *Euphydryas anicia*. Am. Nat. 132: 735–749. (7.1.2, 8.2.3, 8.2.4)
Opler, P. A., and G. O. Krizek, 1984. *Butterflies East of the Great Plains: An illustrated natural history*. xvii + 294 pp. Johns Hopkins Univ. Press, Baltimore, MD. (1.3.1, Ch. 3 Introd., Ch. 4 Introd., 8.6.3, N.0.4, N.7.2, A, T.A.1)
Orive, M. E., and J. F. Baughman, 1989. Effects of handling on *Euphydryas editha* (Nymphalidae). J. Lepidopt. Soc. 43: 244–247. (1.6.2)
Pajunen, V. I., 1966. Aggressive behavior and territoriality in a population of *Calopteryx virgo* L. (Odonata: Calopterygidae). Ann. Zool. Fennici 3: 201–214. (8.5.4)

Palmer, T. N., 2009. Edward Norton Lorenz. Biogr. Mems Fell. R. Soc. 55: 139–155. (N.2.9)

Papke, R. S., D. J. Kemp, and R. L. Rutowski, 2006. Multimodal signalling: structural ultraviolet reflectance predicts male mating success better than pheromones in the butterfly *Colias eurytheme* L. (Pieridae). Anim. Behav. 73(1): 47–54. (2.3.6)

Pascual, M. A., and P. Kareiva, 1996. Predicting the outcome of competition using experimental data: maximum likelihood and Bayesian approaches. Ecology 77: 337–349. (N.9.2)

Peleg, O., and L. Mahadevan, 2016. Optimal switching between geocentric and egocentric strategies in navigation. R. Soc. Open Sci. 3: 160128 (7 pp.). (N.3.6)

Pericet-Camara, R., M. K. Dobrzynski, R. Juston, S. Viollet, R. Leitel, H. A. Mallot, and D. Floreano, 2015. An artificial elementary eye with optic flow detection and compositional properties. J. R. Soc. Interface 12 (20150414): 1–7. (2.3.3, 10.6.2)

Platt, A. P., 1969. A lightweight collapsible bait trap for Lepidoptera. J. Lepidopt. Soc. 23: 97–101. (1.6.2, 9.3.4)

Platt, A. P., 1983. Evolution of North American admiral butterflies. Bull. Entomol. Soc. Am. 29(3): 10–22. (2.2.5)

Platt, A. P., 1984. Stubby-winged mutants of *Limenitis* (Nymphalidae)—their occurence in relation to photoperiod and population size. J. Res. Lepidopt. 23: 217–230. (5.1.2)

Platt, A. P., and J. F. Allen, 2001. Sperm precedence and competition in doubly-mated *Limenitis arthemis-astyanax* butterflies (Rhopalocera: Nymphalidae). Ann. Entomol. Soc. Am. 94(5): 654–663. (1.3.4, T.1.3)

Platt, A. P., and L. P. Brower, 1968. Mimetic versus disruptive coloration in intergrading populations of *Limenitis arthemis* and *astyanax* butterflies. Evolution 22: 699–718. (1.3.4)

Platt, A. P., R. P. Copinger, and L. P. Brower, 1971. Demonstration of selective advantage of mimetic *Limenitis* butterflies presented to caged avian predators. Evolution 25: 692–701. (2.2.5)

Platt, J. R., 1964. Strong inference. Science 146: 347–353. (0.2, 7.3.1)

Pollock, K. H., J. D. Nichols, C. Brownie, and J. E. Hines, 1990. Statistical inference for capture-recapture experiments. Wildlife Monogr. 107: 1–97. (1.6.4)

Prete, F. R., ed., 2004. *Complex Worlds from Simple Nervous Systems*. xx + 436 pp. MIT Press, Cambridge, MA. (N.0.4)

Putnam, R. J., 1995. Ethical considerations and animal welfare in ecological field studies. Biodivers. Conserv. 4: 903–915. (0.3)

Pyle, R. M. 1984. Rebuttal to Murphy and Ehrlich on common names of butterflies. J. Res. Lepidopt. 23: 89–93. (A)

Railsback, V., and S. F. Grimm, 2012. *Agent-Based and Individual-Based Modeling: A practical introduction*. Princeton Univ. Press, Princeton, NJ. 448 pp. (3.3.2, N.1.13)

Real, L. A., 1991. Animal choice behavior and the evolution of cognitive architecture. Science 253: 980–986. (N.8.15)

Reed, J. M., and A. P. Dobson, 1993. Behavioural constraints and conservation biology: conspecific attraction and recruitment. Trends Ecol. Evol. 8: 253–256. (7.2.1)

Riddle, T. W., A. J. Wadcock, J. Tso, and R. M. Cummings, 1999. An experimental analysis of vortex trapping techniques. J. Fluids Eng. 121: 555–559. (2.2.3)

Root, R. B., and P. Kareiva, 1984. The search for resources by cabbage butterflies (*Pieris rapae*): ecological consequences and adaptive significance of markovian movements in a patchy environment. Ecology 65: 147–165. (Ch. 3 Introd., N.8.8)

Rosenberg, R. H., 1989. Behavior of the territorial species *Limenitis weidemeyerii* (Nymphalidae) within temporary feeding areas. J. Lepidopt. Soc. 43: 102–107. (6.2.3)
Ross, G. N. 1963. Evidence for lack of territoriality in two species of *Hamadryas*, Nymphalidae. J. Res. Lepidopt. 2: 241–246. (6.4)
Rutowski, R. L., 1978. The form and function of ascending flights in *Colias* butterflies. Behav. Ecol. Sociobiol. 3: 163–172. (6.4)
Rutowski, R. L., 2000. Variation of eye size in butterflies: inter- and intraspecific patterns. J. Zool. Lond. 252: 187–195. (2.3.1)
Rutowski, R. L., and J. Alcock, 1989. Insect mating systems in the Sonoran Desert of North America. J. Arid Environ. 17: 157–165. (0.2)
Rutowski, R. L., and G. W. Gilchrist, 1988. Male mate-locating behavior in the desert hackberry butterfly *Asterocampa leila* (Nymphalidae). J. Res. Lepidopt. 26: 1–12. (5.2)
Rutowski, R. L., and E. J. Warrant, 2002. Visual field structure in the Empress Leilia, *Asterocampa leilia* (Lepidoptera, Nymphalidae): dimensions and regional variation in acuity. J. Comp. Physiol. A 188(1): 1–12. (2.3.1)
Rutowski, R. L., C. E. Long, and R. S. Vetter, 1981. Courtship solicitation by *Colias* females. Am. Midl. Nat. 105: 334–340. (3.1)
Rutowski, R. L., G. W. Gilchrist, and B. Terkanian, 1988. Male mate-locating behavior in *Euphydryas chalcedona* (Lepidoptera: Nymphalidae) related to pupation site prefereces. J. Insect Behav. 1: 277–289. (5.2)
Rutowski, R. L., L. Gislén, and E. J. Warrant, 2009. Visual acuity and sensitivity increase allometrically with body size in butterflies. Arthopod Struct. Devel. 38: 91–100. (2.3.1, N.10.1)
Salsburg, D., 2001. *The Lady Tasting Tea: How statistics revolutionized science in the twentieth century*. xi + 340 pp. (N.1.11)
Saunders, A. A., 1932. *Butterflies of the Allegany State Park*. 270 pp. N. Y. State Mus. Handbook 13. Albany. (N.2.4)
Scott, J. A., 1968. Hilltopping as a mating mechanism to aid the survival of low density species. J. Res. Lepidopt. 7: 191–204. (8.2.2)
Scott, J. A., 1972–3. Mating of butterflies. J. Res. Lepidopt. 11: 99–127. (2.4.1)
Scott, J. A., 1973. Down-valley flight of adult theclini (Lycaenidae) in search of nourishment. J. Lepidopt. Soc. 27: 283–287. (2.3.3, 8.2.2)
Scott, J. A., 1974. Mate-locating behavior of butterflies. Am. Midl. Nat. 91: 103–117. (1.2, 1.4, 1.7, 4.1.3, 7.3.2, N.1.1)
Scott, J. A., 1986. *The Butterflies of North America: A natural history and field guide*. xiii + 581 pp. Stanford Univ. Press, Stanford, CA. (1.2, 1.4, 1.6.2, 2.2.1, 2.5, Ch. 4 Introd., Ch. 5Introd., 5.1.2, 6.2.1, N.0.4, N.1.1, N.1.4, F.1.1, T.1.1)
Scudder, S. H., 1889. *The Butterflies of the Eastern United States and Canada with Special Reference to New England*, in three volumes. Published by the author, Cambridge, MA. 1958 pp. and 89 plates. (Ch. 4 Epigraph, Ch. 6 Epigraph, 8.6.3, 9.3.5, N.2.11, N.5.4)
Scudder, S. H., 1895. *Frail Children of the Air: excursions into the world of butterflies*. 279 pp. Houghton Mifflin, Boston. (N.0.4)
Scudder, S. H., 1899. *Everyday Butterflies: A group of biographies*. Houghton Mifflin, Boston. (2.2.4, 9.3.5)
Seeley, T. D., 2010. *Honeybee Democracy*. 273 pp. Princeton Univ. Press, Princeton, NJ. (N.0.4)

Shapiro, A. M., 1974. Butterflies and skippers of New York State. Cornell Univ. Exp. Sta. Search (Agriculture: Entomology: Ithaca 6) 4(3): 1–60. (0.2, Ch. 3 Introd., N.1.1, N.4.1, N.7.2, A, B.3, T.A.1)

Shields, O. 1967. Hilltopping. J. Res. Lepidopt. 6: 69–178. (2.3.3, 8.2.2, T.1.3)

Shreeve, T. G., 1981. Flight patterns of butterfly species in woodlands. Oecologia 51: 289–293. (9.2, T.9.1)

Shreeve, T. G., 1984. Habitat selection, mate location, and microclimatic constraints on the activity of the speckled wood butterfly *Pararge aegeria*. Oikos 42: 371–377. (7.2)

Sibatani, A., 1973. Taxonomic significance of reflective patterns in the compound eye of live butterflies: a synthesis of observations made on species from Japan, Taiwan, Papua New Guinea, and Australia. J. Lepidopt. Soc. 27: 161–175. (2.3.2, 2.3.6)

Silberglied, R. E. 1977. Communication in the Lepidoptera. Pp. 362–402 in T. A. Sebeok, ed., *How Animals Communicate*. Indiana Univ. Press, Bloomington. (2.1.2, 7.3.2)

Simonsen, T. J., N. Wahlberg, A. Z. Brower, and R. de Jong, 2006. Morphology, molecules and fritillaries: approaching a stable phylogeny for Argynnini (Lepidoptera: Nymphalidae). Insect Syst. Evol. 37: 405–418. (A)

Sims, E. H. 1972. Fighter tactics and strategy 1914–1970. Second ed. 1980. Aero Publishers, Fallbrook, CA. (6.3.4)

Sims, S. R. 1984. Reproductive diapause in *Speyeria* (Lepidoptera: Nymphalidae) J. Res. Lepidopt. 23: 211–216. (5.1.2)

Singer, M. C., and P. Wedlake, 1981. Capture does affect the probability of recapture in a butterfly species. Ecol. Entomol. 6: 215–216. (1.6.2)

Skellam, J. G., 1951. Random dispersal in theoretical populations. Biometrika 38: 196–218. (1.7, Ch. 3 Introd., 3.2.5)

Smith, G. C., 1995. An evaluation of the methods used to construct life tables in capture-mark-recapture studies. Theor. Popul. Biol. 47: 180–190. (1.6.2, 1.6.4)

Smith, J. N. M., 1974. The food searching behaviour of the European thrushes. II. The adaptiveness of the search patterns. Behaviour 49: 1–61. (8.2.4)

Southwood, T. R. E., 1977. Habitat, the templet for ecological strategies? J. Anim. Ecol. 46: 337–366. (1.7, Ch. 9 Introd., Ch. 10 Introd.)

Southwood, T. R. E., 1978. *Ecological Methods*, 2nd edition, revised. Chapman & Hall, London. (1.6.4)

Southwood, T. R. E., and H. N. Comins, 1976. A synoptic population model. J. Anim. Ecol. 45: 949–965. (9.3.5, N.5.4)

Srinivasan, M. V., M. Lehrer, and G. A. Horridge, 1990. Visual figure-ground discrimination in the honeybee: the role of motion parallax at boundaries. Proc. R. Soc. Lond. B 238: 331–350. (2.3.4)

Srinivasan, M. V., S. Zhang, M. Altwein, and J. Tautz, 2000. Honeybee navigation: nature and calibration of the "odometer." Science 287: 851–853. (2.3.4)

Srygley, R. B., 1999. Locomotor mimicry in *Heliconius* butterflies: contrast analyses of flight morphology and kinematics. Phil. Trans. R. Soc. Lond. B 354: 203–214. (2.2.5)

Srygley, R. B., and E. G. Oliveira, 2001. Orientation mechanisms and migration strategies within the flight boundary layer. Pp. 183–206 in I. P. Woiwod and D. R. Reynolds, eds., *Insect Movement Mechanisms and Consequences*. Proc. R. Entomol. Soc. 20th Symposium, CABI Publ., Wallingford, UK. (8.2.1)

Srygley, R. B., and A. L. R. Thomas, 2002. Unconventional lift-generating mechanisms in free-flying butterflies. Nature 420: 660–665. (2.2.2, 2.2.3)

Stamps, J., 1995. Motor learning and the value of familiar space. Am. Nat. 146: 41–58. (9.4)
Stark, W. S., K. L. Frayer, and M. A. Johnson, 1979. Photopigment and receptor properties in *Drosophila* compound eye and ocellar receptors. Biophys. Struct. Mechanism 5: 197–209. (2.3.2)
Stavenga, D. G., 1979. Pseudopupils of compound eyes. Pp. 357- 439 in H. Autrum, ed., *Comparative Physiology and Evolution of Vision in Invertebrates, A: Invertebrate Photoreceptors, Handbook of Sensory Physiology*. Springer-Verlag, Berlin. (2.3.2, 2.3.6, N.2.21)
Stavenga, D. G., M. Kinoshita, and E.-C. Yang, 2001. Retinal regionalization and heterogeneity of butterfly eyes. Naturwissenschaften 88: 477–481. (2.3.1, 2.3.3)
Stewart, F. J., M. Kinoshita, and K. Arikawa, 2015. The butterfly *Papilio xuthus* detects visual motion using chromatic contrast. Biology Letters 11: 20150687: 1–4. (2.3.4)
Stokes, M. A., and T. L. Smiley, 1968. *An Introduction to Tree-Ring Dating*. Univ. Chicago Press, Chicago. 73 pp. (1.6.1)
Swearingen, J. M., and S. Adams, 2007. Fact sheet: Japanese stiltgrass. Pp. 1–4. Plant Conservation Alliance, www.nps.gove/plants/alien. (N.7.2)
Swihart, S. L., 1967. The neural basis of colour vision in the butterfly, *Heliconius erato*. J. Insect Physiol. 18: 1015–1025. (2.3.1, 2.3.4)
Tang, Y., C. Zhou, X. Chen, and H. Zheng, 2013. Foraging behavior of the dead leaf butterfly, *Kallima inachus*. J. Insect Sci. 13: 58 (16 pp.) . (N.2.23)
Tennyson, A. Lord, 1842, *Poems*. E. Moxton, London. (Ch. 10 Epigraph)
Tinbergen, N., 1951. *The Study of Instinct*. Oxford Univ. Press, Oxford, UK. vii + 228 pp. (N.8.15)
Tinbergen, N., B. J. D. Meeuse, L. K. Boerema, and W. W. Varossieau, 1942. Die Balz des Samtfalters, *Eumenis (= Satyrus) semele* (L.). Z. Tierpsychol. 5: 182–226. English translation: N. Tinbergen, 1972. The courtship of the grayling *Eumenis (= Satyrus) semele* (L.) (1942). Chapter 5, pp. 197–249, in *The Animal in its World: Explorations of an ecologist 1932–1972, vol. 1 Field Studies*. Allen & Unwin, London. (1.6.2, 2.1.2, 7.3.1)
Treusch, H. W., 1967. Bisherunbekanntes gezieltes Duftanbieten paarungsbereiter *Argynnis paphia*-Weibchen. Naturwissenschaften 54 :592. (2.4.1)
Tsai, C-C., N. Shi, J. Pelaez, N. Pierce, and N. Yu, 2017. Butterflies regulate wing temperatures using radiative cooling. Proc. Conf. on Lasers and Electro-optics (CLEO) @ Opt. Soc. Am. FTh3H.6.pdf (2 pp.). (2.2.6)
Tsuji, J. S., J. G. Kingsolver, and W. B. Watt, 1986. Thermal physiological ecology of *Colias* butterflies in flight. Oecologia (Berl.) 69: 161–170. (4.4, 6.3.1)
Turchin, P., 1998. *Quantitative Analysis of Movement: Measuring and modeling population redistribution in animals and plants*. ix + 396 pp. Sinauer, Sunderland, MA. (1.6.4, 7.1.2, 7.3.2, 8.2.3, 8.2.4)
Turchin, P., F. J. Odendaal, and M. D. Rausher, 1991. Quantifying insect movement in the field. Environ. Entomol. 20(4): 955–963. (N.8.8)
Turner, J. D., 1990, Vertical stratification of hilltopping behavior in swallowtail butterflies (Papilionidae). J. Res. Lepidopt. 44(3): 174–179. (8.2.2)
Turner, J. R. G., 1963. A quantitative study of a Welsh colony of the large heath butterfly, *Coenonympha tullia* Muller (Lepidoptera). Proc. R. Entomol. Soc. Lond. (A) 38: 3–111. (Ch. 3 Introd.)

Tutty, O. R., M. Buffoni, R. Kerminbekov, R. Donelli, F. De Gregorio, and E. Rogers, 2013. Control of flow with trapped vortices: theory and experiments. J. Flow Control Meas. Vis. 5: 89–109. (2.2.3)

Vane-Wright, R. I., and P. R. Ackery, eds., 1984. *The Biology of Butterflies.* xxiv + 429 pp. Academic Press, Orlando, FL, reprinted 1989, Princeton Univ. Press, Princeton, NJ. (N.0.4)

Vickers, N. J., and T. C. Baker, 1997. Flight of *Heliothis virescens* males in the field in response to sex pheromone. Physiol. Entomol. 22: 277–285. (2.4.2)

Vogel, S., 1994. *Life in Moving Fluids: The physical biology of flow.* Princeton Univ. Press, Princeton, NJ. xiii + 467 pp. (2.2.3, 2.2.7)

Waage, J. K., 1974. Reproductive behaviour and its relation to territoriality in *Calopteryx maculata* (Beauvois) (Odonata: Calopterygidae). Behaviour 47: 240–256. (6.1)

Waage, J. K., 1988. Confusion over residency and the escalation of damselfly territorial disputes. Anim. Behav. 36: 586–595. (8.5.4)

Wakakuwa, M., D. G. Stavenga, and K. Arikawa, 2007. Spectral organization of ommatidia in flower-visiting insects. Photochem. Photobiol. 83: 27–34. (2.3.3)

Watt, W. B., 1991. Biochemistry, physiological ecology, and population genetics—the mechanistic tools of evolutionary biology. Funct. Ecol. 5: 145–154. (Ch. 9 Introd.)

Watt, W. B., P. A. Carter, and K. Donohue, 1986. Females' choice of "good genotypes" as mates in promoted by an insect mating system. Science 233: 1187–1190. (5.2)

Webb, B., 2002. Robots in invertebrate neuroscience. Nature 417: 359–363. (2.3.3)

Weed, C. M., 1917 (1923). *Butterflies Worth Knowing.* 289 pp. Doubleday, Page & Co., New York. (2.2.4, 2.4.2, 2.4.3, Ch. 5 Introd.)

Wehner, R., 1992. Arthropods. Ch. 3, pp. 45–144 in F. Papi, ed., *Animal Homing.* Chapman & Hall, London. (5.1.4)

Weis-Fogh, T., 1973. Quick estimates of flight fitness in hovering animals, including novel mechanisms for lift production J. Exp. Biol. 59(1): 169–230. (2.2.2, 2.2.3)

Weiss, S. B., and D. D. Murphy, 1988. Fractal geometry and caterpillar dispersal or how many inches can an inchworm inch. Funct. Ecol. 2: 116–118. (N.8.8)

Wickman, P.-O., 1986. Courtship solicitation by females of the small heath butterfly, *Coenonympha pamphilus* (L.) (Lepidoptera: Satyridae) and their behavior in relation to male territories before and after copulation. Anim. Behav. 34: 153–157. (3.1)

Wickman, P.-O., 1992. Sexual selection and butterfly design—a comparative study. Evolution 46: 1525–1536. (2.2.1)

Wickman, P.-O., and C. Wiklund, 1983.Territorial defence and its seasonal decline in the speckled wood butterfly (*Pararge aegeria*). Anim. Behav. 31: 1206–1216. (1.7, 2.3.3, 7.2, 7.3.2, 7.3.3, F.7.6)

Wickman, P.-O., E. Garcia-Barros, and C. Rappe-George, 1995. The location of landmark leks in the small heath butterfly, *Coenonympha pamphilus*: Evidence against the hot-spot model. Behav. Ecol. 6: 39–45. (3.1)

Wiens, J. A., T. O. Crist, and B. Milne, 1993. On quantifying insect movements. Environ. Entomol. 22: 709–715. (N.8.8)

Wiernasz, D. C., 1983. Range expansion and rapid evolution in *Coenonympha tullia* (Lepidoptera): Ecological and genetic change in a new environment. Ph.D. Dissertation, Department of Biology, Princeton University, Princeton, NJ. xii + 202 pp. (0.2, 0.5, 1.56.4, 8.6.1, N.9.4)

Wiernasz, D. C., 1989. Ecological and genetic correlates of range expansion in *Coenonympha tullia*. Biol. J. Linnean Soc. 38: 197–214. (0.2, 8.6.1, A)
Wilbur, H. M., and J. M. Landwehr, 1974. The estimation of population size with equal and unequal risks of capture. Ecology 55: 1339–1348. (1.6.4)
Wilensky, U., 1999. NetLogo http://ccl.northwestern.edu/netlogo/. Center for Connected Learning and Computer-Based Modeling, Northwestern University, Evanston, IL. (0.5, 1.6.5, 7.4, N.0.12)
Wilkerson, R. C., and J. F. Butler, 1984. The Immelmann Turn, a pursuit maneuver used by hovering male *Hybomitra hinei wrighti* (Diptera: Tabanidae). Ann. Entomol. Soc. Am. 77: 293–295. (6.3.4)
Williams, E. H., 2009. Lifestyles of the scaled and beautiful: Pearl and northern crescents. Am. Butterflies, Summer 2009: 4–13. (A)
Worth, C. B. 1980. An elegant harness for tethering large moths. J. Lepidopt. Soc. 34: 61–63. (F.1.8)
Yagi, N., and N. Koyama, 1963. *The Compound Eye of Lepidoptera: Approach from organic evolution*. 319 pp. Shinkyo Press, Maruzen, Tokyo. (2.3.6, F.2.15)
Yeung, W. W. H., 2006. Lift enhancement on unconventional airfoils. J. Mekanikal 22: 17–25. (2.2.3)
Yokoyama, N., K. Senda, M. Iima, and N. Hirai, 2013. Aerodynamic forces and vortical structures in flapping butterfly's forward flight. Physics Fluids 25: 021902: 1–24. (2.2.3)
Zheng, L., T. L. Hedrick, and R. Mittal, 2013. Time-varying wing-twist improves aerodynamic efficiency of forward flight in butterflies. PLOS ONE 8(1): e35360: 1–10. (2.2.3)

# Index

Note: Page numbers in *italic* refer to illustrations; "*t*" indicates information in tables.

age: and behavior, 143–44; wing damage and estimation of, 20–21, 46, 157, 193, *196, 197*
aggregation, *xv,* 162; and cues-and-rules responses, 166–67; at food sources, *xii,* 7, 25, 29, 138–40, 158, 203; and landscape features, 165–66
Allen, A. P., 9
Anderson, D. J, 24, 115, 119, 121
Anthony, G. Scott, *xii*
*The Aurelian* (Harris), 110

Bekoff, M., *xiv*
Bhushan, B., 40
biases, *xiii–xv,* 12, 17
Bixler, G. D., 40
boundary behaviors, *xiii,* 8–9, 162, *172,* 181, 188, 206–7; and cue-and-rule responses, 164–72, 184; and defense of site, 133, 136; of Eyed Browns, 162, 170, 184; of Fritillaries, 170; ignoring boundaries, 84; perches and, 133, 136; and random walk models, 83–85, *84, 85t,* 94–95, 164, 166; of Ringlets, 74, 75–78, 83–85, *84, 85t,* 87, 164–65, 177, 184; and site-fidelity, 105; and site recognition, 184; of Viceroys, 133, 136, 162, 170
bourgeois strategy (home-court advantage), 25, 145–47, 150, 158, 204
Brackenberry, J., 34
Burger, J., 110, 182
"Butterfly Cyclotron" experiment, 102–4, *103, 104*
"Butterfly Rotisserie" lure, *viii,* 18–19, 30, 148, *149, 152,* 152–53, 227–28, *228*

Cabbage Whites *(Pieris rapae), 3,* 5, 34–37, *35;* aerodynamic parameters for, *50t;* maladaptive flight in, *47;* wingbeat and flight of, 41, *42*
Cartwright, B. A., 115, 177, 178

Cech, R., 7, 110, 182
Chai, P., 34
Checkerspot *(Euphydryas chalcedona),* 118
clamps to hold butterflies, 16, *16,* 221, *222, 227*
Clark, A. H., 45, 65–66, 110, 115, 118, 128
*Coenonympha tullia inornata. See* Ringlets, Inornate *(Coenonympha tullia inornata)*
collections, scientific reference, *xiii–xiv*
Collett, T. S., 115, 177, 178
complex behaviors: cues-and-rules as explanation for, *xv,* 1–2, 26, 27, 29, 32, 70, 157–61, 164–73, 184, 187–88, 210, 212 (*See also* cues, environmental; rules, simple internal); sensory capabilities and, 12, 29, 32, 69–70. *See also* decision making
"conga lines," 25, 141, 145–50, *147,* 204, 208
Cordero, C., 9, 17
Corner, E. J. H., 93–94
cues, environmental, *xv,* 1, 10–11, 23, 25, 184; and adaptation to habitat, 204–5; and aggregation, 166–67; and boundary behaviors, 170–72, 184; "channels" in landscape as, 28, 97, *98,* 100, 102–4, 105, 125, 176, 179, 208; and complex behaviors, 25–27, 28, 29, 104–5, 148–50, 157–59, 177–78, 179, 202–12; decision making *vs.* cue-and-rule responses, 29, 160, 175–76, 179–80, 184–85, 201, 208; and experimentation, 18–19; and female behavior, 2, 25, 68, 148–50; for finding and staying in habitat, 102, 125, 164–76 (*See also* boundary behaviors); and male-male interactions, 150–51, 204–5, 208; and mate-finding, 25, 29–32, 53, 72, 149–50, 212; and mechanical behavior, 58, 157–58, 160, 176, 184–85; and oviposition, 2, 68; and perch preferences, 175; and return to previous behavior, 150; sensory cues (*See* hearing; olfaction; vision); and site-fidelity, 104–5 (*See also* Fortuitous Site-Fidelity Model). *See also* horizon; rules, simple internal

Dalton, S., 14–15, 18, 34
*Danaus plexippus. See* Monarchs *(Danaus plexippus)*
data-generated narcosis, *xiii*
Davies, N. B., 25, 145, 147, 150
decision making, 12, 25–26, 174–80, 184, 185, 201, 205, 208–10; *vs.* cue-and-rule responses, 156–60, 175–76, 179–80, 184–85, 201, 208; and Eyed Browns, 174, 177, 179, 181–82, 184; and Fortuitous Site-Fidelity Model as null, 171; and Fritillaries, 115, 174–75, 179, 182, 184, 205, 208; and individual behavior and, 176–77, 178, 208–11; and memory, 177; and Pearly Eyes, 29, 159, 175–76, 178, 179–80, 183, 185, 205, 208–9; and Ringlets, 174, 177–78, 181, 184; and sensory capability, 29, 115; and site-fidelity *vs.* site recognition, 176–78, 184–85, 205, 209–10; and strategic *vs.* mechanistic behavior, 156–59
defense: and boundary behaviors, 133, 136, 185; of "dynamic territories," 129; in Eyed Browns, 11, 24; and "home-court advantage," 25, 145–47, 150, 158, 204; in Pearly Eyes, 25, 100; of perches, 118, 128–29, 136; of specific site, 10, 25, 100, 108, 129, 133–36, 178, 205; and territoriality, 10, 25, 100, 108, 129, 136, 178; in Viceroys, *xii*, 11, 25, 128–37, *134, 135,* 178, 185, 203, 205
devices. *See* machinery
Dickinson, Emily, 74

Ehrlich, P. R., 117
Ellington, C. P., 34
*Enodia anthedon. See* Pearly Eyes *(Enodia anthedon)*
equipment. *See* machinery
Eyed Browns *(Satyrodes eurydice),* 5, *6t,* 7, 96–97, *98,* 216–17; boundary behavior in, 162; "Butterfly Cyclotron" experiment, 102–4, *103, 104;* contest displacement in male-male interactions, 106–7, *107;* and decision making, 174, 177, 179, 181–82, 184; female behaviors in, 11, 96, 99, 218; and Fortuitous Site-Fidelity Model, 22–23, 108–9, 177; habitat of, *7, 15,* 96–97, *97, 98,* 100, 108, 161, 162, *163,* 174; lifespan of, 11, 99; local persistence of, *192, 194, 219t;* male-male interactions in, *30t, 31,* 99–100, *99t,* 106–7, *107, 107t;* mate-finding, 11, 98–99; oviposition in, 11, 96, 218, *220t;* as

patrollers, 28, *97,* 98–99, 203; as perchers, *3,* 100–101; and predator exposure, 199; resightings of, *101, 101t;* and site-fidelity, *xii,* 24, 28, 96–97, 101–2, 104–9, 203; and territoriality, 11, 100–102, 108; thermal ecology of, 106; Viceroy behavior compared with, 97–98, 100–102; vision in, *62, 73;* wingbeat and flight of, 70
Eye-Portrait Studio, 227

FANCYNEXTKISS.NLOGO, 234
female behavior: androgenic bias and study of, 2, 28, 207–8; and environmental cues, 2, 25, 68, 148–50; in Eyed Browns, 11, 96, 99, 218; in Fritillaries, 11, 68, 110–21; and male-female chases, 148–49, *152,* 153–55, *155,* 156, 204, 208; and male-male-female conga lines, 149, 208; and mate-finding, 28, 72, 75, 93–95, 117, 125–28, 178, 179, 207–8; and monandry, 9, 17, 94, 207–8; patrolling, 11; in Pearly Eyes, 67–68, 138–39, 154; ranging, 201; in Ringlets, 67, 75, *79t,* 81, 87, 94, 218; and traplining, 111–12, 117–18; Viceroy, 125–28, 136, 220. *See also* oviposition
Fight-or-Follow Model (NetLogo), 23, 25, 151–53, 158–59, 204–5
flight: and aerodynamic parameters, 49–52, *50t;* flapping modes, 41–45; hindwing and forewing functions in, 49–52; and interactions between butterflies, 23; maladaptive, *47;* and mate-finding, 27, 32 *(See also* patrollers; perchers); and mimicry, 45, 71; of perchers *vs.* patrollers, 191; photographing butterflies in, 18; and predation, 33–34; scales and aerodynamics, 37–41, 47, 195, 207; simple environmental cues and, 70; "slithering" mode, 33, *35,* 36–38, 40–41, 43, *43,* 70–71, 207; and vision, 61, 69–70; vortices and flight dynamics, 34–41, 50–52, 71; and wingbeat cycles, 34–36, *37,* 40–45, *42. See also* wings
flyways: and "channels" in landscape, 28, 97, *98,* 100, 102–4, 105, 125, 176, 179, 208; landscape and orientation of, *xiii,* 24, 70, 72, 114–15, 165 (*See also* horizon); method for mapping, 19; perches as nodes along, 173, *173, 174,* 175; random, 81–93; shared flyways, 115, 117–18, 123, 177; and territorial contests, 24, 106–7, 128–29
food sources: aggregation at, *xii,* 7, 25, 29, 138–40, 158, 203; habitat and, *161t;* larval, 4, 24, 98–99, 111–12, 117–18, *120,* 122–23,

126–27, 136, *161t*, 174–75; nectaring, 68, 117–18, 161, 199; olfaction and finding, 66–67, 73
Fortuitous Site-Fidelity Model (NetLogo), 23, 24, 28, 104–9, *105,* 174, 199, 205, 208, 210; and circuits, 121–23, 174; and fidelity vs. recognition, 108–9, 176–78, 205; as null, 171
Forward Vagrancy Model. *See* Tactical Forward Vagrancy Model (NetLogo)
Fritillaries *(Speyeira cybele),* 3, 5, *6t,* 7–8, 217; aerodynamic parameters for, 49–52, *50t;* and decision making, 115, 174–75, 179, 182, 184, 205, 208; female behaviors in, 11, 68, 110–21, 220; and Fortuitous Site-Fidelity Model, 177; habitat of, *8t, 14, 15,* 110, 124, 161, *163;* horizon and flyways of, *114;* larval food of, 73, 110, 111, 117–18, *120,* 122–23; lifespan of, *112, 119;* local persistence of, *192, 194, 219t;* maladaptive flight in, *47;* and male-male interactions, *30t, 31t;* and mate-finding, *9t, 10t,* 57–58, 111–21, 117 *(See also* traplining); olfaction in, 65–66, 68, 73; oviposition in, 67–68, 110, 111, 118, 123, 220; pheromones and odor of, 110–11, 117; and protandry, 11, 80, 110, 112, *112,* 118, 122; and repeated closed-circuit patrols, *xii,* 5, 11, 24, 29, 110, 114–17, *115, 116,* 120–24, 162, 177, 182, 185, 203, 208, 211; re-sightings of, *101t;* rules for, 174–76; scales of, *39;* and traplining, *xii, 7,* 24, 108, 111–13, 117–18, *120,* 122, 123, 190; and "traveling salesfolk" problem, 24, 118–21, 123–24; vision of, 57–58, *62,* 63, 73; wings and flight of, *43,* 43–44, *44*

game theory, 25, 147, 150
Georgia, Jeff, 96, 110
Gilbert, L. E., 117
Glanville, Eleanor, *ix*
Glassberg, J., 7
Gochfeld, M., 110, 182
Gosse, Philip Henry, 138
GPS tracking, 20–21
Grimm, S. F., 87
"guarding" behavior, *153*

habitat, *7,* 8, *14, 15,* 245n1; adaptation and cues and rules for behavior, 204–5; boundaries and behavior, *xiii,* 8–9, 75–78, 83–85, *84, 85,* 162, 166–72, 181, 184, 188, 206; day-to-day sightings of studied species, *8t;* of Eyed Browns, *7, 15,* 96–97, *97, 98,* 100, 108, 161, 162, *163,* 174; finding, 28, 69–70, 102, 125, 164–76 *(See also* boundary behaviors); of Fritillaries, *8t, 14, 15,* 110, 124, 161, *163;* and landscape features, 165–66, 245n1; and mate-finding, 28–29, *161;* orientation in and navigation of, 70, 104–6 *(See also* horizon); of Pearly Eyes, *7, 14,* 138–39, 141, 161, 162; postagricultural landscape as, 181, 185–86; and predators, *161t;* and ranging-restrictions, *xii;* resources and conditions provided by, *161t;* of Ringlets, *7, 14,* 28, *76, 77,* 161, 162, *163,* 174; summary and speculations on behavior and, 211; Turning Bias Model for Habitat Selection, 167–70, *168, 169;* of Viceroys, *7, 15, 97,* 100, 136, *163;* vision and adaptation to, *64,* 204–5
Hackberries *(Asterocampa liela),* 118
Hairstreaks, 3, 4–5, *6t, 18*
Harris, Moses, 110
hearing, 69, 73
Heinrich, B., 74–75, 80
Heinz, W. C., 28, 76
hill-topping, 165
"home-court advantage," 25, 145–47, 150, 158, 204
horizon: artificial horizon created with mirrors, 19, 24, 55, *56, 130,* 130–31, *131,* 165, 221, 227–28, *229;* and mate-finding, 136; navigation and "V-notches" in, 17, 28–29, 32, 46, 55–57, 65, 69–70, 75, 97, 102, 114, *114,* 117, 125, 164–66, 174–79, 184, 204; and perching areas, 128–31; vision and orientation to, 70, 204
Horn, Charles M., *x*
Horn, David J., *x*
Horn, Elizabeth, *x–xi*
Horn, Eric, *xi*
Horn, Henry E., *x*
Horn, Jennifer, *xi*
Horn, William M., *x*
Horridge, G. A., 55, 58
Hu, H., 39–40

identification: and ambiguity, 7; naming of individuals, 7; number marking of individuals, 5, 15–17, *16,* 143–44, 213–14
individual behavior, 12, 25, 29, 153, 200; and decision making, 176–77, 178, 208–10, *209;* and learning, 201, 208, 210
inference, 1, 12, 22

Koyama, N., 61–65
Krizek, G. O., 182

Large Heaths. *See* Ringlets
larval food, 4, 24, 73, 98–99, 110, 111, 117–18, *120,* 122, 123, 126–29, 136, 160–62, *161t,* 174
learning, 140, 187, 188, 193, 201, 205, 208, 210
Lederhouse, R. C., 129
Lévy Walk, 83, 85–90, 95, 204
lifespan: of Eyed Browns, 11, 99; female *vs.* male, 80, *80,* 93, 207–8; of Fritillaries, 11, *112, 119,* 190, 192, 207; individual variability and, 20; "local lifespan" as term, 21, 79; and reproductive opportunities, 24, 80; of Ringlets, 24, 74, 80, *92,* 93, 94, 190, 192, 207–8; of Viceroys, 11, 124, 192; waiting and, 207–8; wing damage and estimation of age, 20–21, 46, 157, 193, *196, 197*
*Limenitis archippus. See* Viceroys *(Limenitis archippus)*
Lombardi, V., 28, 76
lures: artificial baits, 25, 178, 229, *230;* "Butterfly Rotisserie" device, *viii,* 18–19, 30, 148, *149, 152,* 152–53, 227–28, *228;* responsiveness to, 25, 30–31; tethered butterflies, 18, *18,* 30, 148

MacGregor, J. N., 121
machinery: "Butterfly Rotisserie" lure, *viii,* 18–19, 30, 148, *149, 152,* 152–53, 227–28, *228;* clamps to hold butterflies, *xi,* 16, *16,* 221, *222, 227;* compound-eye simulator, 225–26; Eye-Portrait Studio, 221, 227, *227;* marking and mapping gear, 221; portable strobe flash, 224–25; timer, behavioral, 221–24, *223*
male-female chases, 148–49, *152,* 153–55, *155,* 156, 204, 208. *See also* conga lines
male-male interactions, *155;* and "conga lines," 25, 141, 145–50, *147,* 149, 204, 208; as contests, 106–7, 126, 145, 153–56, 180, 208 (*See also* home court advantage *under this heading*); and cues-and-rules responses, 150–51, 204–5, 208; and decision making, 179; and escalation, 25, 145, 150, 158, 204; in Eyed Browns, *31,* 99–100, *99t, 103t,* 106–7, *107;* in Fritillaries, *30t, 31t;* and "home-court advantage," 25, 145–47, 150, *151t,* 158, 204; in Pearly Eyes, 25, 100, 104–5, 208; and perch displacement, 99–100, *102t,* 106–7, *107,* 126, 129, 136, 151; in Ringlets, *30t, 31t;* "spinning wheel flight," 59, 145, *146,* 148–49, 152, 156, 180, 205; and territoriality, 10, 100, 126, 136; in Viceroys, *30t, 31t,* 126, 129–31, 133–34, *135,* 136–37
mapping, 19–20, 87, 112–13, 180, 221; maps as geometric models, 23
Matchett, W.H., 74
mate-finding: androcentric bias and assumptions about, 2, 3–4, 28, 207–8; courtship behavior, 28, 58, 66, 67, 72, 75, 91, 95, 144–48; environmental cues and, 25, 29–32, 53, 72, 149–50, 212; in Eyed Browns, 11, 98–99; female behaviors, 28, 72, 75, 93–95, 117, 178, 179, 207–8; and flight, 32, 71–72; in Fritillaries, *9t, 10t,* 57–58, 111–21, 117 (*See also* traplining); and habitat, 28–29, *161t;* interaction distances for male butterflies, *30t, 31;* mating statistics, *10t;* monandrous and polyandrous categories of, 17, 94; and olfaction, 32 (*See also* pheromones *under this heading*); "passive" *vs.* "active," 93–94; in Pearly Eyes, *xii,* 6–7, 11, 30–31, 66–67, 139, *139,* 148–50, *154,* 156, *157,* 158; and pheromones, 66–67, 73, 110, 117, 179; and physiology, 27; reproductive strategies and social systems, 11; resource costs of, 24 (*See also* thermal ecology); in Ringlets, 28, 81, 204, 207–8; rules for, 29–32; "single's bar" (lek) congregation, *xii,* 6–7, 11, 139, *139;* traplining and, 117–18, 123; and "traveling salesman problem," 24, 118–21, 123–24; vagrancy as efficient and effective, 94–95; in Viceroys, *xv, 31–32,* 125–28, 136; and visual cues, 28–32, 29–32, 53, 72, 128–32, 205
memory, 177, 209
methods, 1, 27; baiting (*See* lures); "Butterfly Cyclotron" habitat modification, 102–4; capture methods and impact individual life cycle, 20–21; of data analysis, 20–21; equipment (*See* machinery); errors in collection strategies, *xiii–xiv;* and ethical considerations, *xiii–xv;* GPS tracking, 19–20, 20–21; for identification of individuals, *xiv,* 5, 15–17, *16;* for mapping flight paths, 19; models (*See* NetLogo programs); netting techniques, 14–15; Observer Effect and, *xiv–xv;* selection of species for study, 4–7; site selection, *xi–xii;* "Strong Inference," *xiii;* video recording, 18–19
mimicry, 14, 45, *46,* 71, *214*

monandry, 9, 17, 94, 207–8
Monarchs *(Danaus plexippus), 3,* 14; aerodynamic parameters for, 49–52, *50t;* flight mechanics and aerodynamics of, *47,* 50–51, *50t, 51;* flight of, 33, *42,* 43–44, *44,* 46; maladaptive flight in, *47;* and Viceroy mimicry, 14, 45, 71; wingbeat and flight of, 42, *44, 51*

Nachtigall, W., 37–38, 40
naming conventions, 7, 213–15
NetLogo programs, 22–23; FANCYNEXTKISS.NLOGO, 234; Fight-or-Follow Model, 23, 25, 151–53, 158–59, 204–5; Fortuitous Site-Fidelity Model, 23, 24, 28, 104–9, *105,* 121–23, 171, 174, 199, 205, 208, 210; PEARLYEYEMOVIETRAIL.NLOGO, 234–36; RINGLETFANCYHOTCOLDEC18.NLOGO, 230–33; Tactical Forward Vagrancy Model, 22, 24, 59, 74, *89,* 91–95, *92,* 164, 177, 188–89, 204–5

Observer Effect, *xiv–xv,* 5
olfaction: characteristic odors of butterflies, 65–67; "follow your nose" rule, 166; and food finding, 66–67, 73; and interaction between butterflies, 73; and mate-finding, 32, 58 *(See also* pheromones *under this heading);* and oviposition, 66–67; in Pearly Eyes, 66–68, 73, 140, 158, 176, 180, 184, 1140; and pheromones, 66–67, 110–11, 117, 179; *vs.* vision, 68, 73
Oliveira, E. G., 165
Opler, P. A., 182
optic flow, *42,* 52, 58–59, 72–73, 172, 175, 204, 209
Ormerod, T., 121
oviposition, 211, 218–20; delayed, 111; dispersal of eggs, 111, 118, 200, *220, 220t;* and larval food sources, 118, 123, 126, 160; olfaction and, 66–68; and parasitism, 111, 200; site selection for, 2, 11, 111, 118, 123, 127, 136; and traplining, 11

parasitism, 111, 118, 121, 127, 200, 201
patrollers, 2–3, *3,* 10–11, 29; aerodynamic parameters for, 49–52; "Butterfly Cyclotron" experiment and flyways of, 102–4, 107; Eyed Brown as "perchers" who also patrol, *3,* 28, *97,* 98–99, 203; female patrolling, 11; Fritillaries and repeated closed-circuit patrols, 5, 11, 24, 29, 110, *113, 115,* 115–17, *116,* 120–24, 162, 182, 185, 203, 208, 211, 217; Pearly Eyes as "perchers" who also patrol, *3,* 33, 153; *vs.* perchers, 11–12, 191; and predation, 34, 71; Ringlets as, 28, 74–75, 81–83, 94–95; scales and slow flight of, 37–41, 70–71; and site fidelity, 102–7, 178; and sustained flight, 2, *3,* 33, 71–72; and territoriality, 10, 24, 29; and traplining, *xii, xv,* 10, 11, 24, 111–12, 117–18, 117–21, *120,* 123, 178; Viceroys as "perchers" who also patrol, *3, 97,* 128, *133,* 178; wing morphology of, *3,* 33–34, 71–72
PEARLYEYEMOVIETRAIL.NLOGO, 234–36
Pearly Eyes *(Enodia anthedon), xii, 3, 6t,* 7, *39,* 217; aerodynamic parameters for, 49–52, *50t;* and aggregation at food sources, 7, 25, 41, 138–40, *139, 142, 154,* 158–59, 162, 175; artificial baits as lures for observation of, *140,* 140–44, *142, 143;* and Butterfly Rotisserie experiment, 148, *152,* 152–53; "courtship" behavior in, 144–45; and cue-and-rule responses, 148–51, 158–59, 175–76, 180; and decision making, 29, 158–59, 208–9; distribution over habitat, 141 (*See also* aggregation *under this heading);* female behaviors, 67–68, 138–39, 154; Fight-or-Follow Model (NetLogo), 23, 25, 151–53, 158–59, 204–5; and Forward Vagrancy Model, 151; and "guarding" behavior, *152;* habitat of, *7, 14,* 138–39, 141, 161, 162; and hearing, 69, 73; and home court advantage (bourgeois strategy), 25, 145–47, 150, 158–59; and individually variable behavior, 29, 141–44, 153; life span of, 139; local persistence of, *192, 194, 219t;* male-female chases, *146,* 148–49, *152,* 153–55, *155,* 156, 208; male-male-female conga lines, 149, 208; male-male interactions, *30t, 31t,* 145, *146, 152,* 153–56, *154, 155, 155t,* 156, 180, 208 (*See also* home court advantage *under this heading);* and mate-finding, *xii,* 6–7, 11, 30–31, 66–67, 139, *139,* 148–50, *154,* 156, *157,* 158; olfaction in, 66–68, 73, 140, 158, 176, 180, 184, 1140; oviposition in, 67–68, 139, 158, *220t;* as patrollers, 29, 178; and predator exposure, 199–200; re-sightings of, *6t, 101t;* and "single's bars" (leks), *xii,* 6–7, 11, 139, *139* (*See* aggregation *under this heading);* and "spinning-wheel" flights, 136, 145, *146,* 148–49, 152, 156, 180, 204; and Tactical Forward Vagrancy Model, 59; and

Pearly Eyes (*continued*)
  thermal ecology, 25, 153; vision of, 30–31, *63, 73,* 156; wingbeat and flight of, 41, *42, 44,* 44–45, 70
perchers, 2–3, *3,* 10–11, 49; boundaries and perches, 133, 136; and "dynamic territories," 129; and "guarding" behavior, *153;* larval food sources and locations of, 118; and male-male contests, 126, 136, 147; and male-male interactions, 106–7, *107,* 151–53, *151t,* 180; *vs.* patrollers as category, 11–12, 191; and patrolling, *3,* 28, 33, *97,* 98–99, 128, *133,* 153, 178, 203; Pearly Eyes as, *3,* 33, 153; perch displacement in male-male interactions, 99–100, *102t,* 106, *107;* perches as nodes on flight paths, 173, *173, 174,* 175; preferred perches, 29, 70, 125, *127,* 128–33, *129t,* 136, 175; and territoriality, 100–102, 126, 129, 136; time-of-day and perching, *153,* 153–56; Viceroys as, *3, 97,* 128, 132, *133,* 178; wing morphology of, 33–35, 203
pheromones: as detectable by humans, 66–67, 117; and mate-finding, 66–67, 73, 110–11, 117, 179; vortices and concentration of, 36, 58, 73
Platt, A. P., 15, 111
Platt, John Rader, *xiii,* 9, 15
population sizes, calculation of, 20
predation, 29–30; behavior and exposure to, 195–202, 206; by birds, *47,* 69, 198–99; evasion of predators, 34, 131, *198;* flight and escape from, 34; habitats and predator frequency, *161t;* hearing and evasion of, 69, 73; and mimicry, *46;* oviposition and, 121, 127; and patrolling, 71; and perching sites, 131; vision and recognition of predators, 29; by wasps or hornets, 69, 199
protandry, 11, 80, 110, 112, *112,* 118, 122, 207–8
pseudopupils, 61–65, *62, 63*
"pseudo replication," 22

Railsback, V., 87
randomization, 20
random movement: Brownian, 81–83, *82,* 86, 88–90, *89, 168;* comparison of random walk models, *89;* and Fight-or-Flight Model, 151, 204–5; and flight distance, *189;* forward-biased random walk, 88–90, 167, *168, 188–89* (*See also* Tactical Forward Vagrancy Model); Lévy flight, 83, 86, *86,* 88–90, 95, 204. *See also* Tactical Forward Vagrancy Model (NetLogo)
ranging: and adaptive social systems, 188, 202; female behaviors, 201; and habitat finding, 200–202; and individual behavior, *xiii,* 187, 190–91; and landscape features (*See under* cues, environmental); patterns of (*See* territoriality; traplining; vagrancy); and percher *vs.* patroller categories, 11–12; and population genetics, 187–88, 195, 200–201. *See also* boundary behaviors; flyways; territoriality
reproduction. *See* mate-finding; mating statistics; oviposition; spermatophore counts
re-sighting and re-sighting statistics, 4–6, *6t,* 100–101, *101, 113,* 190–91, *191t*
Reynolds number, 49
RINGLETFANCYHOTCOLDDEC18.NLOGO, 230–34
Ringlets, Inornate (*Coenonympha tullia inornata*), *xii, xii–xiii,* 5, *6t,* 7, *16,* 216; and boundary behaviors, 74, 75–78, 83–85, *84, 85t,* 177; and cue-and-rule responses, 174, 176 (*See* "run to daylight" rule *under this heading*); and decision making, 177–78; female behaviors in, 67, 75, *79t,* 81, 87, 94, 218; habitat of, *7, 14,* 28, *76, 77,* 161, 162, *163,* 174; lifespan of, 24, 74, 78–80, *92, 93, 94,* 190, 192, 207–8; local persistence of, *192, 194, 219t;* local population demography, *78,* 78–79; male-male interactions, *30t, 31t;* and mate-finding, 28, 81, 204, 207–8; oviposition in, 67, 218, *220t;* as patrollers, 28, 74–75, 81–83, *82,* 94–95; and predation, 199; and random walk models, 74–75, *82, 86,* 88–89 (*See also* Tactical Forward Vagrancy Model); re-sightings of, *101t;* and "run to daylight" rule, 28, 75–76, *77;* sex and phenology of, *80;* thermal ecology of, 24, 28, 75, 80, *81, 93;* and vagrancy, *xii, 7,* 24, 81–83, *82,* 94–95, 177, 190, 203, 204 (*See also* Tactical Forward Vagrancy Model); vision of, *62, 63;* wingbeat and flight of, 70
rules, simple internal, *xv;* and adaptation to habitat, 204–5; and complex behaviors, 202–12; and decision making, 29; "follow your nose," 166; for Fritillaries, 174–75; for mate-finding, 29–32; for Pearly Eyes, 175; and return to previous behavior, 150; for Ringlets, 174; "run to daylight," 28, 75–76, 164–65, 184; for Viceroys, 175

"run to daylight" rule, 28, 75–76, 164–65, 184
Rutowski, R. L. et al., 31, 51–53, 118

sampling: impacts on population, 20–21; inference and sample size, 22
*Satyrodes eurydice. See* Eyed Browns *(Satyrodes eurydice)*
scales, 37–41, *39,* 47, 66, 67, 70–71, 195, 207
Scott, J. A., 2–3, 10–11, 21, 25, 33, 126, 191
Scudder, Samuel Hubbard, 111, 125, 127, 182
sensory capabilities of butterflies, 12, 25, 27, 65–68, 73, 166–67, 260; and complex behavior, 12, 29, 32, 69–70; and edge detection, 17. *See also* hearing; olfaction; vision
Shreeve, T. G., 147–48, 190
Sibatani, A., 54
Sims, S. R., 111
site-fidelity, *xv*; benefit to local populations, 205–6; and boundary behaviors, 105; and environmental cues, 104–5, 205; and Eyed Browns, *xii,* 24, 28, 101–2, 104–9, 203; and "home" point, 104–5, 205; and patrollers, 102–7, 178; *vs.* site recognition, 176–78, 184–85, 205, 209–10; *vs.* territoriality, 104–5, *105*. *See also* Fortuitous Site-Fidelity Model (NetLogo)
site recognition: and boundary behaviors, 184; and decision making, 176–78, 184–85, 205, 209–10; and "home" point, 104–5; *vs.* site-fidelity, 108–9, 176–78, 184–85, 205, 209–10; and territoriality, 10, 24, 97–98, 100, 108, 126, 176–78
Skelam, J. G., 24, 74, 81–82
Smith, G. C., 17, 20–21
Smith, Maynard, 25, 141, 145, 147, 150
Southwood, T. R. E., 25, 202
Speckled Woods *(Pararge aegeria),* 25, 145, 147–48, 150
spermatophore counts, *xiv–xv,* 9, *10t,* 17, 79, *79t,* 99, 127, 157
*Speyeira cybele. See* Fritillaries *(Speyeira cybele)*
"spinning wheel flight," 59, 145, *146,* 148–49, 152, 156, 180
Srygley, R. B., 35–36, 38, 165
Stark, W. S. et al., 54
statistical analysis, 20–21
Stewart, F. J. M. et al., 58
stroboscopic flash, 18, *224,* 224–25
"Strong Inference," *xiii*
study site: aerial photograph, *13*; description of, 12–13; habitats pictured, *14, 15*;

postagricultural landscape as, 181, 185–86; selection of, *xi–xii*; and speculations about habitat constraints, 205–6
survival (local survival), 21
Swihart, S. L., 52, 58

Tactical Forward Vagrancy Model (NetLogo), 22, 24, 59, 74, *89, 92,* 164, 177, 188–89, 204–5; Ringlet behavior simulation and, 91–95, *92,* 204
Tamai, M., 39–40
taxonomy, 213–15
territoriality, *xv,* 10, 29; behavior criteria for, 108, 126; benefit to local populations, 205–6; and boundaries, 136; and defense, 10, 25, 100, 108, 126, 129, 136, 178; "dynamic territories," 129; and Fortuitous Site-Fidelity Model, 107, 108, 148, 177–78, 205; and perching, 126; *vs.* site-fidelity, 24, 28, 97–98, 108–9, 176–78; and site recognition, 10, 24, 97–98, 100, 108, 126, 176–78; *vs.* traplining, 10–11; in Viceroys, *xii,* 24, 29, 100, 128, *128,* 129–31, 132–36, 137, 208
tethering, 18, *18,* 30, 58, 148
thermal ecology, 25, *81,* 147–48; of Pearly Eyes, 25, 153; of Ringlets, 24, 28, 75, 80, 93; scales and thermoregulation, 47
Thomas, A. L. R., 35–36, 38
Tiger Swallowtails *(Papilio glaucus), ix,* 4, 43, *43,* 245n6; aerodynamic parameters for, 49–52; flight of, 43–44; wingbeat and flight of, *44*
timer, behavioral, 221–24, *223*
traplining, *xv,* 10; female behavior and, 111–12, 117–18; and Great Spangled Fritillary *(Speyeira cybele), xii,* 24, 117–21; and larval food sources, 111, 117–18, *120,* 123; and mate-finding, 117–18, 123; and oviposition, 11; and Pearly Eyes, 178
"traveling salesfolk" problem, 24, 118–21, 123–24
Tudor, G., 7, 110, 182
Turchin. P, 150
Turner, J. R. G., 165
Turning Bias Model for Habitat Selection, 167–70, *168, 169*

vagrancy, *xii, xv,* 25, 190, 200, 202, 211, 218; and constant flight, 33, 95; and efficient mate-finding, 94–95; and flight distances, *189*; and home court advantage, 145, *151t*;

vagrancy (*continued*)
  and resighting of individuals, 5, *6t*. *See also* random movement
valley-bottoming, 55, 166
veins of wing, 70–71
Viceroys (*Limenitis archippus*), 6t, 7, *16,* 217; and artificial horizon experiment, *130,* 130–31, *131t*; boundary behaviors in, 133, 136, 162; and decision-making, 175, 177, 178, 179, 182–83, 185, 205, 208; and defense of site, 24, 100, 126, 128–37, *135*; Eyed Browns compared with, 97–98, 100–102; female behaviors in, 125–28, 136, 220; habitat of, *7, 15, 97,* 100, 136, *163*; landscape cues and, *127, 129t,* 136 (*See also* artificial horizon experiment *under this heading*); lifespan of, 11, 124, 192; local persistence of, *192, 194, 219t*; male-male interactions, *30t, 31t,* 126, 129–37, *132,* 133–37, *135*; mate-finding by, *xv, 31–32,* 125–28, 136; and mimicry, 14, 45, 71; olfaction in, 67; oviposition in, 125, 126–27, 136, 220; and patrolling, *128,* 128–29, *133, 135,* 175; as "perchers" who also patrol, *3, 97,* 128, 132, *133,* 178; and perches, 29, 125, 126, *127,* 128–31, *129t,* 132–33, *133,* 136; population density and behavior of, 128; and predation, 199; re-sightings of, *101t*; and territoriality, *xii,* 24, 29, 100, 126, 128, *128,* 129–31, 132–36, *134t,* 137, 208; thermal ecology of, 131; vision in, 31–32, 62–63, 131–32
videotape recording, 17–18
vision: and adaptation to habitat, *64,* 204–5; artificial compound eye, 17, 55–58, 225–26, *226*; chromatic (color), 53–55, 58, 70; compound eyes, described, 52–53; and finding habitat, 28, 164; and flicker, 52, 55, 61, 70, 72–73, 164, 204, 207, 225 (*See also* optic flow *under this heading*); horizon (V-notches) and navigation in landscape, 17, 28–29, 32, 46, 55–57, 65, 69–70, 75, 97, 102, 114, 117, 125; and mate-finding, 28–31, 32, 72, 128–32, 136, 205; *vs.* olfaction, 68, 73; ommatidia of compound eye, 32, 52–56, 61–65, 204, 207; and optic flow, 52, 58–59, *60,* 72–73, 172, 175, 204–5, 209; and orientation in landscape, 58–61, 207 (*See also* horizon *under this heading*); and predator recognition, 29; pseudopupils, 61–65, *62, 63*; "run-to daylight" rule, 28, 75–76, 164–65, 184; simulated compound-eye view, *56,* 225–27; visual acuity, 31–32, 52–53, 63
vortices: and butterfly interactions, 58, 73; and concentration of pheromones, 36, 58, 73; and flight dynamics, 34–41, *35,* 50–52, 71, 207; veins and management of, 50–51, 70–71, 207

Weed, C. M., 67
Weis-Fogh, T, 34
White Admiral, 4
Wicklund, C., 25, 59, 147, 150
Wickman, P.-O., 25, 59, 147, 150
Wiernasz, Diane, *xiii,* 15, 20, 28, 75, 81, 91
wingbeat patterns, 34–36, 41–45, 70; cycles and vortices, 34–36; maladaptive, *47*
wings: age estimates and damage to, 20–21, 46, 157, 193, *196*; forewing and hindwing functions, 49–52, 72; impaired flight and damaged, 40, 46–49, *48*; morphology of, 32–34, 32–35, *47,* 71–72; of patrollers, *3,* 33–34, 37–41, 70–72; of perchers, 33–35, 34; scales of, 37–41, *39,* 47, 66, 67, 70–71, 195, 207; veins, 50–51, 70–71, 207; veins of, 70–71
Wood Nymphs, *3,* 5, 6t, 41, *42, 44,* 44–45, 70, 211, *215t*; aerodynamic parameters for, 49–52; and color vision, 54; maladaptive flight in, *47*

Yagi, N., 61–65

MONOGRAPHS IN POPULATION BIOLOGY
SIMON LEVIN, ROB PRINGLE, AND CORINA TARNITA, SERIES EDITORS

1. The Theory of Island Biogeography by Robert H. MacArthur and Edward O. Wilson
2. Evolution in Changing Environments: Some Theoretical Explorations by Richard Levins
3. Adaptive Geometry of Trees by Henry S. Horn
4. Theoretical Aspects of Population Genetics by Motoo Kimura and Tomoko Ohta
5. Populations in a Seasonal Environment by Steven D. Fretwell
6. Stability and Complexity in Model Ecosystems by Robert M. May
7. Competition and the Structure of Bird Communities by Martin L. Cody
8. Sex and Evolution by George C. Williams
9. Group Selection in Predator-Prey Communities by Michael E. Gilpin
10. Geographic Variation, Speciation, and Clines by John A. Endler
11. Food Webs and Niche Space by Joel E. Cohen
12. Caste and Ecology in the Social Insects by George F. Oster and Edward O. Wilson
13. The Dynamics of Arthropod Predator-Prey Systems by Michael P. Hassel
14. Some Adaptations of Marsh-Nesting Blackbirds by Gordon H. Orians
15. Evolutionary Biology of Parasites by Peter W. Price
16. Cultural Transmission and Evolution: A Quantitative Approach by L. L. Cavalli-Sforza and M. W. Feldman
17. Resource Competition and Community Structure by David Tilman
18. The Theory of Sex Allocation by Eric L. Charnov
19. Mate Choice in Plants: Tactics, Mechanisms, and Consequences by Nancy Burley and Mary F. Wilson
20. The Florida Scrub Jay: Demography of a Cooperative-Breeding Bird by Glen E. Woolfenden and John W. Fitzpatrick
21. Natural Selection in the Wild by John A. Endler
22. Theoretical Studies on Sex Ratio Evolution by Samuel Karlin and Sabin Lessard
23. A Hierarchical Concept of Ecosystems by R. V. O'Neill, D. L. DeAngelis, J. B. Waide, and T.F.H. Allen
24. Population Ecology of the Cooperatively Breeding Acorn Woodpecker by Walter D. Koenig and Ronald L. Mumme
25. Population Ecology of Individuals by Adam Lomnicki
26. Plant Strategies and the Dynamics and Structure of Plant Communities by David Tilman

27. Population Harvesting: Demographic Models of Fish, Forest, and Animal Resources by Wayne M. Getz and Robert G. Haight
28. The Ecological Detective: Confronting Models with Data by Ray Hilborn and Marc Mangel
29. Evolutionary Ecology across Three Trophic Levels: Goldenrods, Gallmakers, and Natural Enemies by Warren G. Abrahamson and Arthur E. Weis
30. Spatial Ecology: The Role of Space in Population Dynamics and Interspecific Interactions edited by David Tilman and Peter Kareiva
31. Stability in Model Populations by Laurence D. Mueller and Amitabh Joshi
32. The Unified Neutral Theory of Biodiversity and Biogeography by Stephen P. Hubbell
33. The Functional Consequences of Biodiversity: Empirical Progress and Theoretical Extensions edited by Ann P. Kinzig, Stephen J. Pacala, and David Tilman
34. Communities and Ecosystems: Linking the Aboveground and Belowground Components by David Wardle
35. Complex Population Dynamics: A Theoretical/Empirical Synthesis by Peter Turchin
36. Consumer-Resource Dynamics by William W. Murdoch, Cheryl J. Briggs, and Roger M. Nisbet
37. Niche Construction: The Neglected Process in Evolution by F. John Odling-Smee, Kevin N. Laland, and Marcus W. Feldman
38. Geographical Genetics by Bryan K. Epperson
39. Consanguinity, Inbreeding, and Genetic Drift in Italy by Luigi Luca Cavalli-Sforza, Antonio Moroni, and Gianna Zei
40. Genetic Structure and Selection in Subdivided Populations by François Rousset
41. Fitness Landscapes and the Origin of Species by Sergey Gavrilets
42. Self-Organization in Complex Ecosystems by Ricard V. Solé and Jordi Bascompte
43. Mechanistic Home Range Analysis by Paul R. Moorcroft and Mark A. Lewis
44. Sex Allocation by Stuart West
45. Scale, Heterogeneity, and the Structure of Diversity of Ecological Communities by Mark E. Ritchie
46. From Populations to Ecosystems: Theoretical Foundations for a New Ecological Synthesis by Michel Loreau
47. Resolving Ecosystem Complexity by Oswald J. Schmitz
48. Adaptive Diversification by Michael Doebeli

49. Ecological Niches and Geographic Distributions by A. Townsend Peterson, Jorge Soberón, Richard G. Pearson, Robert P. Anderson, Enrique Martínez-Meyer, Miguel Nakamura, and Miguel Bastos Araíjo.
50. Food Webs by Kevin S. McCann
51. Population and Community Ecology of Ontogenetic Development by André M. de Roos and Lennart Persson
52. Ecology of Climate Change: The Importance of Biotic Interactions by Eric Post
53. Mutualistic Networks by Jordi Bascompte and Pedro Jordano
54. The Population Biology of Tuberculosis by Christopher Dye
55. Quantitative Viral Ecology: Dynamics of Viruses and Their Microbial Hosts by Joshua Weitz
56. The Phytochemical Landscape: Linking Trophic Interactions and Nutrient Dynamics by Mark D. Hunter
57. The Theory of Ecological Communities by Mark Vellend
58. Evolutionary Community Ecology: The Dynamics of Natural Selection and Community Structure by Mark A. McPeek
59. Metacommunity Ecology by Mathew A. Leibold and Jonathan M. Chase
60. A Theory of Global Biodiversity by Boris Worm and Derek P. Tittensor
61. Time in Ecology: A Theoretical Framework by Eric Post
62. Fish Ecology, Evolution, and Exploitation: A New Theoretical Synthesis by Ken H. Andersen
63. Modeling Populations of Adaptive Individuals by Steven F. Railsback and Bret C. Harvey
64. Scaling in Ecology with a Model System by Aaron M. Ellison and Nicholas J. Gotelli
65. Social Butterflies by Henry S. Horn